You Can't Make It Rain

THE STORY OF THE NORTH AUSTRALIAN PASTORAL COMPANY 1877-1991

MARGARET KOWALD • W. ROSS JOHNSTON

First published in 1992 by Boolarong Press with North Australian Pastoral Company

Second published in 2015 by Boolarong Press
655 Toohey Road
Salisbury Qld 4107
Australia
www.boolarongpress.com.au

National Library of Australia Cataloguing-in-Publication entry:

Creator:	Kowald, Margaret, 1950- author.
Title:	You can't make it rain : the story of the North Australian Pastoral Company 1877-1991 / Margaret Kowald & W. Ross Johnston.
ISBN:	9781925236064 (paperback)
Subjects:	North Australian Pastoral Company--History.
	Cattle trade--Northern Territory.
	Cattle trade--Queensland.
	Farm corporations--Northern Territory.
	Farm corporations--Queensland.
Other Creators/Contributors:	
	Johnston, W. Ross (William Ross), 1939-
	North Australian Pastoral Company.
Dewey Number:	338.1762130994

Front cover: *Mustering on Alexandria Station*, photo by John Thompson

Back cover: *NAP properties, past and present*.

Printed and bound by Watson Ferguson & Company, Salisbury, Brisbane, Australia.

CONTENTS

Preface

As a young boy I had always been fascinated with the early history of the outback of Queensland and the Northern Territory and with the exploits of our early explorers and pioneers in opening up and developing these vast areas. My father had often told me of his friendship and association with the Forrest and Collins families, who had played an important part in taking up huge areas in the west and in particular in the founding and development of the partnership 'The North Australian Pastoral Company'.

During my early school days in Sydney I was associated with many 'country boys' who came from western grazing families and would speak of their families' pastoral background and history. This interest in history also stemmed from my mother's background as a member of the Belfield family, one of the oldest New England (NSW) grazing families.

After I joined the Board of The North Australian Pastoral Company Pty. Limited, more than thirty years ago, I spent a considerable amount of time associating with people in the pastoral industry and visiting many of the well-known grazing properties in New South Wales, Queensland and the Northern Territory. At times I became quite embarrassed by my inability to answer many of the questions put to me about the early history and background of the company and of some of its founders.

For many years it has been my ambition to have sufficient research carried out for an accurate account of these matters to be recorded, not only for the information and enlightenment of present generations but also for posterity.

The late Mr. W.F. Alexander, a Director of the company for many years, spent a considerable amount of time and energy in compiling a detailed account of most company matters covering the period to 1967. But despite his efforts, and the time he had taken in research, it was felt that because of the lack of resources then available to him, a more detailed and thorough account should now be attempted.

With this in mind, in August 1989, the Board of the North Australian Pastoral Company Pty. Limited engaged the services of Margaret Kowald and Associate Professor Dr. Ross Johnston, of the History Department, The University of Queensland, to write a history of the company. The authors have spent a tremendous amount of time in researching and compiling this account of the early days of the company (originally a partnership) to the present time. The research has meant visits to the various company properties, to the archives kept in Brisbane, Darwin and Adelaide, and interviews with station managers and other personnel, both past and present, and with the descendants of those who were involved with the company many years ago. It has meant a thorough research of all the old company records that have been preserved over the years.

The company is proud of its record. Few pastoral companies, private or public, can claim to have retained their basic identity for over 113 years. Indeed, few of the early pastoral companies exist today and only a small number of the old and well-known properties have not changed hands many times over the last fifty years.

Many descendants of the original founders, directors and shareholders of the company are still shareholders today. Over all these years they have believed in and remained loyal to the company despite attractive opportunities to sell out.

Their loyalty and faith has been rewarding to them and gratifying to the directors.

The North Australian Pastoral Company Pty. Limited has been a good tenant of those pastoral leases it has held in both Queensland and the Northern Territory; this fact has been acknowledged and documented. The company has not 'traded' in the properties that it has acquired and, apart from the stations which were sold for sound commercial reasons — the Burdekin properties in 1910 and Islay Plains in 1980 — it has retained and substantially improved all its other properties.

The company has never been flamboyant; neither has it sought publicity. It has always quietly consolidated its position before moving further ahead. Many pastoral companies which have expanded too rapidly have eventually failed and collapsed, leaving the properties overgrazed with broken-down improvements and generally in a neglected state.

The company has faith in its future and in the future of the pastoral industry in Australia. It believes in the integrity of those on the land and in their commitment to the industry and to the future of Australia.

I hope those who read this history will find it interesting and informative and that it will give them some insight into the background of one of Australia's oldest private pastoral companies that has withstood the stresses of droughts, floods, depressed prices and increased costs for well over 100 years and can rightly claim to be a 'quiet achiever'.

E.M.Crouch
DEPUTY CHAIRMAN AND DIRECTOR
THE NORTH AUSTRALIAN PASTORAL COMPANY PTY. LIMITED
Brisbane, December 1991.

Edward Michael Crouch
Director 1955-
Managing and Executive Director 1972-1981
Chairman of Directors 1967-1990.

Acknowledgments

This commissioned history presented a rare opportunity to trace the history of one of Australia's oldest cattle companies in the context of the Australian cattle industry for that period. The task was facilitated at all times by the cooperation of, and encouragement from, the company. The most deserved thanks must go to Michael Crouch, immediate past Chairman of Directors and present Deputy Chairman, for his perseverance in ensuring that a history of NAP be written. William Foster, a Director, and Ross Brunckhorst, the General Manager, have also been unstinting in the time and helpful advice they have given to the project.

There were gaps in the research material — some early partnership and station records have not survived and the 1974 Brisbane floods had destroyed files. There was sufficient material, however, to allow the varied and fascinating features of NAP's history to be explored — the stations, the cattle, and the people of the company; those who own it, those who administer it and those who work on the properties.

As well as using the company records and the family papers of those associated with NAP, research in Brisbane was undertaken at the Queensland State Archives, John Oxley Library, The University of Queensland and the Lands Department. Northern Territory records were researched in Adelaide at the Public Record Office and Mortlock Library and in Darwin at the State Archives, the State Library and the Lands Department. Gordon Hoffer did a splendid job in compiling and drawing the maps.

Interviews have been conducted with past and present employees and shareholders. Some also kindly donated photographs, especially those valuable early photographs of Alexandria taken by James Broadbridge. As well as the people and organisations already mentioned, we wish to thank these people for their time, hospitality and contribution towards the finished product.

Bill and Rhondda Alexander	Manager, Marion Downs
David Alexander	Director
Harry Barnes (Jnr)	Son of former manager
Barbara Beswick	Head Office
David Brown	Head Office
Malcolm and Elizabeth Debney	Manager, Glenormiston
Cam Doglioni	Former road train driver
Jim Dwyer	Former manager Soudan and Glenormiston
Henry Foster	Director
Bill Fraser	Director
Nancy Grace (née Forrest)	Daughter of former managing director
Jean Haughton-James	Shareholder
Tony Johnston	Son of former manager
Bob Kirk	Manager, Herbert Downs
George MacQueen	Former manager Gallipoli, Monkira and Alexandria
Ken McGuire	Former manager Soudan, Marion Downs and Alexandria
Joan McKean	Shareholder
Peter and Lenore McLaren	Manager, Monkira
Lee McNicholl	Former company veterinarian
Steve Millard	Head Office, former manager Gallipoli, Soudan and Coorabulka
Ken Moore	Former general manager and director
Philip Myers	Manager, Wainui
John and Wendy Ohlsen	Former manager Gallipoli, Islay Plains and Alexandria
Richard Purvis-Smith	Former shareholder and manager of road trains
Ron Spedding	Former bookeeper Alexandria, head office staff
Bill Young	Former manager Alexandria

Margaret Kowald, Ross Johnston.
BRISBANE, DECEMBER 1991.

Notes on Measurement

		IMPERIAL				*METRIC*
Length						
12	inches	=	1 foot			
3	feet	=	1 yard	=		91.44 centimetres
1760	yards	=	1 mile	=		1.61 kilometres
1	inch	=				25.4 millimetres
39.37	inches	=	1 metre			
Area						
4840	sq yards	=	1 acre	=		0.405 hectares
640	acres	=	1 square mile	=		2.6 square kilometres
2.47	acres			=		1 hectare
Weight						
16	ounces	=	1 pound	=		0.453 kilograms
112	pounds	=	1 cwt (hundredweight)			
20	cwt	=	1 ton	=		1.02 tonnes
Rainfall						
4	points			=		1 mm
100	points	=	1 inch	=		25 mm

Abbreviations

ADB	*AUSTRALIAN DICTIONARY OF BIOGRAPHY*
CPP	*COMMONWEALTH PARLIAMENTARY PAPERS*
JOL	JOHN OXLEY LIBRARY
MCNT	MINISTER CONTROLLING NORTHERN TERRITORY
NAPR	THE NORTH AUSTRALIAN PASTORAL COMPANY PTY. LIMITED RECORDS, BRISBANE
NTAS	NORTHERN TERRITORY ARCHIVES SERVICE
OMCNT	OFFICE OF MINISTER CONTROLLING NORTHERN TERRITORY
PROSA	PUBLIC RECORD OFFICE OF SOUTH AUSTRALIA
QPP	*QUEENSLAND PARLIAMENTARY PAPERS*
QVP	*QUEENSLAND VOTES AND PROCEEDINGS*
SAPD	*SOUTH AUSTRALIAN PARLIAMENTARY DEBATES*
SAPP	*SOUTH AUSTRALIAN PARLIAMENTARY PAPERS*

GLOSSARY

Agistment	Grazing cattle on other people's land under an agreement.
Bangtail muster	A muster which gives an accurate count of cattle. The hair on the tail of a beast is cut square so the animal will not be counted twice.
Bookkeeper	A person on large stations who attends to clerical and accounting tasks and also usually manages the station store. NAP has a bookkeeper on Alexandria.
Bos indicus	Tropical breeds of cattle e.g. Brahman and Sahiwal popular for their resistance to heat and ticks. They have been crossed successfully with *Bos taurus* (European) breeds e.g Shorthorn.
Boss drover	Man in charge of a cattle droving camp, contracted to deliver cattle to a certain destination.
Botulism	Disease in cattle caused by a toxin. Vaccination is a very effective method of preventing losses.
Boxing	Mixing of stock.
Branding	The act of burning a station's distinctive registered mark into the hide of a beast to show ownership.
Breeders	Female cattle of breeding age.
Bronchoing	To broncho a beast is to put a catching rope on it and pull it up to a broncho panel.
Broncho horse	A strong horse which pulls the animal to the broncho panel.
Broncho panel	A fixture of post and rails to which a beast is pulled when it is being branded, ear-marked, dehorned, inoculated and in the case of male animals, castrated.
Brucellosis	Contagious abortion in cows resulting in low calving percentages. Can be prevented with a vaccination.
BTEC	Brucellosis and Tuberculosis Eradication Campaign.
Bullock or Ox	Castrated adult bovine animal over two years of age with dressed weight greater than 225 kilograms.
Calabashes	Also referred to as 'shin plasters'. Private money orders used by outback stores and some stations instead of cash. Often of thin paper, they would sometimes disintegrate in the holder's pocket and so never be cashed or used.
Cowboy	Man employed at the station homestead to work as gardener and general hand.
Crush	A narrow race or passage in a stockyard through which animals can only pass in single file.
Cull, Culls	Unwanted stock removed from the herd.
Cutting out	The act of separating selected cattle from a mob by stockmen on horseback.
Drafting	The process of separating an animal or groups of animals from the main herd.
Dressed weight	Weight of a carcass after slaughter.
Droving	Walking cattle 'on the hoof', usually over long distances.
Duffing	Stealing cattle.
Earmark	Distinguishing mark cut in the ear of an animal.
Fat cattle, fats	Cattle considered to be in prime condition for slaughter
Forward selling	The selling of cattle in a specified condition for a predetermined price.
Galah session	Two-way radio communication between properties which can be heard by everyone.
Gibber	Round stone
Hawker	Travelling shopkeeper selling a wide variety of goods to station residents.
Heifers	Young female cattle.
Jackaroo	Cadet stockman.
Jap Ox	Bullocks (or Oxen) which meet carcass specifications required by the Japanese market.
Jillaroo	Female equivalent of a jackaroo.

K, KB wagons Particular classes of railway wagons which transport cattle. K wagons are approximately 10 metres long and each carry around 17 bullocks or 20 cows depending on the size of the animals. A KB wagon includes compartments to accommodate the drover and the guard.

Killer Cattle killed for food on the station and during droving trips.

Lick blocks Supplementary feeding blocks containing specific nutrients.

Licker drums Method of feeding molasses to stock. Drums rotate as cattle lick the mixture.

Mustering Gathering cattle into a group.

Off side Refers to the right side of the animal, e.g. off ribs or off rump, the position where the brand is placed.

Pleuro-pneumonia Contagious cattle disease which can be treated with a vaccine.

Pumper Also referred to as a boreman; a person who is responsible for the maintenance of bores.

Punkah Large swinging, screenlike fan hung from the ceiling; of Indian origin.

Rager An old untamed and aggressive bullock or cow.

Reactor A beast showing a positive reaction to a disease test.

Redwater Cattle disease also referred to as tick fever, the name being derived from the blood-coloured urine of cattle severely affected by ticks.

Ringer A stockman. Derived from the drover's stockman who had to 'ring' or ride around the cattle at night.

Road train Means of transporting cattle by road, consisting of a prime mover with two or three trailers ('dogs').

Rush Stampede; sudden movement of cattle through fright.

Sandy blight An acute conjunctivitis usually infectious in which the eyes of infected persons feel as though they have sand in them.

Spaying The rendering infertile of female cattle by the removal of their ovaries.

Steers Young castrated cattle. Dressed weights are usually less than 225 kilograms.

Stockman Man who herds and handles stock; the Australian equivalent of the American 'cowboy'.

Stock routes Designated public routes for the movement of stock on foot between properties or from properties to market.

Store cattle, stores Young cattle which are not fat.

Tailing To 'tail' is to mind. Thus a horse-tailer in a stock camp looks after the horses. The end of a travelling mob is also referred to as the 'tail'.

Teamster One who drives a team of horses, often bringing supplies to remote stations.

Tendering Derived from the word attending, tendering occurred at a tender muster more particularly before fences were common when stockmen from neighbouring stations met during musters to sort out and claim stock.

Tuberculosis (Bovine) Contagious disease affecting the lungs of animals. With no known treatment, destroying affected stock is the only way to eradicate the disease.

Turkey's nest A circular wall of earth above ground to store water.

Turnoff Cattle leaving stations for fattening properties or sale.

Weaner Calf 6-12 months old no longer feeding from its mother.

Wet, the wet Period in northern Australia when monsoonal influences bring rain, typically from November to April.

INTRODUCTION

The North Australian Pastoral Company Pty. Limited[1] is, by any standards, a major player in Australia's cattle industry. It presently leases 56,462 square kilometres or 0.7 per cent of the Australian continent, a land mass greater than Switzerland and Denmark and almost three times the size of Wales. Its chain of properties extends from the breeding properties in the north — Alexandria (Northern Territory) and Boomarra (Queensland) — to the fattening properties in south-west Queensland — Marion Downs, Monkira, Coorabulka, Glenormiston, Connemara, Kynuna and Vergemont. The company also operates a feedlot, Wainui, much closer to Brisbane on the Darling Downs. The company owns an office in Mount Isa and head office premises in Brisbane from which the total operation is managed. A depot in an industrial area of Brisbane facilitates the dispatch of a wide range of essential materials and foodstuffs to the properties. Staff numbers vary with the cattle season but currently average around 150. With a herd of some 120,000 head the company is one of Australia's largest cattle producers[2], whose economic importance can be gauged from the strong role beef plays in Australia's exports, an importance which should only increase with expanding opportunities in Japan and Korea.

Underlying the company's strong contemporary position is a fascinating story of growth and development since 1877 when the first leases were acquired by the founding partnership. Much of this story has been shaped by the changing fortunes of the pastoral industry itself and how the company responded to broader crises, opportunities and change. Devastation from drought is an ever-present threat, for example, in the pastoral industry. While different NAP properties have been affected by drought to varying degrees, the company's herds were particularly vulnerable during the droughts of 1897-1903, 1926-1928, 1952, the late 1950s, 1960s and the mid to late 1980s. Stock disease has also been a vital issue for pastoralists. Redwater disease associated with ticks was a major problem until early this century with brucellosis and tuberculosis the main concerns in more recent years. Market and price swings have been a continuing source of destabilisation in the cattle industry, with price slumps in the 1920s and mid-1970s having a particularly detrimental effect.

On the other hand, technological advances in areas such as veterinary science, machinery, transport and communication have been of great benefit to the industry. The widespread drilling of sub-artesian bores since early this century has opened previously inaccessible country for pastoral use and helped to cope with drought. Herd and station management practices have improved, with one catalyst for change being the large-scale Brucellosis and Tuberculosis Eradication Campaign (BTEC) of the last two decades. The introduction of the more environmentally suitable *Bos indicus* breeds with the traditional British bloodlines has lifted industry productivity. There have also been positive developments in

---

1 The company is usually referred to as NAP although NAPCO has also been used. The term NAP will be used throughout this book.

2 Other large cattle companies include Stanbroke Pastoral Company, Australian Agricultural Company, Kidman Holdings and Heytsbury Pty. Limited.

overseas marketing with the export of chilled as well as frozen meat in earlier years, and the broadening of Australia's markets in North America and Asia more recently.

But beyond the influences of common industry-wide factors, the long-term survival and development of NAP is a tribute to the people involved; in particular the character, ability, foresight and often quite extraordinary dedication of those who have contributed both on and off the properties throughout NAP's existence.

At one level, the strong family base of ownership has been a most important factor in sustaining the company. Its origins date from 1877 when Queenslanders William Collins, William Forrest and Sir Thomas McIlwraith were joined by Englishmen John Warner and Sir William Ingram to form a partnership, the North Australian Pastoral Company. This marked the founding of one of Australia's oldest — if not the oldest — family-based private cattle companies still in existence. The company is justly proud of its heritage. At the time of writing, NAP retains an identity and feeling akin to that of an extended family. Descendants of the first Collins, McIlwraith and Ingram partners are still shareholders, although members of the Foster and Launder families now have major holdings also.

Indeed, the ownership structure of the company has significantly altered at only three stages in its 114-year history. The first stage was in the very early years of this century when the McIlwraith interest in the partnership was greatly reduced. The second was in the late 1930s, when the then remaining Forrest shares were sold to the Foster family, while the third was in the mid-1980s, when the Fosters moved to majority control, the Launders entered the company and the Warners sold their remaining shares. The long-term continuity of ownership has resulted in a corresponding stability in decision making. The company has had only eight managing partners or general managers, and since incorporation in 1931 only twenty-five men have served on the board and only four of these have been Chairman.

The company's development can be seen in terms of four distinct eras, and, indeed, this present account is structured very much in those terms. The initial period takes the partnership through its earliest years to about 1910. In 1877 the first leases were taken up for Alexandria Station, in the Northern Territory, which remains central to the operation today. In the following year, two properties, Inkerman and Woodstock in the Burdekin district of Queensland, were purchased primarily to stock Alexandria. The Burdekin properties were sold at a good price for sugarcane land in 1910, during an otherwise unsettled time for NAP, with the deaths of original partners and other family members. The first three chapters of this book focus on the founding of the partnership, the original partners, the taking up of the Alexandria leases and the Burdekin properties.

The second period runs from about 1910 to the late 1930s. These were difficult years, with World War I followed by severe drought conditions and market downturn. For most of that time, Alexandria was the company's only station, although the first Channel country property, Marion Downs, was purchased in 1934. The partnership was incorporated as a private company in 1931, with Philip Forrest, manager of the company for most of this era, becoming the first chairman of directors. The end of this period is marked by the departure of Forrest from the company in 1936 with the sale of his shares to Francis Foster, and by the acquisition in 1939 of two further Channel properties, Monkira and Coorabulka, which served to more firmly establish NAP's subsequent style of operation — breeding cattle on Alexandria and fattening on, and selling from, the Channel country. Chapter 5 deals with general managerial and administrative issues through these decades, while chapters 4 and 6 describe the early improvements and station life on Alexandria from its founding to World War II.

NAP's third era extends for another three decades to the late 1960s. This was an important period of consolidation and internal development for the company. No further properties were added, but the way was opened for the expansion that followed in more recent years. The leases were negotiated favourably and, although labour and materials were scarce during the war, major improvement programs were undertaken on the properties in the post-war years. This era saw the managerial and administrative dominance of Douglas Fraser and the strengthening of the Foster influence in the company. The end of the era is marked by Fraser's death in 1968. Chapters 7 to 10 focus specifically on this period, dealing in particular with the parts played by Fraser and Foster, on general administrative issues, the next stages in the development of Alexandria, the three Channel properties and the road train fleet established by NAP.

The modern era, which starts in the late 1960s, has been a period of expansion and increasing sophistication for the company. A new Channel property, Glenormiston, was purchased in 1968 and Islay Plains, in the Clermont district, was held for ten years from 1970. The most striking period of acquisition, however, has been since the mid-1980s, with Wainui Feedlot (1985), Connemara (1986), Kynuna (1986), Boomarra (1989) and Vergemont (1990) purchased in that time, partly in response to prevailing drought conditions but also to add greater flexibility in general herd disposition. Management practices on the properties have significantly advanced in the modern era, stimulated in particular by the demands of the disease eradication program, BTEC. Head office has responded to the increasing complexity of the company's operations with new levels of sophistication in scientific approach, computing and communications.

Dominant figures in this era have been Michael Crouch, chairman until 1990, and the two general managers, Ken Moore (whose contribution straddled both the Fraser years and the modern era) and Ross Brunckhorst. Chapters 11 and 12 cover administrative and general managerial issues through this period, and the properties, both old and new. The final two chapters focus on two particularly important historical aspects: the BTEC program that dominated property management for much of that time, and the takeover period of the mid-1980s, when the company lost some of its innocence in the broader corporate world, and which brought the Foster family to formal control.

The company has always maintained a policy of quiet determination and fairness, whether in dealing with its employees, with other members of the industry, with government officials or in the handling of larger affairs such as during the takeover period. In the process, it has accumulated an impressive list of achievements: the first artesian bore in the Northern Territory was drilled on Alexandria; the company has been in the forefront of disease control; it was the first company in Queensland to establish a road train fleet to move cattle; and NAP is now achieving recognition for its composite breeding program.

It is worth stressing again the one factor that emerges above all others — the people of the company. While some key persons have been mentioned, in fact the combined, dedicated efforts of three generations of partners, shareholders and employees have generated the present sound position of The North Australian Pastoral Company Pty. Limited with its commitment to the pastoral industry.

This publication is a tribute to these men and women who have served the company so faithfully and so well for over a century — the early partners, the directors, administrative staff in Brisbane and Mount Isa, station employees and generations of shareholders. If the company can continue to inspire the same dedication, its future must be assured.

CHAPTER 1

A Partnership in the Making

Pessimism and despair have often enough marked men's strivings in the [Northern] Territory. Yet it has not always been so. The pastoral pioneers entered upon their domain with buoyant hearts.[1]

The Barkly was in every sense a study in immensity. Immense cattle runs, and the measureless land-ocean dwarfed everything.[2]

THE birth of the colony of Queensland on 10 December 1859 gave an impetus to the northern and westward march of pioneers to settle and stock new country. By the mid-1870s, with much of Queensland taken up, there was pressure from pastoralists for more land. Some saw the Northern Territory as the next frontier to be settled, and in 1877 it was here that leases were taken up by three men from Queensland: William Collins, Thomas McIlwraith and William Forrest. They formed a partnership called the North Australian Pastoral Company, and their amalgamated Northern Territory leases were called Alexandria Downs.

The historical significance of this pastoral undertaking went unnoticed initially since other capitalists from the eastern States also took up land in the Northern Territory. The impact was only later fully apparent. Now, over a century later, Alexandria is still owned by NAP with descendants of McIlwraith and Collins among the present shareholders. The company has expanded with the purchase of other cattle properties, but most importantly it has survived, despite the vicissitudes of harsh environments, fickle markets, the effects of disease and attempted corporate takeover.

The earliest documentary evidence of the company is a telegram dated 9 February 1877 from William Collins to the South Australian Chief Commissioner of Lands, asking the rate per square mile for land to be leased for pastoral purposes in the Northern Territory.[3] Collins formally applied a week later, with Forrest and McIlwraith also applying for leases during 1877. By the end of 1882 they had taken up a total of 44,408 square kilometres (17,080 square miles) of land, 39,156 square kilometres (15,060 square miles) on the Barkly Tableland in the vicinity of the Playford River and Buchanan Creek, and a further 5,252 square kilometres (2,020 square miles) east of the Victoria River.[4]

Regulations required stocking the lands within three years. With this in mind, the three men looked to Queensland as a source of cattle. Rather than purchase stock to be overlanded, they acquired properties in the Burdekin and Haughton River areas of north Queensland. William Collins travelled to north Queensland by steamer in late 1877 to inspect the area[5] — freehold land as well as the leasehold cattle properties of Woodstock and Inkerman. The land was purchased from the executors of the will of Robert Towns of Sydney early in 1878.[6]

1 Ross Duncan, *The Northern Territory Pastoral Industry 1863-1910*, Melbourne, Melbourne University Press, 1967, p.1.

2 V.C. Hall, *Outback Policeman*, Melbourne, Rigby, 1970, p. 122.

3 William Collins, Melbourne, to South Australian Chief Commissioner of Lands, 9 February 1877, GRS 1, Docket 40, PROSA.

4 *SAPP*, 1882, No. 119, pp. 13-6.

5 Harry C. Perry, *Pioneering: the life of R.M. Collins MLC*, Brisbane, Watson Ferguson, 1923, p. 197.

6 Stock Mortgage between Thomas McIlwraith and Sophia Towns, 5 March 1878, SCT/CD 24, BK 13, QSA.

Sir Thomas McIlwraith: original partner 1877-1900.

John Oxley Library

The partnership looked for other members to share the financial burden. In 1878 they had promises to join from John Henry Boyer Warner, a London barrister, and in February 1879 Warner paid more than £6,000 towards the partnership.[7] In early 1880, while in London, McIlwraith made arrangements for Sir William Ingram, the proprietor of the *Illustrated London News*, to join the partnership. Collins, McIlwraith and Forrest each held a quarter share and Warner and Ingram one eighth each. Ingram paid £12,500 in 1880 for his share, making the partnership by that time worth £100,000.[8] The partnership was sound, with McIlwraith the entrepreneur and influential politician, Forrest the businessman who saw to administration, Collins the pastoralist who was an expert in stock matters, and Warner and Ingram financially secure partners who kept an interest, albeit a distant one.

Sir Thomas McIlwraith was born on 17 May 1835 at Ayr, Scotland. He was educated at the University of Glasgow, and emigrated to Victoria in 1854. After a time on the Mount Alexander goldfields, he prospered as a government railway engineer and private railway contractor while retaining family shipping interests. In 1864 he contested, unsuccessfully, a seat in the Victorian Legislative Assembly. Meanwhile he had taken up, in partnership with Joseph C. Smyth, eight runs in the Maranoa district of Queensland.

After moving to Queensland in the early 1870s he embarked on a vast range of speculative and developmental investments of which the Queensland National Bank, Queensland Investment and Land Mortgage Company, the Darling Downs and Western Land Company, the North Australian Pastoral Company and McIlwraith, McEachern and Company were the main elements.[9] Entering parliament in 1870, he was Premier of Queensland from 1879 to 1882 and again in 1888 and in 1893 during the crucial period when NAP was in its infancy. He was a director of the Queensland National Bank until he became Premier in 1879; he received a knighthood in 1883. With such connections he could provide such a favourable political and business environment for his companies that historian Duncan Waterson has commented, 'These enterprises intermingled public and private business concerns to an extraordinary degree'.[10]

As an entrepreneur, McIlwraith was ahead of his time; it was claimed after his death that 'Queensland never fully appreciated him'.[11] He was very much at one with the enthusiasm and optimism of the 1880s when loan money was being spent lavishly and immigrants were streaming into Queensland. McIlwraith endeavoured to develop Queensland by exploiting and managing the total resources of the colony. He did not narrowly conceive the frontier as 'an advancing wave of Queensland pastoral settlement propelled by

7 John Collins to William Collins, 19 February 1879, Collins Family Papers held privately by Joan McKean, Brisbane; McIlwraith's Statement of Account, 1881, Sheet 15, McIlwraith-Palmer Papers, OM64.19, JOL.

8 William Forrest, Brisbane, to Thomas McIlwraith, 21 May 1880, Sheet 7, OM64.19, JOL.

9 For McIlwraith's involvement with the Queensland National Bank see Tony Gough, 'Tom McIlwraith, Ted Drury, Hugh Nelson and the Queensland National Bank 1896-1897', *Queensland Heritage*, Volume 5, No. 9, November 1978, pp. 3-13.

10 Duncan Waterson, 'Pastoral capitalism and the politician Thomas McIlwraith and two land companies, 1877-1900', *Royal Historical Society Queensland Journal*, 12, 6, November 1986, p. 402.

11 R. Spencer Browne, *A Journalist's Memories*, Brisbane, Read, 1927, p. 67.

British industrial needs and . . . British capital'.[12] Rather, he saw expansion in terms of interlocking pastoral, processing, mining and transport enterprises with finance from British, Victorian and New South Wales capital, and direction from Brisbane entrepreneurs, politicians, merchants and pastoralists.[13]

A groundswell of support was built up through McIlwraith's ability to persuade and convince governments, companies, banks and individuals that opportunities for financial enterprise in Queensland were almost limitless and certainly worth pursuing. As Waterson has summarised, his aim was to be

> not only the catalyst but the magnet; not only the exploiter but the developer and the satisfier; not only the entrepreneur but the complete colonial man operating on a vast territorial scale and ultimately on an empire basis.[14]

But inevitably his commercial and political operations were heavily dependent on the financial and institutional climate, and when the Queensland economy began to deteriorate in the late 1880s, so too did McIlwraith's success.

William Collins: original partner 1877-1909.

John Oxley Library

In 1890 he astounded everyone by joining his old political enemy Griffith to form the 'Griffilwraith' coalition. He resigned from the Ministry on 9 December 1897 and after spending his last years in broken health, trying to piece together his shattered fortunes resulting from the economic crisis of 1893, he died in London on 17 July 1900. Described as 'an able bully with a face like a dugong and a temper like a buffalo',[15] McIlwraith certainly had a grand vision for Queensland and for nearly twenty-five years was one of its greatest personalities. His descendants, the Alexanders, are shareholders in the company, with David Alexander, a great grandson, presently on the NAP board. A lasting reminder of McIlwraith's involvement with NAP is the Burdekin sugar town of Ayr, surveyed in 1882 while he was Premier and named after his birthplace in Scotland.

William Collins, second son of John and Anne Collins, née Martin, was born on 26 April 1846 at Mundoolun,[16] Beaudesert and died in Brisbane on 22 January 1909. Situated on the Albert River, Mundoolun had been taken up by Collins' parents in 1844.[17] It became and remained the hub of the family's business activities, though the Beaudesert properties were subsequently increased with Tamrookum[18] in 1878, Rathdowney in 1884 and Nindooinbah in 1906.[19] Educated at Brisbane and Sydney, he joined his father and brother Robert in 1863 in pastoral pursuits. Together with another brother John George, the family formed the partnership John Collins and Sons. The brothers undertook many western trips to inspect the country and stock the partnership's increasing pastoral holdings which included Westgrove (1862), Box Vale (1872), Whitula (1875), Morney Plains (1875), Mount Leonard (1875), Mount

12 D. Waterson, *Personality, Profit and Politics: Thomas McIlwraith in Queensland 1866-1894*, The John Murtagh Macrossan Lecture 1978, St Lucia, University of Queensland Press, 1984, p. 31.

13 For discussion on the indebtedness of the north Queensland pastoral industry see Dawn May, *From Bush to Station*, Townsville, James Cook University, 1983, pp. 12-21.

14 D.Waterson, *Personality, Profit and Politics: Thomas McIlwraith in Queensland 1866-1894*, p. 31.

15 *ADB*, Volume 5, 1974, p. 164.

16 Originally spelt Moondoolan, it is an Aboriginal word meaning 'death adder'.

17 Until their full ownership of Mundoolun in 1847, the Collins family had a share in the property with William Humphreys.

18 Tamrookum became the family home for another son, Robert Martin Collins.

19 Nindooinbah became the home for the William Collins family. It had initially been owned by William Duckett White who passed it to his son Ernest White in 1867.

Merlin (1877), Springvale (1877), Warenda (1877), Coorabulka (late 1890s), Chatsworth (1903), Noranside (1908), and Babbiloora and Carnarvon. John Collins and Sons rented Waterton, a property on the upper waters of the Dawson River which they later owned as part of Gwambegwine. To this day the NAP company uses the name Waterton as the code for cables and telegrams.[20]

The Collins family had other business interests including the partnership Collins, White and Company, formed in 1883 between Simon Fraser and Sir Malcolm McEachern of Melbourne, Albert W.D. White of Bluff Downs, north Queensland, and John Collins and Sons. Adjoining properties near McKinlay — Eulolo, Beaudesert and Strathfield stations — were taken up by the partnership, and Glenormiston in south-west Queensland was purchased in 1899.

While it was William Collins who appeared in the initial NAP partnership, the involvement with NAP was very much a Collins family affair. As early as 1882, when Forrest was suggesting they go into sugar-growing, he wrote to McIlwraith that if William was absent, to 'discuss this with his father or Rob at once'.[21] Indeed, by 1910 William's share in the company was distributed three ways between himself and his two brothers, in the name of John Collins and Sons. The family worked closely together, Robert Collins commenting in 1877 that

> a worthier motive than that of the acquisition of wealth lay at the bottom of our general policy, and amongst other things it required of us was the exercise of a good deal of forbearance.[22]

Robert Martin Collins was born in Sydney on 17 December 1843, four years after his parents arrived from Ireland. He joined them at Mundoolun following his education at Brisbane and Sydney, and after his marriage in 1879 he settled at Tamrookum on the Logan River. Being the eldest son, he took the lead in family concerns and possessed those qualities said to be common to all the Collins family, 'sound judgment' and 'steadfastness of purpose'.[23] He represented the electoral district of Albert in the Legislative Assembly from 1896 to 1898, and in June 1913 he was nominated to the Legislative Council but died two months later. A lasting legacy is the Lamington National Park in Queensland, an area which he had consistently argued should be reserved as a national park.

Robert Martin Collins. John Oxley Library

Two sisters, Jane and Anna Bertha Collins, also contributed to the family enterprise. But when the three Collins brothers died within four years (1909 to 1913), the tradition of male domination within pastoral families ensured that the unmarried Jane Collins, despite being a most capable woman, did not assume the most influential position.

The onus of responsibility fell instead for a short time to Christopher John Collins, only son of Robert Martin Collins. Born 15 September 1881, he took his father's place as director of the Queensland Meat Export Company and was prominent in the affairs of NAP until his early death in the influenza epidemic of 1919. The mantle of responsibility then passed to Douglas Martin Fraser who was to play a most prominent part in NAP's later development. Son of Anna Bertha Collins and Simon Fraser (knighted in 1918), Douglas married his cousin Marion Dorothea Jane Collins, a daughter of R.M. Collins. While the Collins name disappeared, the family's association with NAP continued, as two of William's three daughters married into the Persse family and those of Robert married into the Bruxner, Philp, Ralston, Persse and Harris families.

20 *ADB*, Volume 3, 1969, pp. 442-3; M. Fox (compiler) *The History of Queensland; its people and industries*, Volume 1, Brisbane, States Publishing Company, 1919, pp. 298-305.

21 William Forrest, Adelaide, to Thomas McIlwraith, 15 February 1882, Sheet 18, OM64.19, JOL.

22 R.M. Collins in Perry, *Pioneering*..., p. 205.

23 Fox, *The History of Queensland*..., p. 300.

Daughters of R.M. Collins on the occasion of Annie Bruxner's (née Collins) 80th Birthday, 1960. Left to Right: Bertha Harris, Katharine Philp, Dorothea Fraser, Annie Bruxner, Mabel Ralston. Front: Lesley Bruxner. Jean Haughton-James

William Forrest was born on 11 January 1835 at Ballykelly, Ireland. He studied engineering at Glasgow, and arrived in Victoria in 1853. After success in goldmining at Bendigo, where he met McIlwraith, he joined the rush to pastoral Queensland in 1860, managing an out-station in the Dawson district and in 1867 purchasing the head station Mount Hutton with Simon Fraser.

McIlwraith appointed Forrest to the Legislative Council in 1883 and twice offered him a portfolio. In 1893 Forrest was gazetted Queensland's Agent-General in London but the financial problems of the mercantile and pastoral community held him back. He stayed to restructure the mercantile and stock and station agency B.D. Morehead and Co, of which he was managing director. This ensured the progress of the refrigerated meat export trade.

Business was his passion. 'Dogged perseverance and thrift leads . . . to prosperity',[24] he once said. Indeed, with William Collins absent in Europe and North America from May 1878 to February 1879, it fell to McIlwraith and Forrest to see to the NAP properties. As historian Duncan Waterson has commented, Forrest was the key figure in 'shrewdly negotiating with conservative South Australians worried about their virgin incubus, the Territory'.[25]

Forrest was a tireless worker and the most prolific correspondent of the three. Letters written from the Queensland, Melbourne and Adelaide Clubs went in all directions: to the other partners, to the station managers, to the South Australian and Queensland Governments, to the agents for the partnership in Adelaide, Sanders and Packard, and to the company secretaries at the head office at 21 Queen Street, Brisbane (F.M. Hannington, and after 1900 George Oakes Beardmore). While consulting with the others on major decisions, William Forrest preferred to attend to the administration himself. For instance, he wrote in 1882, after William Collins had put through a transaction incorrectly, 'I wish to heavens Collins would do as I asked him . . . viz. leave the ac/s

24 William Forrest in *ADB*, Volume 8, 1981, p. 552.

25 D. Waterson, *ADB*, Volume 8, 1981, p. 551.

William Forrest: original partner 1877-1901.

Nancy Grace

alone until I go back and then I'll put them in order'.[26] He never married, and with failing health passed his shares to his brother John in August 1901, two years before his death on 23 April 1903.

John Forrest succeeded William as managing partner of NAP and chairman of Moreheads in 1902 and had an important role in negotiations for the sale of NAP's Burdekin properties. He was born on 18 October 1848 at Ballykelly, Ireland. After joining his brother William at Mount Hutton on the Dawson in 1868, he managed McIlwraith's Gin Gin Station between 1875 and 1880 and became McIlwraith's local political organiser in Bundaberg. In 1880 with financial assistance from McIlwraith, he purchased the Avoca sugar plantation at Bundaberg where he gained recognition for his humane treatment of Kanakas who worked on Avoca, allowing them a holiday on the coast with decent food and shelter in order to prevent them succumbing to dysentery and other diseases.[27] A period (1889-1902) as a pastoral inspector for the Queensland National Bank followed, until he took up his positions as managing director of Moreheads and managing partner of NAP. Like his brother, he was offered but declined the Agent-Generalship in 1909. He died in Brisbane on 29 September 1911.[28]

As well as Thomas McIlwraith, William Collins and William Forrest, another member of the original partnership was Sir William James Ingram, Baronet, of The Strand, London. Born in 1847, he attended Winchester College and Cambridge University, graduating in law. His father Herbert Ingram had begun the *Illustrated London News* in 1842 and William followed as managing director and proprietor of the newspaper. He was a Member of Parliament in 1874-80, 1885-86 and 1892-95. In 1874 he married Mary, the eldest daughter of Edward Stirling, a businessman and member of the Legislative Council of South Australia, who had lived in Australia between 1839 and 1865. Two of Mary's brothers, Edward and John, returned to Adelaide to pursue successful careers. As prominent members of the Adelaide Club, they kept their brother-in-law well informed of NAP's progress from their conversations with William Forrest who always stayed at the club on his frequent trips to Adelaide.

Sir William Ingram was made a Baronet on 9 August 1893 and after his death in 1924, his title passed to his eldest son Sir Herbert[29] who had received his father's NAP shares in 1920. Collingwood Ingram, another son of Sir William, shared his father's interest in ornithology.[30] Indeed his family's connection with NAP enabled him to study birds of the Alexandria district, the results of which were published in 1907 and 1909 editions of *The Ibis*, a British ornithological magazine.[31]

Naturalist William Stalker spent 1905-1906 on the Barkly on a trip arranged by Sir William Ingram. Stalker, the first to undertake such a study in the area, collected mammals and 105 species of birds which were dispatched to England for identification. They were shot mainly along the watercourses during the wet season and included

26 W. Forrest, Melbourne Club, to T. McIlwraith, 22 July 1882, OM64.19, JOL.

27 Newspaper cutting, 9 January 1884, Newspaper Cutting Book, Book A1, 1880s-1900s, Bundaberg and District Historical Museum, Bundaberg.

28 *ADB*, Volume 8, 1981, p. 552.

29 *Dictionary of National Biography*, Volume 10, Oxford University Press, 1917, pp. 450-1; *Debrett's Peerage, Baronetage, Knightage and Companionage*, 1963, pp. 470-1; Sir B. Burke, *History of the Landed Gentry of Great Britain and Ireland*, sixth edition, Volume 1, London, Harrison, 1879, p. S50.

30 William Ingram wrote the article 'On the Display of the King Bird-of-Paradise', *The Ibis*, April 1907, pp.225-9.

31 Collingwood Ingram, 'On the birds of the Alexandra District of North Territory of South Australia', *The Ibis*, July 1907, pp.387-415; 'Supplementary list of the birds of the Alexandra District, North Territory of South Australia', *The Ibis*, October 1909, pp.613-8.

doves, pigeons, dotterels, curlews, ducks, cormorants, eagles, owls, corellas, woodswallows, wrens, finches and cockatoos. Collingwood named one bird *Mirafra rufescens* which was a sub-species of the horsfield bushlark[32] and another, a sub-species of the singing bushlark, he named *Ptilotis forresti* after the Forrest brothers, William and John.[33] Stalker also collected a small mouse, commonly known as Forrest's Territory Mouse, which was named *Leggadina forresti* after the Forrests.[34]

The Ingram family has retained a real interest in NAP over the years. Several family members have served on the company's board since incorporation in 1931 — Sir William's son Sir Herbert Ingram, Sir Herbert's two sons Sir Herbert and Michael, and Michael's son Andrew. They have been represented at board meetings in Brisbane by Edward Robert Crouch, his son Michael and the present chairman of directors, Christopher Lyndon.

The fifth member of the partnership, John Henry Boyer Warner, was born in 1849 and met William Ingram while both were studying law at Trinity College, Cambridge. For a short time Warner was private secretary to the Governor of New South Wales[35] and his interest in NAP probably arose through a variety of factors: his period in Australia, a genuine interest in cattle (his home being in Leicestershire, a grazing district of England), and his association with the Ingram family. A cousin of the wife of William Ingram, Archibald Stirling, wrote *The Never Never Land*, published in 1884, in which he thanked his friend J.H.B.Warner for sketches in the book.[36] During research for the book, Stirling stayed at the Queensland Club in Brisbane and renewed his friendship with William Forrest.

Warner's entry into NAP, however, was not without incident, although the early misgivings were later overcome. In 1878 he tried to go back on his promise to join the partnership and McIlwraith, Collins and Forrest threatened legal action. In a letter to McIlwraith, Collins complained 'he [Warner] treated us shamefully . . . , his past conduct proves that he is not to be trusted'.[37] Forrest also commented, 'It is not an unmixed pleasure having him in'.[38] In Warner's defence it should be mentioned that in 1878 he was only 29, the youngest of the partners, and at that age to find sufficient capital to risk would have been somewhat difficult. Following his death in 1891, John Warner's shares passed to his brother Captain William Pochin Warner who was captain of Queen Alexandra's own eighteenth Hussars.[39] The Warner family involvement in NAP continued through the descendants of Edward Handley Warner, another brother of John and William, until 1987.

Communication between England and Australia was facilitated by overseas trips made by the partners. John Warner visited Queensland at least once and in 1899 Robert Collins met Captain Warner and Sir William Ingram in London.[40] From early 1895 until his death in 1900, McIlwraith lived in Scotland and was visited by other partners.

Overall, the activities of the Queensland partners complemented each other to maintain 'a careful balance of political influence and administration lubrication'.[41] They also protected each other's interests. In 1883, in offering Forrest a seat in the Legislative Council, McIlwraith was assured of Forrest's support in lobbying Melbourne capitalists to support his proposed transcontinental railway. Forrest and his brother John also lobbied for McIlwraith's re-election for Mulgrave in 1883. For NAP, McIlwraith used his influence to make sure that critical areas of NAP's Burdekin sugar holdings were freeholded.[42] William Forrest provided detailed instructions for McIlwraith to present to Patrick Perkins, McIlwraith's Minister for Lands, in order that the land could be secured before the Griffith Administration assumed power. This lobbying was not difficult since at that time Perkins was a member of the NAP partnership bringing the number of partners to six. He had joined on 24 November 1882 following discussions about the possible purchase of an adjoining station

32 It is also now known as *Mirafra javanica rufescens*. Bulletin British Ornithologists Club, volume xvi, 10 July 1906, p.116 in Collingwood Ingram, 'On the birds of the Alexandra district of North Territory of South Australia', *The Ibis*, July 1907, p.414.

33 It is also known as *Meliphaga virescens forresti*. C.Ingram, 'On the birds of the Alexandra district of the North Territory of South Australia', pp.412-3.

34 Ellis Troughton, *Furred Animals of Australia*, second edition, Sydney, Angus and Robertson, 1943, p.294.

35 Warner Family Papers, held privately by Mr.C.F.Warner, Halesworth, Suffolk, England.

36 A.W.Stirling, *The Never Never Land*, London, Sampson Low, Marston, Searle and Rivington, 1884, p. vii.

37 W. Collins, Inverness, to T. McIlwraith, 8 October 1878, Sheet 2, OM64.19, JOL.

38 W. Forrest, Melbourne, to T. McIlwraith, 25 December 1879, OM64.19, JOL.

39 Warner Family Papers.

40 The members of the Collins family who travelled overseas included Robert, his wife Belle, daughter Annie and Robert's brother John George Collins. Letters from their trip are found in the Fraser Family Papers, Mundoolun. Queensland.

41 Waterson, 'Pastoral capitalism . . .', p. 407.

42 T. McIlwraith to W. Forrest, 14 July 1883, OM64.19, JOL; *ADB*, Volume 8, pp. 551-2; Waterson, 'Pastoral capitalism . . .', p. 404.

Patrick Perkins: a partner 1882-1899. John Oxley Library

in the Burdekin area, Woodhouse.[43] The purchase fell through, but Perkins had other business interests in common with McIlwraith and joined the NAP partnership anyway. His share of the partnership was one-fifth, the same as McIlwraith, Collins and Forrest. Ingram's and Warner's shares decreased to one-tenth each.

Patrick Perkins was born at County Tipperary, Ireland on 10 October 1838. He migrated to Victoria in 1855 and after pursuing brewing interests at Castlemaine, he opened a brewery at Toowoomba in 1869. Operations extended to Brisbane with Castlemaine XXXX ale. Settling in Brisbane in 1876, he speculated in real estate, mining, brewing and hotel businesses. He was elected to the Legislative Assembly from 1877 to 1884 and from 1888 to 1893, and was appointed to the Legislative Council in 1893.

His period as Minister for Lands from 1879 to 1883 can be criticised for what seems to be an abuse of his ministerial powers. For instance, he would appear to have acted in the interests of the NAP partnership when the company successfully applied for an additional selection of some 290 hectares on the Burdekin in 1883. William Forrest requested McIlwraith to 'ask Mr.P.[Perkins] to accept the applications. The land is of considerable importance to us'.[44] Perkins' friendship with McIlwraith ended in 1888, after he publicly criticised 'McIlwraith, Orangemen and Psalm-singers'. He was troubled by financial difficulties after the 1893 economic crash, having speculated heavily in Mount Morgan shares and hotel properties. By 1898 he was unable to pay a call of £8,000 on his share in NAP and his association with the partnership ended in a deed of dissolution of 21 August 1899 when he was paid out with £1,500.[45] In his last years he was almost impoverished and died at Hawthorn, Victoria on 17 May 1901.[46]

With the departure of Perkins, the partnership reverted to the original five and remained with the descendants of these partners for another four decades. But by 1910 Sir William Ingram was the only surviving member of the original partnership, and the composition of NAP was changing as deceased partners passed their shares on to family members. Following McIlwraith's death in 1900 his shares passed to a trust operated by his brother Andrew McIlwraith and his wife's brother-in-law, Arthur Palmer. As Collins and Forrest were then the only resident and active partners, they received £150 each per year for managing the affairs of the partnership.[47]

Following the death of William Collins in 1909, there was a redistribution of his shares among the Collins family. At the same time, the trustees of the McIlwraith estate decided to dispose of the shares held by the estate, receiving £38,000 in the transaction.[48] Mrs Blanche Alexander, the youngest daughter of Thomas McIlwraith and a beneficiary under her father's will, decided to retain her one-sixteenth share of NAP by re-acquiring her 54 shares from the McIlwraith estate, so the McIlwraith involvement with the company was not completely severed. William Tyler Forrest, son of John Forrest, was also regarded as a partner from this time, the 54 shares acquired from the McIlwraith Estate going to him instead of to John Forrest under a family arrangement. At the same time, John intimated that at a later stage he would transfer some of the 216 shares standing in his name to his son, Philip.

---

43 W. Collins, Inkerman, to T. McIlwraith, 31 May 1882, OM64.19, JOL.

44 W. Forrest, Inkerman, to T. McIlwraith, 26 August 1883, OM 64.19, JOL.

45 *ADB*, Volume 5, 1974, pp. 431-2; Deed of Dissolution, 21 August 1899, NAPR (NAP Records held in The North Australian Pastoral Company Pty. Limited Offices, Brisbane).

46 *ADB*, Volume 5, 1974, p. 431.

47 Deed of Partnership of NAP, 7 October 1901, NAPR.

48 Receipt and Acknowledgment, Messrs McIlwraith and Palmer to NAP, 6 July 1910, NAPR.

By 1911, as a result of these various transactions, the arrangement of shares was:[49]

	Holding before acquisition from McIlwraith's Trustees	Acquired from McIlwraith's Trustees	Acquired from Administratrix late Wm. Collins	New holding
Mrs. Alexander (Blanche Margaret)	–	54	–	54
Captain Warner (William Pochin)	108	27	–	135
Sir Wm. Ingram (William James)	108	27	–	135
John Forrest	216	–	–	216
W.T. Forrest	–	54	–	54
R.M. Collins	72	24	24	120
J.G. Collins	72	24	24	120
Mrs.Wm.Collins (Mary Adelaide Gwendoline)	–	6	24	30
			TOTAL	864

The overall effect was that the interests of the Collins and Forrest groups had increased from 25 per cent each after Perkins' withdrawal, to around 31 per cent each. The Ingram and Warner interests had also increased, but were still only half those of the others. The McIlwraith connection was retained through Mrs Alexander, but at a much reduced level of some 6 per cent. Over the next two decades this general configuration would be altered with a decrease in Forrest shares to 19 per cent. The next major wave of changes, however, occurred in the 1930s.

49 Letter by R.M. Collins: Memorandum re The North Australian Pastoral Company, Tamrookum, 14 March 1911, NAPR.

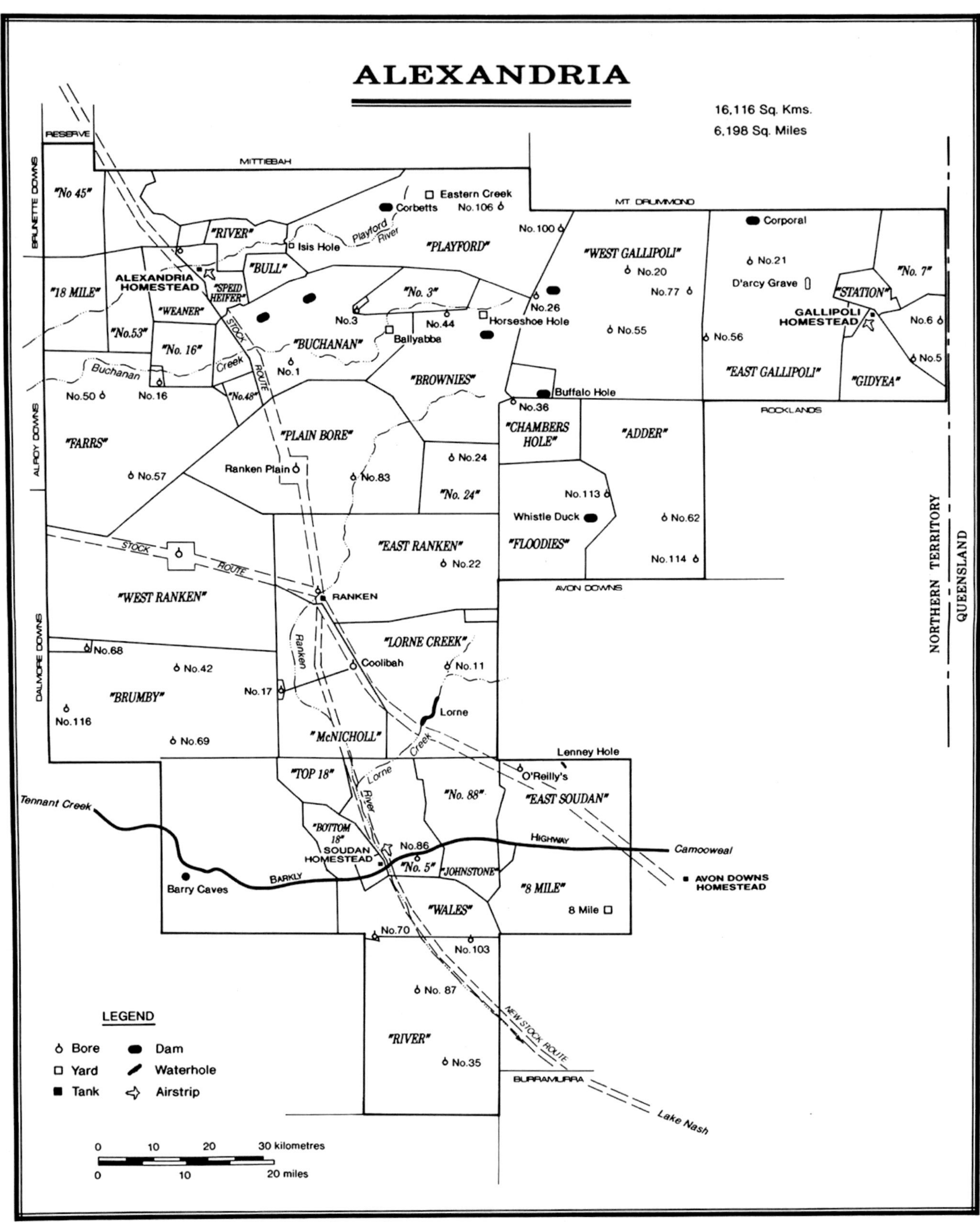

Alexandria, 1991

Chapter 2

Alexandria: Taking Up the Leases

The Territory is a hard mistress and it is surprising how many people who have come up determined to conquer, have departed without having made the grade.[1]

Alexandria is without doubt the most historically significant of the NAP properties. The leases were first taken up in 1877 by the original NAP partners and have remained with the company ever since, a rare situation in the Australian cattle industry. Being the company's largest property and its most significant for breeding, Alexandria tends to overshadow the contribution of the remaining properties.

It is a magnificent property and can rightly claim to be the 'jewel in NAP's crown'. In a good season the vast plains of Flinders and Mitchell grass resemble an extensive wheat field and the herd can number more than 60,000 head. But the Barkly Tableland can expect a drought approximately once every ten years[2] and poor seasons produce the inevitable water and feed problems. Devoid of permanent water, except for the Ranken and Playford rivers and Buchanan Creek, the property depends on its 120 sub-artesian bores to service the huge area.

The climate is monsoonal with well defined wet and dry seasons. Nearly all the rainfall — an annual average on Alexandria of 400 millimetres (16 inches) — is received during the hot summer months, November to April, with the heaviest falls during January and February. The seasonal nature of the rainfall imposes physical constraints on cattle work such as mustering, branding, disease control and turnoff which must be done in the dry.

Alexandria has always been the company's major breeding property with store cattle being sold off the station to others until 1934 when Marion Downs, a fattening property in the Channel country of south-west Queensland, was purchased. The breeding/fattening policy is based on the fact that the Cooper, Diamantina and Georgina river systems of the Channel country flood out over fertile, alluvial flats which allow deep water penetration and produce an abundance of pasture which includes herbage and legumes rich in protein. Water penetration on the Barkly Tableland is not as deep, and pasture growth, mainly of Flinders and Mitchell grasses, while of high quality when young, is not comparable with the pasture of the flooded Channel country. Also, the flooded Channel country is less deficient in the essential minerals for growth, calcium and phosphorus. The limestone soils of the Barkly, on the other hand, while high in calcium, are low in phosphorus. This complicates the phosphorus problem in animal production.

Because of Alexandria's enormous size, Gallipoli and Soudan are essential out-stations. Before resumptions in 1965 the property was 28,085 square kilometres in area or nearly the size of Belgium. It was reputedly the largest cattle station in Australia if not the world,[3] although this was not strictly true. Until 1915, Victoria River

1 Rowand H.Murray, manager, Alexandria, to Douglas M.Fraser, managing director and general manager of NAP, 24 April 1953, R.H.Murray File, 1951-55, NAPR.

2 R.M. Holt and J.D. Bertram, *The Barkly Tableland Beef Industry 1980*, N.T. Department of Primary Production, Technical Bulletin, No. 41, June 1981, p.9.

3 *North Queensland Register*, 3 March 1979; E.Hill, *The Territory*, Sydney, Angus and Robertson, 1951, p.172.

Downs Station in the Northern Territory was the biggest and after that, the Kidman family's Anna Creek Station in South Australia was and still is the biggest in Australia at 30,005 square kilometres.[4] Alexandria, however, has a greater carrying capacity — 60,000 head compared with Anna Creek's 25,000. Even after resumption Alexandria is huge. Its area of 16,116 square kilometres is bigger than Northern Ireland and it is the largest station in the Northern Territory. The external boundary extends for approximately 730 kilometres which is about the distance from Melbourne to Adelaide. From the most northerly to the southerly boundaries it is 177 kilometres and the corresponding maximum distance between eastern and western boundaries is 164 kilometres.

This vast area was not inhabited by Europeans until NAP took up the Alexandria leases. The first European to explore the broader region was William Landsborough in 1861, who named the Barkly Tableland after the Governor of Victoria, Sir Henry Barkly. Leading one of four separate expeditions to discover the fate of Burke and Wills, Landsborough followed the Albert River to a tributary, the Gregory, which he named after the Queensland Surveyor General and explorer, Augustus Charles Gregory.[5] The Plains of Promise (as Stokes in 1841 had called Queensland's Gulf country) were crossed to the head of the Herbert River which Landsborough named after the first Colonial Secretary of Queensland, Robert G.W.Herbert. Following confusion with another Herbert River in Queensland, also named after the Colonial Secretary, the Northern Territory Herbert was changed to the Georgina after Georgina Kennedy, the daughter of Sir Arthur Kennedy who was Governor of Queensland from 1877 to 1883.[6]

Landsborough traversed the Mitchell grass plains to within a few miles south of the present town of Camooweal, but turned back because of the threatening attitude of the Aborigines.[7] Although he was unsuccessful with the primary object of his travels, he demonstrated more than any other explorer the existence of excellent pastoral country. His very favourable reports of land 'destined to rank as a first-class sheep country' gave impetus to settlement.[8]

The first settlers moved on to the Barkly in the mid-1860s and by 1865 George Sutherland was at Rocklands.[9] In the same year John Logan Campbell Ranken and his cousin Lorne abandoned Afton Downs in Queensland for want of surface water and took their 4,000 sheep west to establish Avon Downs, declaring it stocked in 1867. The Ranken River[10] and Lorne Creek, both of which are important features of Alexandria, are named after these men. The James River is presumably named after James Brown, who, in 1866, stocked the Alexander holdings in the vicinity of the James and Herbert rivers with his brother Alexander.[11] Other early settlers in the area included Thomas Nash and the Steiglitz brothers.[12]

At this time, however, wages and transport costs were high, station hands unavailable, and wool prices low. The schooners which carried the wool leaked and destroyed the cargo. These adversities and Australia's financial crisis of the late 1860s forced the pastoral pioneers to abandon their runs and for the next ten years the Barkly was deserted. The Australian economy improved with the gold rushes of the 1870s, and with the building of the overland telegraph line from Darwin to Adelaide in 1872, pastoralists were again hungry for land. Burketown was reborn and became a significant

4 *The Guinness Book of Records, 1992*, England, Guinness Publishing, 1991, p.176.

5 L.Leichhardt (1845) and A.C.Gregory (1855-56) had traversed the Gulf region.

6 Marion Downs on the Georgina was purchased by NAP in 1934.

7 G. Bourne, *Journal of Landsborough's Expedition from Carpentaria in search of Burke and Wills*, Melbourne, Dwight, 1962; C. Wagstaff, *Avon Downs NT 1882-1982*, Armidale, Australian Agricultural Company, 1982, p.4.

8 W. Landsborough quoted in Professor Tate's Report on the Northern Territory, 27 June 1882, in SAPP, Volume 3, 63 of 1882, p.6.

9 G. Sutherland, *Pioneering Days: In the early Sixties*, Brisbane, Wendt, 1913, p.9.

10 Mistakes have occurred with the spelling. RANKIN was often used; by 1878 on Frank Scarr's Map (Mortlock Library, Adelaide) and 1885 on the Goyder Map, it was called RANKINE by South Australian surveyors such as David Lindsay and Charles Winnecke who apparently thought it was named after the Rankine family of Strathalbyn in South Australia. The name RANKEN was gazetted 21 May 1947. Reports on Tablelands, Northern Territory, *SAPP*, No.66 of 1898-99, Volume 2, p.1; Records held by the Place Names Committee, Department of Lands and Housing, Darwin.

11 Letter, James and Alexander Brown, Newcastle, to Resident Commissioner, Northern Territory of South Australia, 28 January 1868, Place Names Committee, Department of Lands and Housing, Darwin.

12 G.Sutherland, *Pioneering Days: In the early sixties*, pp.9-22; Lilian Ada Miller, *The Border and Beyond: Camooweal 1884-1984*, Toowoomba, Harrison Printing, 1984, p.7; C. Wagstaff, *Avon Downs NT 1882-1982*, Tamworth, Australian Agricultural Company, 1982, p.4; V.O'Brien, personal papers and his article 'The Nation Builders; Pioneer Pastoral Attempts in the Territory', V.Dixon, (ed.) *Looking Back: The Northern Territory in 1888*, Casuarina, Historical Society of N.T., 1988, pp.60-5.

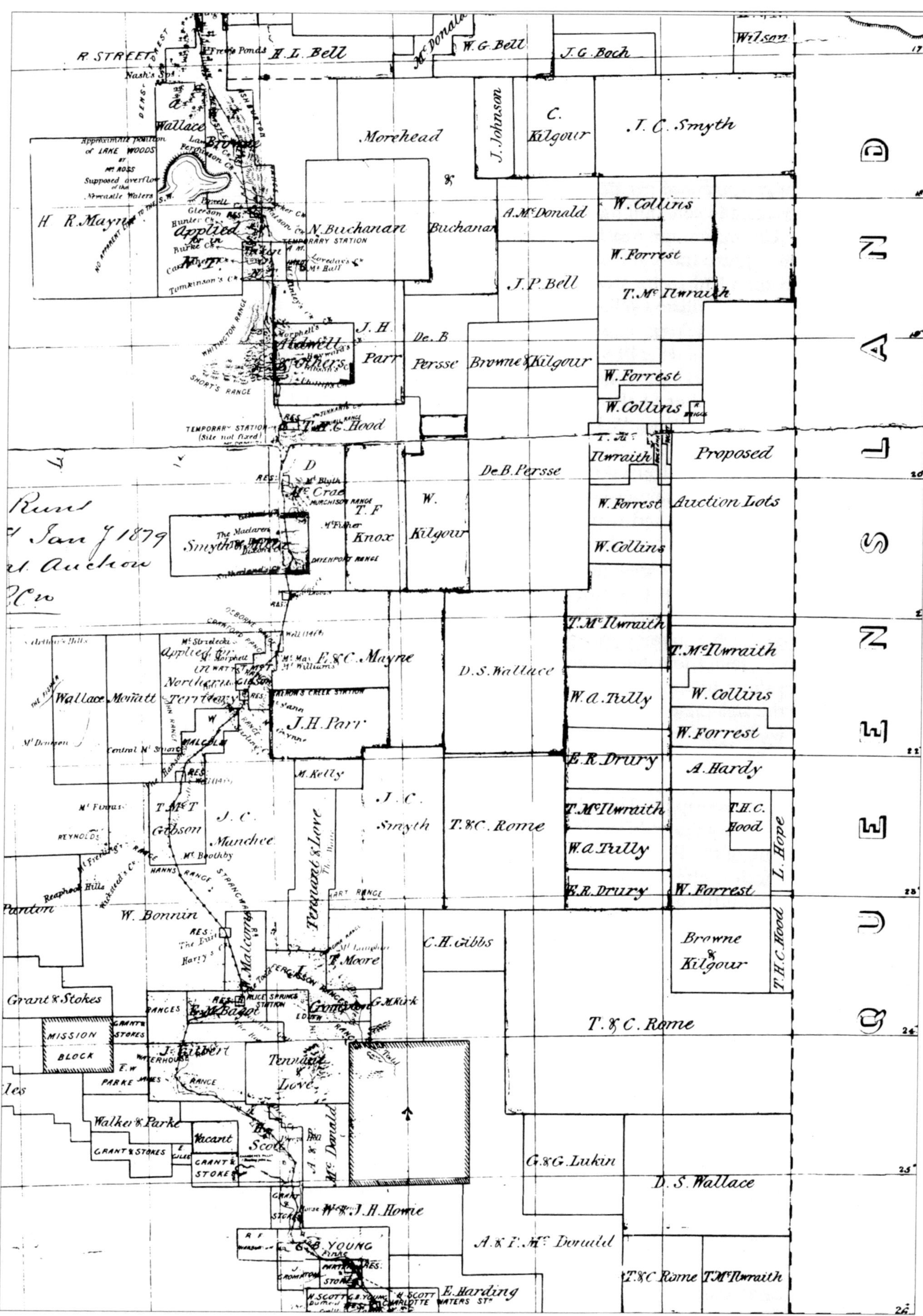

Map A: Pastoral Leases 1878.

(Dept. of Lands and Housing, Darwin)

port for the Barkly as leases were taken up from the mid-1870s and settlement followed.

The NAP partners were keen to be part of this land boom and took up as much as they could even without seeing the country. The Northern Territory Land Act of 1872 stated that twenty-five year pastoral leases could be obtained without competition; rent was sixpence per square mile for the first seven years and ten shillings per square mile thereafter. The maximum area for one lease was to be 400 square miles (1,040 square kilometres) and runs had to be stocked within three years.[13] Following William Collins' enquiry in February 1877 about the taking up of land in the Territory, leases totalling 18,240 square miles (47,424 square kilometres) were applied for in February, June and October of that year in the separate names of W.Collins, W.Forrest and T.McIlwraith[14] (see Map A).

In early 1878, very soon after taking up their leases, the NAP partners and other lessees in the area employed the surveyor and explorer Frank Scarr to make a feature survey of the country. He left Marion Downs Station in Queensland with a well-equipped party, and passing through Glenormiston Station he traversed the Barkly to the telegraph line. Such a journey was hazardous; early in the trip he found four horses and clothes belonging to the Prout brothers who had perished while exploring the area the previous year.[15] Scarr's survey map shows his journey as he attempted to cross due west only to encounter desert. His route basically followed the Herbert and Ranken rivers, with descriptions on his map of Alexandria country as 'perfectly open downs' and 'very open downs . . . blue bush', and of the Playford and Buchanan as being 'grassy, well sheltered'.[16] He crossed the Playford in the vicinity of today's site of the Alexandria homestead. Scarr's journal is a desperate tale of the constant search for water in inhospitable country, but despite these difficulties, McIlwraith was still somewhat dissatisfied that the survey was not more detailed.[17]

The surveyor Charles Winnecke explored the Tableland area between 1877 and 1881. He named the Playford River after the South Australian politician Thomas Playford who had engaged him for the work.[18] Ernest Favenc was another early explorer who reported on Northern Territory/Queensland lands in 1878-1883. He was trusted by both the South Australian Government and the Aborigines, which gave him a unique opportunity to examine new country. In 1878-79 he headed a transcontinental expedition to investigate the possibility of an overland railway line being built on the land grant system between Brisbane and Port Darwin. Sponsored by the *Queenslander* newspaper, he travelled across the Tableland passing in the vicinity of the Herbert and Ranken rivers and Lorne Creek. He set up a base camp on the Buchanan and while camped at the junction of the Herbert and the Ranken he met Frank Scarr who was returning from Tennant Creek.[19]

While the NAP partners would have been interested to read Favenc's reports, they had already employed Scarr to traverse the area and were safe in the knowledge that they held the land over which these explorers travelled. Not that it was all good country; one block was held in the southernmost corner of the Northern Territory, South Australia and Queensland borders, country which was found to be desert.

There is a belief that the famous Nat Buchanan was employed by NAP to report on the country.[20] Certainly in 1881 William Collins had arranged for Frank Hann to inspect and report on the N.T. country for the sum of £200 [21] but it is not clear whether Buchanan had been involved. But Nat Buchanan does come into the NAP story indirectly, in relation to the rush to take up leases.

In 1877, with Sam Croker, Buchanan was the first to cross the Barkly from east to west. He was keen to acquire country for himself, but unfortunately when he arrived at Powell Creek and telegraphed Adelaide to take up country on the Buchanan River, he found he was too late. Others, including the NAP partners, had taken up the area on the map in Adelaide, sight unseen.[22] While Nat Buchanan was a well-known and respected cattle man, taking up leases on the Victoria River in 1880

13 Northern Territory Pastoral Regulations, *SAPP*, Volume 2, 29A of 1881, p.1.

14 Docket 219, GRS 1, PROSA.

15 F.Scarr, Journal of Expedition to the Northern Territory, 1878, privately held by Bill Kitson, Museum of Mapping and Surveying, Brisbane; Ernest Favenc, *The History of Australian Exploration from 1788 to 1888*, Sydney, Turner and Henderson, 1888, pp.273-4.

16 Map of Frank Scarr's journey, 1878, Mortlock Library, Adelaide.

17 Agreement between NAP partners and Frank Scarr, 5 April 1878, Sheet 23, OM64.19, JOL; Gordon Buchanan, *Packhorse and Waterhole*, Sydney, Angus and Robertson, 1933, pp.32-3.

18 Charles Winnecke, Reports on Tablelands, Northern Territory, SAPP, Volume 2, 66 of 1898-99, p.2.

19 *Queenslander*, 26 April 1879, p.531.

20 P.Forrest to W. Alexander, 24 August 1952, NAPR.

21 W.Collins, Mundoolun, to T.McIlwraith, 23 June 1881, OM64.19, Sheet 12, JOL.

22 Ernestine Hill, *The Territory*, Sydney, Angus and Robertson, 1951, p. 167; Gordon Buchanan, *Packhorse and Waterhole*, pp.36-7.

and establishing Wave Hill Station in 1883, it shows the advantage of the city businessman over the outback man in his ability to move quickly to take up leases as soon as they became available.

It is interesting that in the same rush De Burgh Persse applied for 7,400 square miles (19,240 square kilometres) in October 1877 and 1,950 square miles (5,070 square kilometres) in the next month, in an area which later formed part of Brunette Downs Station.[23] Although Lake De Burgh is a reminder that he once held the country, he subsequently forfeited the leases, focusing instead on his Queensland properties including Connemara which was purchased by NAP many decades later. His son De Burgh married Fanny, daughter of Robert Collins, and his grandson Dudley (son of Charles) married Margaret Anne Janette Collins, daughter of William Collins.[24] Their descendants are included among the present NAP shareholders.[25]

As a result of the early exploration and surveys in the area, various wide-ranging ideas about the potential of the Tableland were floated. The Mayor of Adelaide in 1895, for example, saw the Tableland as fit for lavish cultivation:

> Two crops of wheat could be grown in a year, also maize, yams, rice, sugar-cane, pomeloes (sic), dates, mangoes, pineapples, cocoanuts (sic), citrons, bananas, melons . . . .[26]

The Tableland, with its rich black or red loam soil mixed with clay, had excellent grasses — Mitchell and Flinders, and edible bushes: bluebush, roley-poley, munyeroo, acacias, whitewood, Landsborough pines, gidyea and currant bushes. Surveyor David Lindsay's description in 1898 was typical:

> Much of it, especially the country on the Herbert, James, Rankine, Playford, and Brunette creeks and the head of the Creswell Creek, can scarcely be surpassed . . . It is equal to some of the finest country on the Cooper, in Queensland.[27]

Though this was a fine basis for pastoral enterprise, the porous soil and flat treeless plains made water collection difficult.

It was thought by many that the Tableland was destined to be sheep country; as late as 1901 Government Resident Parsons claimed that, with sufficient water, 10 million sheep could be supported.[28] Sheep, however, were stocked only on Austral Downs and Avon Downs. The main factors which prohibited more widespread stocking were that fencing was costly, transport facilities for marketing wool were difficult, cattle prices in the early 1900s were good and the promised railway through the Barkly did not eventuate.[29]

Ernest Favenc first recognised that while the area to the south of the 19th parallel,[30] which divided Alexandria, was excellent for sheep, country to the north in the Nicholson River area was good cattle country.[31] In the case of NAP, despite the Collins family's interest in sheep, the intention from the beginning was to stock Alexandria with cattle from the Burdekin properties.

The naming of the company illustrates the scale of enterprise which the original NAP partners envisaged. The partnership combined pastoral interests in Queensland and the Northern Territory of South Australia (the latter renamed the Northern Territory in 1911 after the Commonwealth Government took control), hence the NORTH AUSTRALIAN PASTORAL COMPANY. Two similar company names have appeared in the Territory's history but neither company survived. The title, NORTH AUSTRALIAN COMPANY, was proposed by geologist J.S. Wilson of the Gregory Expedition in 1855-56 with the aim of establishing northern ports to facilitate English commerce; and the NORTH AUSTRALIAN TERRITORY COMPANY was established with English capital in London in 1887 in order to develop the extensive leasehold and freehold properties held by C.B. Fisher.[32]

Until it was declared stocked in 1884, there was no urgency to give Alexandria a name. The partners referred to it as their 'Northern Territory land' or 'Tableland country', and when the first

23 Pastoral Lease applications, 10 October 1877, Docket 2403; 15 November 1877, Docket A2475, GRS 10, PROSA.

24 De Burgh Persse, in Fox, *The History of Queensland: Its People and Industries*, Volume 1, 1919, pp. 168-72.

25 Mary Connolly, Desmond Persse, Margaret Hockey, Christopher Persse, Jonathon Persse, Michael Collins Persse; and the Estates of Margaret Anne Janette Persse and Dorothea Gwendoline Martin Scott.

26 C. Tucker, 17 January 1895, GRS 9, Volume 4, PROSA.

27 David Lindsay, Reports on Tablelands, Northern Territory, *SAPP*, Volume 2, 66 of 1898, p.2.

28 J.L.Parsons, *Proceedings Royal Geographical Society of Australasia*, S.A.Branch, Volume 5, 1901-02, Appendix 1, p.11.

29 C.Wagstaff, *Avon Downs NT 1882-1982*, p.12.

30 Since resumption in 1965, the 19th parallel forms part of the northern boundary of Alexandria.

31 E. Favenc, 'Reports on Country in the Northern Territory', *SAPP*, Volume 4, 181 of 1883-84, p. 1.

32 Government Resident's Report, 1887, *SAPP*, 53 of 1888, p.1.

manager, Thomas Harding, established a camp on the Ranken River near the Avon Downs boundary in 1882, it was referred to as 'Top Camp'.[33] Later this appears on early maps as 'Harding's Old Station'.[34] But by early 1885 Harding was sinking wells and erecting buildings and yards at the site of the present homestead on the Playford.[35] This was approximately at the centre of the holding, so the move was presumably made in the name of efficiency (although in the wet, the black soil became a quagmire).[36]

The name 'Alexandra Station' appears for the first time in a government report of 1884[37] but the following year the spelling 'Alexandria' is used.[38] While the earliest mention of Soudan out-station, situated on the Ranken River north-west of Harding's Old Station[39] comes in 1892, it was possibly established as early as 1886 at a time when the herd was increasing and a base was needed in the south for the stock camps. The war in the Sudan in north-east Africa provides the geographic and historical context of the naming of Alexandria and Soudan.[40] Military conflict occurred in 1882 at the Egyptian port of Alexandria, and further fighting occurred in the Sudan to the south in 1885. This suggests that Alexandria station could have been named as early as 1882. NAP's Burdekin properties had names associated with the Crimean War and although the partners were not responsible for the Burdekin names, perhaps they wanted to continue the military theme on Alexandria. The geographic/military association was further enhanced with the purchase of Herbert Vale in 1918; it was renamed Gallipoli.

While 'Alexandria Downs' was commonly used for the first few decades, from the early 1900s the 'Downs' was gradually omitted and 'Alexandria' remained. What is confusing is the use of 'Alexandra Downs' on maps and in documentation as late as the 1920s. While a spelling error seems the simplest explanation, there is another interpretation which is worth some consideration. 'Alexandra Land' was the term John McDouall Stuart gave to central and northern Australia following his exploration north from Adelaide in 1861. He named the area after Alexandra, the Princess of Wales, who married the future King Edward VII in 1863. The South Australian Government decided instead that the area should be the Northern Territory of South Australia.[41]

The NAP partners were not afraid to invest on a grand scale and were anxious to take up as much land as possible. The original idea had been to apply for 52,000 square kilometres (20,000 square miles) in the Alice Springs region but this idea was abandoned, perhaps because greater general interest was being shown in the Barkly region.[42] Here they took up their own 'Alexandra Land', which might explain the name 'Alexandra Downs'. In the absence of written proof from any of the original partners, however, the geographic/military interpretation of the naming of Alexandria seems more plausible.

There was considerable fluctuation in NAP's holdings in the early years after acquisition, and in one sense the initial size mattered relatively little because stocking was not eventually required until 1884. The first catalyst for change was the Land Act of 1880 which lowered rents from ten shillings to two shillings and sixpence per square mile, because lessees were finding the rents too high. As a result, leases were forfeited through nonpayment of rent so that they could be taken up again at the lower rents. During 1879, NAP forfeited several blocks in poorer areas especially in the south, and this forfeiture continued in 1880.[43] The total land held by the partners at January 1880 was 24,627 square kilometres.[44] Six months later, this had decreased to 15,834 square kilometres[45] although

33 T.McIlwraith, Brisbane to Minister of Education, Adelaide, 20 December 1882, Docket 734, GRS 1, PROSA.

34 Maps dated 1907, Sheet 2; 1910, Sheet 2; 1915, Sheet 2; Department of Lands and Housing, Darwin.

35 J.S.Little to Government Resident, 1 April 1885, Docket A7811, GRS 10, PROSA.

36 With the resumption in 1965, the homestead is no longer central.

37 Government Resident's Quarterly report on the Northern Territory, 1 July 1884, *SAPP*, 53A of 1884, p.1.

38 Letter from station managers including Thomas Harding of 'Alexandria, Playford River' to the Government Resident, 10 September 1885, Government Resident's Half-yearly report on the Northern Territory, 31 December 1885, *SAPP*, Volume 3, 53 of 1886, p.4.

39 T.Harding to W.Forrest, 24 August 1892, Docket 481, GRS 1, PROSA.

40 Soudan was the spelling used at the time of the war, hence the naming of Alexandria's out-station as Soudan, not the current spelling, Sudan.

41 K.T.Borrow, 'The Northern Territory of Australia, North Australia and Alexandra Land (1862-1900)', *The Australian Law Journal*, Volume 28, 22 July 1954, p.148; M.S.Webster, *John McDouall Stuart*, Melbourne, Melbourne University Press, 1958, pp.263-4; V.O'Brien, 'The Nation Builders; Pioneer Pastoral Attempts in the Territory', p.60.

42 Philip Forrest to Bill Alexander, 24 August 1952, NAPR.

43 John Collins to William Collins, 19 February 1879, Collins Family Papers, held privately by Joan McKean, Brisbane.

44 Docket 570, January 1880, PROSA.

45 Pastoral land applications in the Northern Territory, *SAPP*, Volume 4, 80 of 1880, p.1.

they still retained land around watercourses and along a corridor to the Gulf.

Forrest showed some disquiet about this reduction in area, which was compounded by an unintentional forfeiture through oversight to pay the rent,[46] and he was eager to replace the leases thus forfeited. When the Herbert River Auction Blocks were offered for sale on 21 January 1882, he was anxious to take up as many as possible. An Act in 1881 provided a twenty-one year lease for land in this region which included the area settled in the 1860s and then abandoned.[47] The legalities of the sale of the Herbert River Blocks were not settled, however, until the Northern Territory/Queensland border was surveyed. Consequently, it was established only in 1886, after the survey was completed by J.Carruthers and his party, that Camooweal was in Queensland, Lake Nash Station in the Northern Territory and that Rocklands Station straddled the border. This clarified the situation of John Costello of Lake Nash, for example, who had been paying rent to both the Queensland and the South Australian Governments.[48] To illustrate further the anomalies that can occur in border regions, in 1921 the station manager of Rocklands, John Campbell, pointed out that rentals charged in Queensland were much higher (thirteen shillings and sixpence per square mile), than the three shillings and sixpence per square mile charged in the Territory; yet the land in the Territory part of Rocklands was better than that in Queensland.[49]

By taking up new leases at the lowered rents and securing 1,200 square miles (3,120 square kilometres) of the auction blocks, by the end of 1882, NAP's holdings in the Northern Territory totalled 44,200 square kilometres.[50] The Alexandria out-station of Soudan was established on one of NAP's blocks (block 115), while other blocks became Rocklands, Avon Downs and Austral Downs. Other successful bidders for the auction blocks included Thomas Guthrie, John Costello, Lawrence Munroe, T.L.Richardson, C.J.Sanders, W.Kilgour and Oscar De Satge[51] (see Map B).

In order to encourage the pastoral industry further, the Northern Territory Crown Lands Act 1890 issued pastoral leases for forty-two years. Annual rent was to be not less than sixpence per square mile for the first seven years, not less than one shilling for the next seven years, and not less than two shillings for the third period of seven years. For the remaining twenty-one years, rent was to be fixed by the Minister.[52]

More forfeitures and taking up of new leases followed. So too did criticism. In 1893 the lessees' practice of gradually allowing the leases of the inferior country to be forfeited while taking up the same or other leases at lower prices was noted in the South Australian Parliament. The example was given of William Forrest's forfeiture of 13,133 square kilometres originally rented at £721 15s, which he then took up for £127 2s.[53] Indeed, in 1895 NAP surrendered seventeen leases comprising approximately 4,420 square kilometres[54] and sixteen more leases were cancelled the following year.[55] As a consequence, when in 1898 Forrest requested a list of all NAP leases, the total area had decreased to 23,457 square kilometres.[56] Many blocks in the Nicholson River region remained forfeited because ticks and redwater disease had infected the area. An Administrator later commented:

> the North Australian Pastoral Company forfeited 10,000 square miles in one day owing to the losses due to redwater.[57]

But NAP's surrender of leases was consistent with what was happening generally in the pastoral industry of the Northern Territory. Whereas in 1886, 702,062 square kilometres were held by pastoral lessees, in 1895 this had decreased by well over half to 254,800 square kilometres.[58] Lowered rentals also meant lowered annual revenue for the government; there was a decrease from £16,000 before the 1890 Land Act to £6,980 in 1895.[59]

46 W.Forrest to T.McIlwraith, 21 May 1880, OM 64.19, Sheet 7, JOL.

47 W.Forrest, *SAPD*, 11 December 1891, p.2463.

48 J.S.Little, Austral Downs, to Government Resident, 1 September 1886, Docket A8351, GRS 10, PROSA.

49 John Thomson Campbell, Camooweal, 21 October 1921, Parliamentary Standing Committee on Public Works, *CPP*, 1921, p.85.

50 *SAPP*, 119 of 1882, pp.13-6.

51 1885 Pastoral Leases Map, Department of Lands and Housing, Darwin.

52 Chief Surveyor, Report by Administrator, 1912, CPP, 1913, p.157.

53 *SAPD*, 14 December 1893, pp.3583-4.

54 Packard to MCNT, January 1895, Docket 354, GRS 1, PROSA.

55 Dockets 264, 375, GRS 1, PROSA.

56 W. Forrest, 5 January 1898, Docket 6, GRS 1, PROSA.

57 Report of Administrator for NT for 1912, *CPP*, 1913, p.116; also quoted in E.Hill, *The Territory*, p.179.

58 Half-yearly Report on the Northern Territory for 1886, *SAPP*, Volume 3, 53 of 1887, p.1; Government Resident's Report on the Northern Territory for 1895, *SAPP*, Volume 2, 45 of 1896, p.1.

59 Government Resident's Report on the Northern Territory for 1895, *SAPP*, Volume 2, 45 of 1896, p.1.

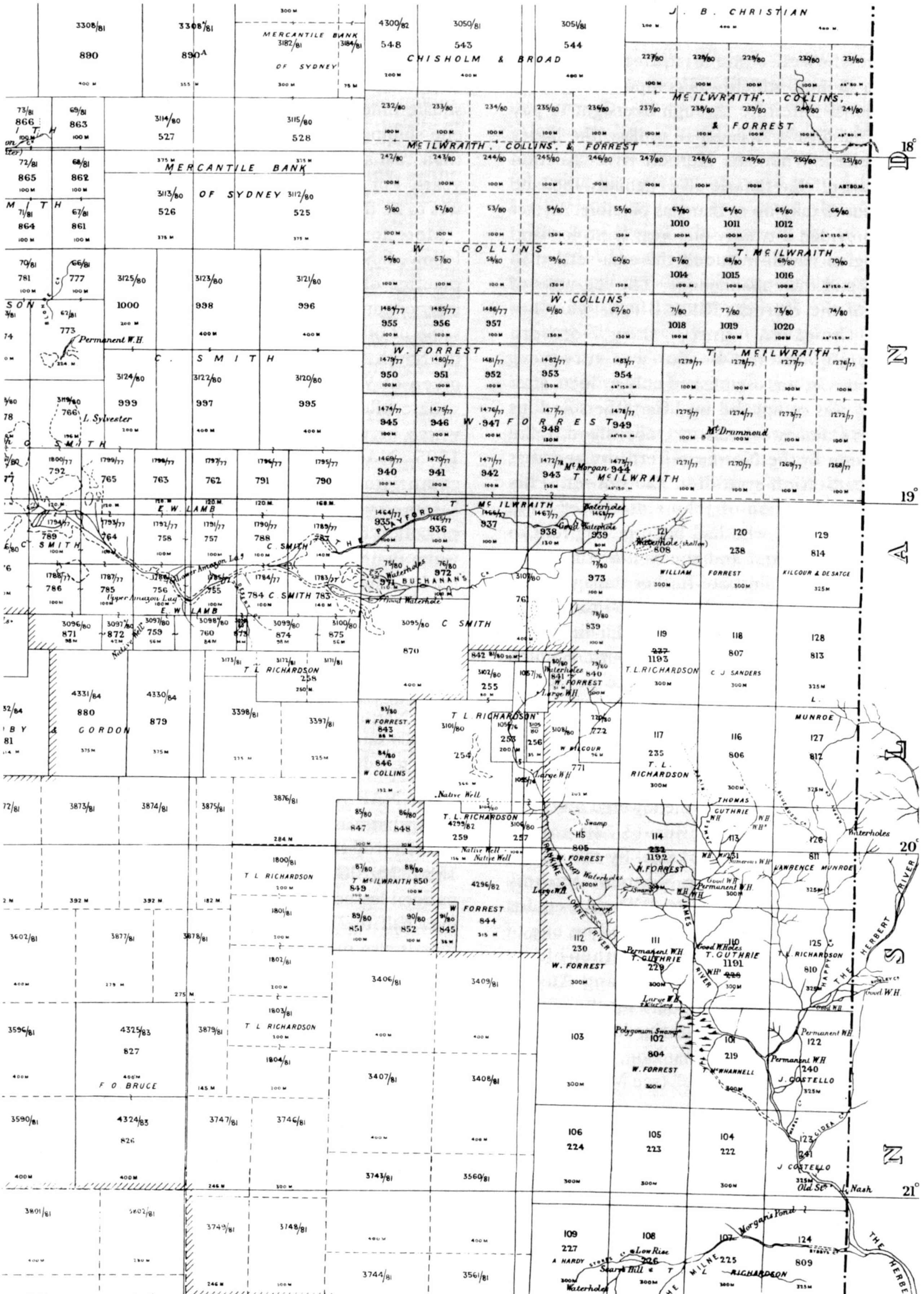

Map B: Pastoral Leases 1885.

Department of Lands and Housing, Darwin

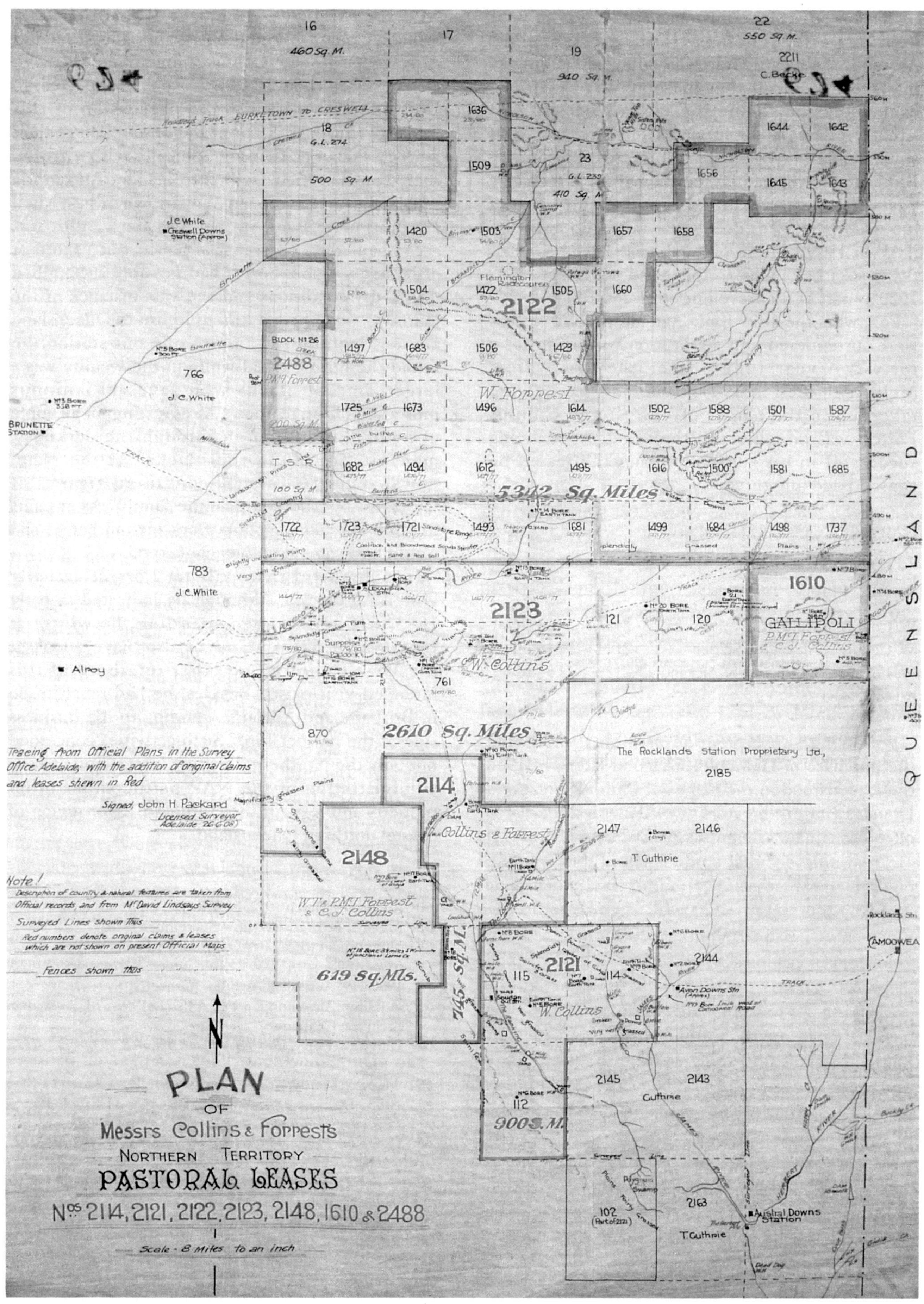
PLAN
OF
Messrs Collins & Forrest's
NORTHERN TERRITORY
PASTORAL LEASES
Nos 2114, 2121, 2122, 2123, 2148, 1610 & 2488
Scale - 8 Miles to an inch
Tracing from Official Plans in the Survey Office, Adelaide, with the addition of original claims and leases shewn in Red
Signed, John H. Packard
Licensed Surveyor
Adelaide 26.6.08
Note!
Description of country & natural features are taken from Official records and from Mr David Lindsays Survey
Surveyed Lines shown Thus
Red numbers denote original claims & leases which are not shown on present Official Maps
Fences shown Thus
N
QUEENSLAND
BURKETOWN TO CRESWELL
460 Sq. M.
500 Sq. M.
940 Sq. M.
550 Sq. M.
410 Sq. M.
G.L. 274
G.L. 239
C. Beckie
J.C. White
Creswell Downs Station (Approx)
Brunette
Brunette Station
BLOCK No 26
2488
P.M.T. Forrest
200 Sq. M.
766
783
870
761
T.S.R.
100 Sq. M.
W. Forrest
2122
Flemington Racecourse
5302 Sq. Miles
2123
W. Collins
2610 Sq. Miles
1610
GALLIPOLI
P.M.T. Forrest & C.J. Collins
Alroy
The Rocklands Station Proprietary Ltd.
2185
2114
Collins & Forrest
2148
W.T. & P.M.T. Forrest & C.J. Collins
619 Sq. Mls.
745 Sq. M.
2121
W. Collins
2147
2146
T. Guthrie
2144
2145
2143
Guthrie
900 S.M.
112
115
114
102 (Part of 2121)
2163
T. Guthrie
Avon Downs Stn (Approx)
Austral Downs Station
Rocklands Stn
CAMOOWEAL
TRACK
1636
1509
23
1644
1642
1656
1645
1643
1420
1503
1657
1658
1504
1472
1505
1660
1497
1683
1506
1423
1725
1673
1496
1614
1502
1588
1501
1587
1682
1494
1612
1495
1616
1500
1581
1685
1722
1723
1721
1493
1681
1499
1684
1498
1737
121
120
Splendidly Grassed Plains
Magnificently Grassed Plains

There were other adjustments made to the total holdings after Thomas Harding established Alexandria. Richard Holt, who followed Harding as manager, reported that in about 1890 Harding exchanged the Connells Lagoon blocks of 400 square miles (1,040 square kilometres) for 300 square miles (780 square kilometres) of country Brunette Downs had on the Buchanan. After a few years, Brunette forfeited three of the Lagoon blocks, which were eventually taken up again by NAP in 1909 to form pastoral lease 2488.[60] One of the blocks belonging to this lease was gazetted in 1920 as part of the travelling stock route.[61]

Following the 1899 Land Act, all the NAP leases were surrendered and application was made for a forty-two year tenure over the same area. McIlwraith died in 1900 so the new leases were taken up in the names of Forrest and Collins. Forrest was prominent in the negotiations, being successful in his suggestion that the leases be consolidated into three.

NAP Leases under 1899 Land Act[62]

Lessee	Situation	Lease	Rental
W. Collins	Herbert River Blocks	2121	3s 6d sq ml
W. Collins	Alexandra Station*	2123	9d sq ml
W. Forrest	North-east of Alexandra	2122	1s 6d sq ml

*Spelling used

Following subsequent adjustments, including the purchase of Gallipoli (discussed later), in 1933 the leases stood in the names of Philip Forrest and Douglas Fraser, directors of NAP, with a total area of 28,085 square kilometres (10,802 square miles).

Alexandria: 1933[63]

	Rental	square miles	square kilometres
PL 2122	9d	5,327	13,850
PL 2488	6s 0d	200	520
PL 189 (1610)	3s 0d	412	1,071
PL 2123	1s 6d	2,610	6,786
PL 2114	6s 0d	742	1,929
PL 2148	7d	610	1,586
PL 2121	3s 6d	901	2,343
	Combined Area	10,802	28,085

Of the original partners, it seems that only William Collins visited the Alexandria leases in the early years. Robert Collins' biographer, Harry C. Perry, states that 'It fell to Robert and George Collins to go out and take possession of this property [Alexandria]'[64] but the description of their journey does not fit geographically with a trip to that area. The first Collins to inspect Alexandria thus seems to have been William when he visited Harding in 1883.[65] On that occasion Collins was accompanied by a sergeant of police, but a shortage of horses — some horses had possibly succumbed to the dry conditions and the vast distance of the journey — compelled him to return to Queensland alone. Planning to camp at one out-station, he found the hut deserted while in the vicinity was a large camp of blacks who gave an 'ominous indication of the reason'.[66] After resting for a couple of hours, he pushed on through the darkness, managing to elude the blacks. At the time, newspapers frequently contained reports of 'outrages' by Aborigines so the family was at pains to keep all such accounts from his mother so she would not become unduly concerned.

It is doubtful that William Forrest actually visited Alexandria. Although he indicated in early 1883 that he intended spending the winter at Inkerman and that he might also 'have a look at the West Country', no other reference of this intended visit exists. Besides, he had much to do in Brisbane and Adelaide, tending to the business side of the partnership.[67] Such activities concerned not only the Northern Territory leases, but also the administration of the NAP partnership's other property interests in the Burdekin River region of coastal northern Queensland.

---

60 R.Holt, 18 March 1922, NAPR.

61 *Northern Territory Gazette*, 4 September 1920.

62 Northern Territory Pastoral Country Stocked, 1902, *SAPP*, 61 of 1902, p.2; NAP Leases under Act of 1899, A1640, 1900/219, Australian Archives, Canberra.

63 Holding: Alexandria, 1933, A452/1, 1952/310, Australian Archives, Canberra.

64 Harry C. Perry, *Pioneering: the life of R.M.Collins MLC*, p.194.

65 Thomas McIlwraith to Minister for Education, 11 May 1883, Docket 280, GRS 1, PROSA; Harry C.Perry, *Pioneering: the life of R.M.Collins MLC*, pp. 199-201.

66 Harry C. Perry, *Pioneering: the life of R.M. Collins MLC*, p.200.

67 W.Forrest to T.McIlwraith, 14 February 1883, OM64.19, Sheet 27, JOL.

Chapter 3

The Burdekin Properties 1878-1910

It is difficult to imagine that the lush green expanse of sugar land of the Burdekin Delta between the Leichhardt Range and the Pacific Ocean in north Queensland was ever used for any other purpose. Yet when it was first taken up in 1861, sheep and cattle grazed on abundant grasses and rested in the shade of ash, gum, bloodwood and acacia trees. By the time NAP bought into the area in early 1878, cattle had replaced sheep, and sugar was becoming profitable. And when William Forrest commented in 1882 'I think our selections are about the best card that we have played. They will soon be very valuable',[1] his predictions were correct. Not only were the cattle to stock Alexandria provided by the Burdekin properties, but the latter were also independently profitable, especially with the expansion of the frozen meat trade. A further boost to profits came in 1910 when the land was sold to the Queensland Government. While no other properties were purchased by NAP as a result of the sale, Alexandria and the partners were assured of financial security.

The potential of the Burdekin had been known since Leichhardt's exploration in 1845 when he described the region in the Upper Burdekin as 'one of the finest we had seen. It was very open, with some plains, slightly undulating or rising into ridges, beautifully grassed with sound ground'.[2] He also predicted how well the country was suited for pastoral pursuits.[3] Expeditions by A.C. and F.T.Gregory in 1856 and George Elphinstone Dalrymple in 1859 further enhanced the prospects so that when the Kennedy District was thrown open for selection in 1861, there was a rush to take up these lands. Bowen was established in the same year; it was the major centre until the early 1870s when Townsville with its better access to mining and pastoral districts became the preferred port.

On the northern bank of the Burdekin River, Jarvisfield was taken up in 1861 by John Graham Macdonald. The following year he took up Inkerman, both runs being transferred to Robert Towns in 1864.[4] Towns, a wealthy Sydney merchant who had extensive holdings in north Queensland, also acquired the neighbouring runs of Springfield, Kilbogie and Tandara which later became part of the consolidated property, Inkerman. He traded with other partners, Alexander Stuart and Charles Cowper, under the name Robert Towns and Company; John Melton Black who was responsible for the early development of Townsville was general manager of the company's northern interests.[5] On the Haughton River, Woodstock, which included the runs Cleveland Plains, Cintra, Wyoming, Magenta and Landsdowne, was taken up

1 W. Forrest, Adelaide, to T. McIlwraith, 12 February 1882, OM 64.19, JOL.

2 Ludwig Leichhardt, *Journal of an Overland Expedition in Australia from Moreton Bay to Port Essington*, Adelaide, Libraries Board of South Australia, 1964, p. 209.

3 *ibid.*, p. 229.

4 Register of Runs under Unoccupied Crown Lands Occupation Act 1860, CLO/N18-19, QSA.

5 Stock Mortgage, R. Towns and Co. and Bank of New South Wales, 12 June 1869, SCT/CD 14, Mortgage 50, Bk 8, QSA; Stevens, John F. 'Townsvale and Townsville: the activities of the Hon. Robert Towns MLC in Queensland 1848-1873', Typescript, North Tamborine, 1966, p. 14, JOL.

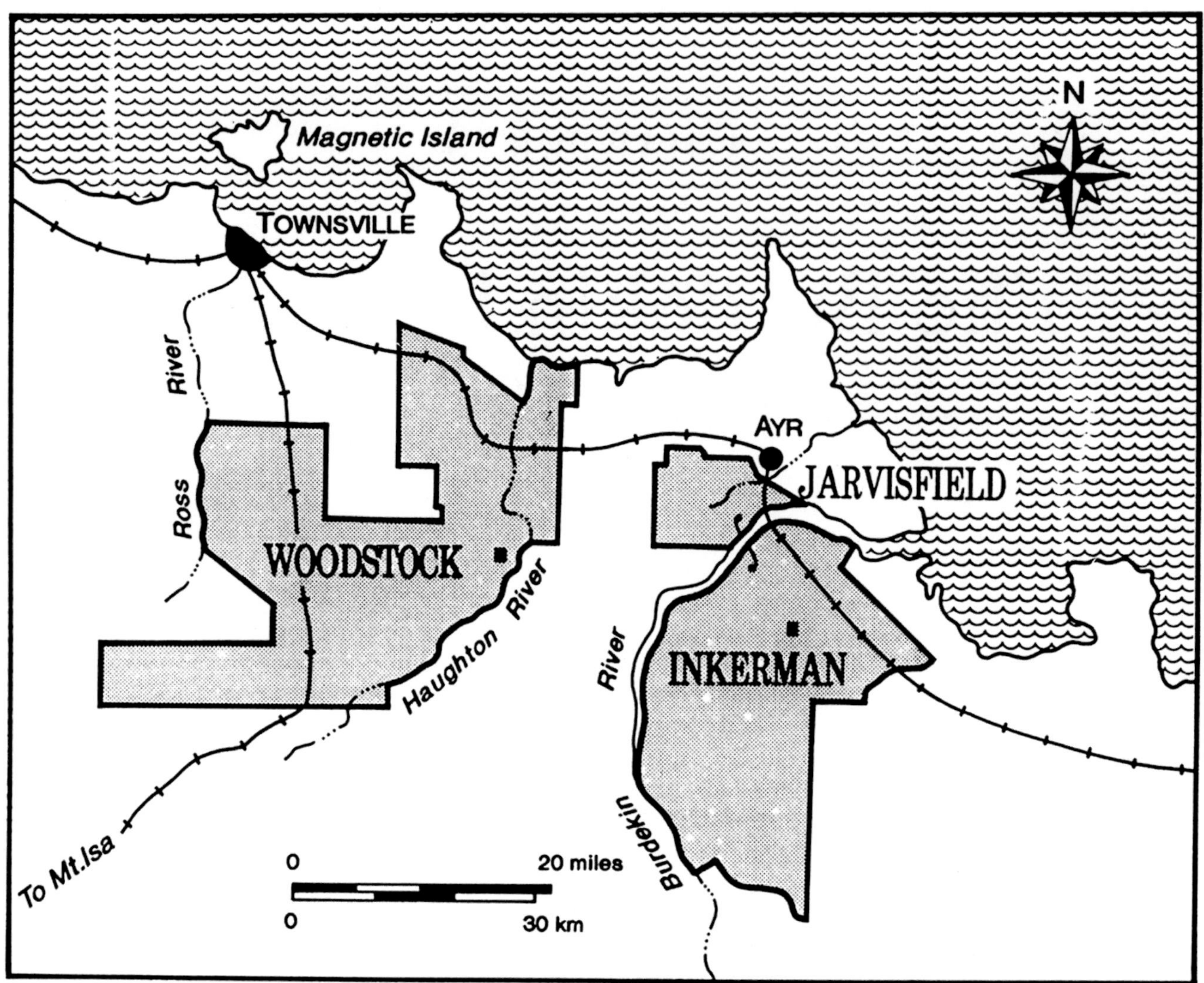

NAP's Burdekin Properties 1878-1910.

by Black in 1863 but transferred to the Towns company the following year.[6]

By the time Towns died in 1873 he had amassed more land. In early 1877, when the executors of his will tried to auction his Burdekin lands, the area had increased to 31 square kilometres (12 square miles) of freehold land, plus 2,314 square kilometres (894 square miles) of the leasehold properties Inkerman, Leichhardt Downs (later part of the Inkerman consolidation), Woodstock and Jarvisfield. No bid was forthcoming because it was felt that the lease of two pounds per square mile was too high.[7] But the partnership of the North Australian Pastoral Company had both the finances to cope with, and the foresight to recognise, the financial gains which could be made by acquiring the land and then selling it for sugar growing. Furthermore, they realised that the properties could provide cattle to stock Alexandria, which in turn was to be used for breeding stock for the Burdekin properties.[8] After negotiation, NAP's purchase of the Burdekin properties owned by Towns was finalised in March 1878. Although the purchase price is unknown, there was a mortgage of £47 18s 6d to be paid over five years.[9] William Collins, the expert in stock matters, took charge of the 16,000 head of Shorthorn cattle and the 350 horses.[10]

Under the 1860 Land Act, runs had been restricted to between 25 and 100 square miles but there was no limit to the number of runs which a person might hold.[11] Robert Towns was therefore

6 Woodstock, CLO/N19, p. 687, QSA.

7 George King (executor of will) to Henry Douglas, Sydney, 29 March 1877, LAN/AF 368, QSA.

8 Waterson, 'Pastoral capitalism . . . ', p. 404.

9 Stock Mortgage between Thomas McIlwraith and Sophia Towns, 5 March 1878, SCT/CD 24, BK 13, QSA.

10 *ibid.*

11 B. Kingston, 'The origins of Queensland's Comprehensive Land Policy', *Queensland Heritage*, Volume 1, No. 2, 1965, pp. 4-8.

able to amass large tracts of land. The Pastoral Leases Act of 1869 extended tenure to twenty-one years and lowered rents, which enhanced the potential of the properties. The first legislation to affect NAP directly was the Crown Lands Act of 1884, which provided for consolidation of adjoining runs, and also for resumptions. The Inkerman consolidation of 738 square kilometres consisted of the runs Inkerman Downs, Leichhardt Downs, Springfield and Kilbogie[12] and the Woodstock consolidation of 881 square kilometres consisted of Woodstock 1, 2 and 3, Cleveland Plains, Cintra, Balaclava, Wyoming, Magenta, Landsdowne, Callandoon 2 and Crimea.[13] Each consolidation was divided into two parts. One continued as leasehold for a term of fifteen years, while the other (up to one-half of the total) was resumed. The lessee could, however, obtain an Occupation Licence, which allowed them to graze stock on the resumed portion.[14] The consolidations were meant to enable more people to settle on the land by gradual resumption, although the Act did not require residency by the owner. So Inkerman and Woodstock continued as before, with resident managers John Townshend and John Carr.

Although NAP was insulated to a degree because of its extensive holdings, it did implement a policy of purchasing more land as it became available, to protect against resumptions by the government. Moreover, the partners saw the value of the land for sugar growing and were waiting for capital gain. In 1884, land leased by Robert Collins and his father John was purchased by the partnership to add to the consolidation; approximately 21 square kilometres from Robert and 8 square kilometres from John.[15] The company continued to purchase land so that by 1890 it was the largest occupier of land on the Lower Burdekin.

In his doctoral thesis on the Burdekin, Peter Griggs explains:

> It [NAP] had acquired just over twenty per cent of the Crown land that had been selected in the region since 1877 and leased almost half of the remaining unresumed portions . . . .[16]

Even when world sugar prices suddenly declined in 1884, NAP kept buying; between 1885 and 1895 it increased its holdings by 99 square kilometres. Indeed in 1900, despite lower profits because of drought and ticks, it spent £42,360 12s on freehold purchases. Balance sheets show further purchases were made in 1903 and 1907 until the amount of freehold land peaked in 1908 at 341 square kilometres.[17] Resumptions by the Crown and conversions to occupation licence accounted for a corresponding decrease of leasehold land.

The land was worth acquiring. Watered by the Haughton, Ross and Reid rivers, Woodstock had the convenience of the railway from Townsville to Charters Towers through its northern region. Francis Heeney, a member of the Land Court, wrote about the Inkerman Consolidation in 1907 describing the excellent well-watered country in that area:

> The country is generally level, gently sloping or undulating, with areas of plains, flats, swampy plains, some flooded country, some lightly and some heavily timbered country. There are large areas of black and brown soils, considerable areas of dense scrub on parts of the frontage to the Burdekin River, some areas of swampy, stony, ridgy, and sandy lands . . . . The country is well grassed . . . The timber consists of ash, gum, bloodwood, acacia, apple. There is a large quantity of timber of commercial use.[18]

Generally on the runs there were chains of lagoons containing good water, and where there was no surface water it could be obtained from wells at a depth of 9 metres. Some of the NAP land was in the sub-artesian area although irrigation was not commenced until after the company had sold.

Individual pastoralists, unlike the companies, were generally unhappy with the 1884 Consolidation Land Act. Robert Collins complained that 'one class of pastoralists was dispossessed of great areas of country to make room for another class that made precisely the same use of the land'.[19] He advocated greater security of tenure by providing longer leases, so that settlers would feel more inclined to invest capital and time in the development of their holdings.[20] The concern of the Collins brothers continued, to the extent that William Collins was a member of a deputation of

12 Inkerman Downs Consolidation, LAN AF 368, QSA.

13 Woodstock Consolidation, LAN AF 891, QSA.

14 Charles A. Bernays, Queensland politics during sixty years: 1859-1919, Brisbane, Government Printer, n.d., pp. 321-2.

15 Declarations of Trust, R.M. Collins (31 May 1884) and J.G. Collins (5 May 1884) in favour of North Australian Pastoral Company, NAPR.

16 Peter D. Griggs, Plantation to small farm: A historical geography of the lower Burdekin Sugar Industry, 1880-1930, PhD Thesis, University of Queensland, 1990, p. 91.

17 Balance sheets, 1900-1908, NAPR.

18 Francis Heeney, 'Report to the Honourable the Secretary for Public Lands on the Inkerman and Woodstock Freeholds Situated in the Townsville Land Agent's District', reproduced in W.J. Scott 'Report by the Under Secretary for Public Lands under *The Closer Settlement Act of 1906* on the Inkerman Estate', QPP, 2 (1911-1912), p. 740.

19 R.M. Collins, in Perry, *Pioneering . . .*, p. 340.

20 *ibid.*, pp. 341-3.

pastoralists to the Premier in 1899 about this matter.[21]

As well as the business acumen of the partners and the excellent quality of the country, the sound management practices on the stations also contributed to the success of the venture. A reading of the correspondence between William Collins and the Inkerman manager John Townshend suggests that as early as 1878

> sound advice, clear directions and firm control were the hallmarks of this enterprise where technical mastery, patience and solid knowledge had brought some measure of stability and success.[22]

Though isolated from the stations, the partners kept in remarkably close contact with the working of the properties. William and John Forrest and William Collins made frequent visits to the Burdekin. Records show that one of the Warner brothers visited from England at least once[23] and Jane Collins travelled to the Burdekin with her brother John George in 1903.[24]

Strategically placed freehold blocks could be acquired on such visits. In 1883 for instance, William Forrest purchased two selections near Woodstock close to the railway, which provided the only permanent water for the back country, and which gave a clear run into the yards.[25]

The role of the manager was a vital one, made difficult through the owners' absence and the distance of the markets, while the tropical climate was trying in a way that many southerners did not appreciate. The partnership's 'inherited' managers, John Townshend[26] at Inkerman and John Carr at Woodstock and Jarvisfield, had to deal immediately in 1878 with the Land Commissioner at Bowen in taking up the leases, completing the leasing arrangements and making payments. The partners had confidence in Carr but not in Townshend; William Forrest commented in 1879

Jane Collins. John Oxley Library

> Townsend is no good, Collins and I agree that he must as soon as convenient be got rid of . . . Carr spends his time in making money Townsend in making excuses.[27]

Townshend remained, however, and following his death in 1887, James Henry Rae was appointed manager of Inkerman. On the other hand, the partners helped Carr when he was keen to secure a selection of his own and 'knock out a living for [his] wife and family'.[28] William Forrest was at Woodstock in 1883 when, instead of drafting a letter for Carr to copy, Forrest wrote directly to McIlwraith requesting that he see if Carr's selection had been confirmed.

The vagaries of location and climate in north Queensland cannot be denied. The isolation was highlighted by a traveller in 1882.

> It is quite the exception to meet men who have travelled into the interior from either Townsville or Bowen . . . . A change of outer garments, a couple of flannel shirts, a revolver, a blanket, a saddle, and a toothbrush — that great emblem of civilisation — together with a few yellow-covered novels in the way of literature, made up nearly the whole of our respective outfits, and quite enough we found it, in a land where the dandy is unknown, and a man need

21 *Reports of two Deputations of Pastoralists in Brisbane*, Brisbane, British Australasian Society, 1900.

22 Waterson, 'Pastoral capitalism . . . ', p. 405.

23 The records do not indicate which brother; merely a reference to 'Warner'.

24 Jane Collins, Inkerman, to Bertha Fraser, October 1903, Fraser Family Papers, Mundoolun.

25 W. Forrest to T. McIlwraith, 8 September 1883, OM 64.19, JOL.

26 Spelling given as Townshend in *QVP*, 2 (1884), 1127 and in letters such as by William Collins to John Collins, 14 July 1879, Collins Family Papers privately held by Joan McKean, Brisbane; also Jane Collins to Bertha Fraser, October 1903, Fraser Family Papers, Mundoolun; also seen in letters as Townsend although Townshend is most probably the correct spelling.

27 W. Forrest to T. McIlwraith, 25 December 1879, Sheet 4, OM 64.19, JOL.

28 J. Carr, Woodstock, to T. McIlwraith, 28 July 1881, Sheet 13, OM 64.19, JOL.

> not fear being cut by an acquaintance because of a crease in his coat, or the want of a pair of gloves.[29]

Yet life at the homesteads on the Burdekin had all the expected comforts. In the late 1870s, the main link with Bowen was the weekly coach which left on Tuesdays, travelling via Salisbury Plains, Inkerman, Leichhardt Downs, Woodhouse and Hillsborough, to arrive at Ravenswood on Thursdays.[30] Each of these places had permanent water, so essential for the horses as well as for the bullocks. From Inkerman, another track went via Salisbury Plains, Hamilton's Crossing and Woodstock to Townsville, a distance of 208 kilometres, although this was impassable when the Burdekin and Haughton flooded. Often at such times, both NAP properties assisted stranded travellers.[31]

A dwelling was erected on Inkerman soon after the leases were taken up; Joseph Hann remarked in his diary in 1863 'Inkerman is improved very much, a nice home, yards and a paddock'.[32] A homestead was built in 1870 on Leichhardt Downs when it was devoted to sheep, and Inkerman was merely an out-station running cattle. Inkerman had become the more important property by the time NAP took control so the homestead was moved there to an area on Plum Tree Creek where it was 'most lavishly endowed with natural water received from large lagoons, springs, and waterways'.[33] Wells equipped with windmills supplemented this supply.

In 1907 the homestead was a nine-roomed dwelling with kitchen attached. There were also three cottages, an office, store and a meat house. A blacksmith's shop and stables were located near a horse yard, beside which was erected a small iron-roofed shed with four small yards attached. These stockyards were used for drafting, branding and spaying. Milking yards were also required for the station's herd of dairy cows.[34] Valued at £929, the homestead at Inkerman was one of the largest in the Burdekin although this was not surprising considering its prominence as the northern headquarters for the company.

Situated on Major Creek near the junction with the Haughton River, the Woodstock homestead and outbuildings were not as extensive, valued in 1907 at only £278 15s. There were a house, kitchen and outbuildings, two huts, storeroom, men's huts and stock and milking yards. Improvements on both properties also included fencing, dips and huts on out-stations.[35] In 1913, after NAP had sold, the Woodstock railway stationmaster offered to

> give £5 for the iron 2 roomed humpy where the blacks used to camp and £10 for the 3 roomed place used as a store and dray shed etc and £30 for the main buildings.[36]

The government refused this offer, saying it should have been higher.

Stores and supplies were purchased at six monthly intervals, which created a constant problem for food storage. To keep ants away, the legs of food safes were painted with creosote or stood in tins of kerosene. To check mice, flour sacks were stacked upright and placed apart on a stand with tin tacked around the legs. To combat weevils, a layer of ti-tree bark was placed underneath the bags, which were then covered with ti-tree bark. Sugar in 34 kilo (75 lb) bags were also kept on a stand.[37] Scurvy and 'Belyando rot' were not uncommon because of a diet lacking fruit and vegetables. The new settlers were unaccustomed to the available tropical fruits and a myth circulated that such exotic fruits were the cause of disease.[38]

Station life encouraged some formalities. Historian Joan Neal describes homestead life on the Burdekin:

> The pastoralist, his wife and family would usually "dress" for dinner: it was said that this practice set the standard for the organization and efficiency of the rest of the station. Afternoon tea was also a formal occasion, not always restricted to women. Many stations built up extensive libraries, subscribed to southern newspapers, and of course, the *North Queensland Register*. Evenings were spent reading, singing around the piano, and most importantly working on accounts. Evening entertainment was held by carbide or kerosene lights with their inevitable associated odours and the accompanying flying insects.[39]

Inkerman and Woodstock would almost certainly have had copies of Ingram's *Illustrated London News*. He would have been impressed had he visited Mrs

29 A.W. Stirling, *The Never Never Land*, London, Sampson Low, Marston, Searle and Rivington, 1884, pp. 61-2.

30 *Port Denison Times*, 15 December 1877.

31 *Port Denison Times*, 23 August 1879.

32 'Joseph Hann and Family, Settlement in North Queensland 1861-1871', Typescript, 15 November 1863, p. 50.

33 M. Fox (compiler), *The History of Queensland: its People and Industries*, Volume 2, 1921, p. 681.

34 Report and Valuation by Crown Lands Ranger Byers on Inkerman and Woodstock Freeholds, 24 January 1907, *QPP*, 2 (1911-1912), pp. 747-8.

35 *ibid.*

36 B.J. Walsh, Railway Station Master, Woodstock, 1 April 1913, LAN AE 674, QSA.

37 Joan Carmichael Neal, *Beyond the Burdekin*, Charters Towers, Mimosa, 1984, pp. 15-6.

38 R. Cilento 'Medicine in Queensland', *Journal Royal Historical Society Queensland*, Volume 6, 1961-62, p. 873.

39 *ibid.*, p. 17.

Fulford, of Lyndhurst Station in north Queensland, for she had 'lined the walls with calico tents that were not good enough for camping out and papered [them] with pictures out of the "Illustrated London News" '.[40]

Because of the extent of the holdings, NAP and its managers had a high profile in the community. James Rae, the manager at Inkerman, took a prominent part in the Ayr Divisional Board, later known as the Ayr Shire Council.[41] He was elected to the board in 1889, a year after its formation, and held the position of chairman from 1896 to 1911. He was also chairman of directors of the Ayr Co-operative Dairy Company Limited between 1901 and 1903 and in 1907; and was the first president of the Lower Burdekin [Ayr] Show Society in 1904, a position he held until 1910.[42] The partnership subscribed to the Townsville and Ayr cyclone fund in 1903; it donated funds to the Townsville and Ayr shows;[43] and it supported the local dairy industry of the Burdekin area with the purchase of shares in the Ayr Co-operative Dairy Association in 1903.[44] It was placed on the list of annual subscribers to the Church of England diocese of north Queensland after Robert Collins, in London in 1878, visited the newly appointed Bishop, the Right Rev. Dr. Stanton.[45]

By the time NAP took up the Burdekin properties, earlier problems between Aborigines and Europeans had dissipated. In 1867 there was a report that Aborigines had been spearing cattle at Woodstock.[46] Two years later, however, the manager of Woodstock reported that this had ceased after the blacks had been 'allowed in' on the property.[47] The success of this venture encouraged other stations to do the same especially when it was found that the use of Aboriginal labour reduced costs.[48]

Initial experiments with sheep on the Burdekin in the 1860s had proved unsuccessful because of footrot, fluke, lungworm, spear grass and attacks by dingoes. Exacerbated by a slump in wool prices, only a few properties persisted with sheep after 1870. There was, however, a remarkable growth in the cattle industry in the 1870s so that by 1880 cattle numbers in north Queensland were estimated at 450,000, a threefold increase from 1874. Despite the droughts of 1884-85 and 1888-89 the herds grew to a peak of 1,390,899 in 1894 before the next disastrous drought at the turn of the century.[49]

In keeping with the general increase, the Inkerman and Woodstock herds grew from 16,000 in 1877 to 25,000 in 1882.[50] Stocking Alexandria during 1883 caused a drop to 14,580 and apart from a low of 12,844 in 1903 (because of drought), the numbers remained at around 16,000.[51] Horses were bred on the stations in significant numbers, which on Woodstock and Inkerman increased from 316 in 1881 to 640 in 1901. They were an essential part of the working of the station, being used for droving, mustering and transport. Horses bred on Inkerman became prominent in northern racing circles after manager Rae introduced the original bloodstock.

Stock was the focal point of the partnership's enterprise, and prizes gained at the Townsville and Ayr shows were testimony to its quality. The Shorthorn cattle on the properties had been bred from good quality stock imported from England during the ownership by Towns. There were newspaper reports during the 1860s which referred to the 'splendid imported cattle . . . of pure Durham breed . . . bred by the Earl of Anglesford'.[52] The NAP partners were just as conscious of good breeding stock. William Collins suggested buying bulls in 1878 when he visited Scotland and Yorkshire. William Forrest wired from Sydney in 1882: 'Inspected today 40 good Chisholm bulls 8 to 10 months think can buy about 12 pounds freight 3 pounds reply promptly leaving tomorrow'.[53] The requirement of the 1884 Act to fence the leases improved breeding methods and resulted in an improved quality of stock. In the absence of land surveys, Admiralty charts had been consulted for subdivision, a practice which sometimes led to quirkish consequences. In 1889, for example, when NAP decided to fence the southern and eastern boundaries of two of the Woodstock runs, it was found that there was insufficient country for all of

40 Mrs. John Fulford, in G.C. Bolton, *A Thousand Miles Away*, Brisbane, Jacaranda, 1963, p. 105.

41 Townsville Bulletin, 11 September 1948, p. 3; Rae resigned from the Ayr Shire Council in 1911, as stated in a letter from James Rae to R.M.Collins, 5 May 1911, Collins Family Papers, privately held by Joan McKean, Brisbane.

42 *Australian Sugar Journal*, 11/10, 1910, p.359.

43 Balance sheets, 1903-1910, NAPR.

44 Balance sheet, 1903, NAPR.

45 Perry, *Pioneering . . .*, p. 245.

46 *Port Denison Times*, 13 July 1867.

47 D. May, *From Bush to Station*, p. 66.

48 D. May, 'The North Queensland Beef Cattle Industry: an historical overview', *Lectures on North Queensland History*, No. 4, James Cook University, History Department, 1984, p. 123.

49 Bolton, *A Thousand Miles Away*, p. 99.

50 W. Collins, Inkerman, to T. McIlwraith, 31 May 1882, OM 64.19, JOL.

51 Balance Sheets 1901-1911, NAPR.

52 Cleveland Bay Express, 13 February 1869; Shorthorn cattle originated from Durham County, England.

53 W. Forrest, Sydney, to T. McIlwraith, 25 April 1882, Sheet 21, OM 64.19, JOL.

Yarding Cattle at Inkerman. A.W. Stirling, *The Never Never Land*, 1884

Woodstock 3 run which at sea level was actually in the mangroves.[54]

One of the major problems of the early cattle breeders was the absence of markets. Some local demand for fresh meat was created in the late 1860s by the discovery of gold in various parts of the north such as Ravenswood (1869) and Charters Towers (1872) and a real boost to the north Queensland cattle industry came with the opening of the Palmer goldfield in 1873. This demand continued into the 1880s, although by 1879 the population of the Palmer was declining and local cattle prices were falling.

Further respite came in the 1880s with an increased demand for store cattle for the Northern Territory, and a price rise on southern markets as some southern pastoralists switched from cattle to sheep. There was also a demand for store cattle for some of the other properties owned by the Collins family.[55] Until 1886, however, north Queensland pastoralists found it more profitable to send their cattle to the boiling-down works in Townsville or to the scattered mining settlements. The Townsville plant produced fertilizer from meat, neatsfoot oil from hoofs and horns, tallow for candles and soap, as well as hides. Of the 1,782 cattle sold in 1881 from NAP's Burdekin properties the carcasses of 1,698 were boiled down.[56]

To provide further markets, in 1882 NAP went into a butchering business with local pastoralists White, Allingham and Philp.[57] The drought of 1883-1886, however, restricted access to markets, causing further oversupply to the local market. After 1886, north Queensland cattlemen began to participate in southern markets, especially Melbourne and Adelaide, until 1892 when a poll tax of five shillings per head on cattle entering New South Wales made interstate trade uneconomic.[58]

The northern pastoralists could see the potential for an enormous export market — if only the beef could be preserved and transported. The NAP partners had great faith in the meat-producing potential of the State and, as a logical extension of the company's philosophy, looked to develop an integrated meat export trade, from breeding to fattening and processing. While meat could be preserved in cans, as meat extract and as essence and tallow, profits and the future lay in freezing.

Although Thomas Sutcliffe Mort made an unsuccessful frozen meat shipment to Britain in 1876,[59] the Collins brothers, William and Robert, were convinced they could succeed after an overseas trip in 1878 where they saw the Bell-Coleman compressed air cooling process. Robert described the system in detail to his father and McIlwraith in letters from England.[60]

Until then their pastoral interests had taken them chiefly on trips to western Queensland to seek and stock new properties, but Robert commented that such expeditions were hindering 'more rational, if not more profitable, employment'.[61] They were eager to learn all they could and while in England they visited Thomas McIlwraith's brother Andrew who later arranged with his shipping firm, McIlwraith, McEachern and Company to fit out the 'Strathleven' with Bell-Coleman machinery. Forty tons of beef and mutton were frozen on board, with the boat leaving Sydney on 29 November 1879 and

54 W. Forrest, Brisbane, to Minister for Lands, Brisbane, 29 September 1889, LAN AF 891, QSA.

55 William Collins to John Collins, 14 July 1879, Collins Family Papers privately held by Joan McKean, Brisbane.

56 Woodstock and Inkerman Returns, 30 June 1881, Sheet 15, OM 64.19, JOL.

57 W. Collins, Westgrove, to T. McIlwraith, 2 March 1882, OM 64.19, JOL.

58 May, *From Bush to Station*, pp. 28-30.

59 Perry, *Pioneering . . .*, p. 67.

60 *ibid.*, p. 236.

61 R.M. Collins in Perry, *Pioneering . . .*, p. 202.

arriving in London on 2 February 1880. The meat which had cost up to 2d per pound in Australia sold in Britain for up to 5.5d for beef and 6d per pound for mutton. A carcass of lamb was sent to Queen Victoria and a sheep to the Prince of Wales.[62] Although the ship left via southern ports, Queenslanders were the moving spirits behind the venture, with William Collins actually accompanying the ship to Melbourne.[63] Commercial success did not follow immediately, however, and the 1880s was a decade of trial and error.[64]

William Forrest, McIlwraith and the Collins brothers were principals in one of the first Queensland companies to export frozen meat, the Queensland Freezing and Food Export Company, registered in 1880.[65] It established works on the Brisbane River and on 20 May 1884 despatched Queensland's first shipment of frozen meat to England on the British East India Company's steamer 'Dorunda'.[66] North Queensland's first meat freezing export company was located at Poole Island just off Bowen, but in January 1884 the works suffered extensive cyclone damage while the first shipment of frozen meat was being loaded. The site was then abandoned.[67]

The NAP partners were particularly keen to develop the frozen meat industry and in 1890 another firm, the Queensland Meat Export and Agency Company Ltd., was formed with Thomas McIlwraith as chairman of provisional directors[68] until William Forrest became chairman. Robert Collins was a director and by 1905 William Collins and John Forrest had also become directors.[69] The company established a freezing and preserving works at Pinkenba and in 1892 works were also opened at Ross River, Townsville. Another company, the Bowen Meat Export and Agency Company, followed in 1895 with NAP holding shares in both.[70]

62 ibid, pp. 238-9.

63 William Forrest, Melbourne, to T. McIlwraith, 25 December 1879, Sheet 4, OM64.19, JOL.

64 J.C.A. Camm, 'The Queensland frozen beef industry 1890-1914', *Australian Geographer*, Volume 16, No.1, May 1984, p. 29.

65 William Forrest to T. McIlwraith, 21 May 1880, Sheet 7, OM64.19, JOL.

66 Noela Corfield, The Development of the Cattle Industry in Queensland, 1840-1890, B.A. (Hons) Thesis, University of Queensland, 1959, p. 112; *Brisbane Courier*, 21 May 1884, p.6.

67 *Queenslander*, 2 February 1884, p. 226.

68 T.McIlwraith, 4 July 1890, Collins Family Papers, held privately by Joan McKean, Brisbane.

69 *Pastoral Review*, 14 (1904-5), p. 986.

70 J.T.Critchell and J.Raymond, *A History of the Frozen Meat Trade*, London, Constable, 1912, pp.47-9.

By 1893, NAP had entered into a contract with the Queensland Meat Export and Agency Company Ltd., to send stock to the Townsville works.[71] Between 1901 and 1904, however, most of the cattle sold went to the Bowen Freezing Meatworks, presumably for better prices, but also for Inkerman, there was no river to cross and cattle could be walked the sixteen kilometres to the railhead at Bobawaba.[72] Thus in 1901, of the 1,744 sold, 1,183 were treated at Bowen, 454 were sold locally and the remaining 107 were killed for rations.[73] By the turn of the century, the frozen meat export trade was flourishing, with 120,000 head being purchased annually by northern meatworks.[74]

The partners saw a dramatic increase in exports as one way out of the economic crash of 1893 which was the result of massive speculation in mining and property in the late 1880s. Commodity prices dropped, interest rates accelerated and debt repayments could not be met. In 1893, many banks, including the Queensland National Bank with which McIlwraith was closely involved, suspended payment. Prominent Queensland professional men such as politicians Sir Arthur Palmer, Robert Philp and B.D. Morehead, Judge Harding, and entrepreneurs such as Perkins and McIlwraith especially, had their fortunes decimated. The survival of NAP was a tribute to the partners and their management practices on the land and financially because they endeavoured to keep partnership matters separate from their other financial affairs.

To pursue export markets, Forrest, as chairman of the Queensland Meat Export Company and by then a member of parliament, made a trip to London in 1893.[75] In September that year the Meat and Dairy Encouragement Act was passed to encourage the establishment of meatworks for the export trade, and William Collins became a member of the Meat and Dairy Board empowered to enforce the Act. For a period of five years a levy was placed on all sheep and cattle depastured in Queensland — fifteen shillings per hundred cattle and one shilling and sixpence per hundred sheep. Capital advances of up to 50 per cent were made from these funds for the erection of meatworks and dairy establishments for which no repayments were due for five years. This co-operative effort was very successful, with forty-

71 W. Forrest to Under Secretary for Lands, Brisbane, November 1893, LAN AF 891, QSA.

72 A railway was built between Bowen and Bobawaba in 1891.

73 Balance Sheets, 1901 and 1905, NAPR.

74 Neal, *Beyond the Burdekin*, p. 63.

75 *Pastoral Review*, 4 (1894-95), p. 345.

seven meatworks listed in Queensland by 1899, twenty-three of these registered as exporters.[76] With the growth of the export trade, the percentage of the colony's cattle herd slaughtered each year by meatworks rose progressively from an annual turnoff of less than 1 per cent in 1890 to 6 per cent a decade later. With the impact of tick disease especially in north Queensland, this saw cattle numbers decrease by 40 per cent to 4,000,000 in 1900. The 1898-1903 drought then had a devastating effect on numbers, and by its end cattle numbers had fallen to 2,500,000. The meat processing industry was likewise affected to the extent that by 1906 only eleven meatworks were killing.[77]

The cattle disease pleuro-pneumonia was a persistent problem faced by pastoralists last century. Pleuro or 'lung disease' was initially introduced into Australia in 1858 by a shipment of five cattle from England.[78] It reached north Queensland in 1864 and in the wake of remedies such as carbolic acid, inoculation was developed by the end of the nineteenth century. The disease was usually more widespread in travelling stock because of stress, and outbreaks continued, especially in incomplete musters.[79] Tuberculosis, often mistaken for pleuro-pneumonia, was also a problem in the 1880s.[80] Cattle duffing, and poisonous plants such as rubber vine, noogoora burr, lantana, and heart leaf poison bush, were other hazards which the pastoralist had to face.

But the north Queensland cattle industry suffered its greatest setback in 1896 with the arrival of the cattle tick (*Boophilus microplus*), which by 1891 had crossed into Queensland from the Northern Territory. It is most probable that the tick was introduced into Australia at Darwin (then called Palmerston) in August 1872 in a cargo of Indonesian cattle. Although intended for slaughter, some of the cattle escaped and mixed with station cattle at Adelaide River.[81] On the appearance of the tick in north Queensland, cattlemen tried to hide the resultant redwater disease, fearing lengthy quarantine would cut them off from markets. But the spread was relentless and in 1897 the Land Commissioner at Townsville advised the Under Secretary of Lands not to open any fresh country. 'At the present time no one desires country, but are waiting to see if the Ticks will die out'.[82]

They didn't die out and 'the stench of dead cattle hung heavy over the stock routes and watercourses of North Queensland'.[83] In 1897 the NAP properties were so affected that Inkerman lost 65 per cent of its cattle.[84] In 1900 William Forrest wrote that of the 26,325 cattle on Inkerman and Woodstock in 1896, a square tail muster[85] the following year showed that only 6,814 remained, a loss of 25 per cent.[86] But the NAP partners were not alone in their loss.

With the financial resources of its partners, NAP was more fortunate than most. The properties were able to be restocked with an additional 15,000 cattle between 1897 and 1899[87] although further losses followed with the 1898-1903 drought. The dry conditions then halted the spread of the ticks, but the movement of stock to the coast on agistment helped to keep the menace in the area.[88] Losses on Woodstock and Inkerman occurred until 1904 and although the cattle developed an immunity to redwater, the ticks remained, with dipping an added expense for the northern pastoralist. The sale of hides was also adversely affected by the tick plague, and ended another source of income for the pastoralist. But NAP's strong financial base again helped to protect it and stock numbers were boosted in 1903 and 1904 with the purchase of 1,000 and 891 head for £3,150 and £2,807 respectively, from Fanning Downs, a station in the area.[89]

William Collins and Dr. J. Sidney Hunt visited the United States in 1896 as Commissioners of the Queensland Government to make 'expert inquiry' into ticks, their subsequent comprehensive report providing additional knowledge on the subject.[90] The following February two young steers from Inkerman which had recovered from an acute attack of fever, were transported to Brisbane to be used in the first inoculation experiment in Queensland which was carried out successfully at the Indooroopilly Experiment Station and Mundoolun, establishing the procedures not only of quarantine

76 Perry, Pioneering . . . , pp. 299-302; Camm, 'The Queensland Frozen Beef Industry 1890-1914', p. 31.
77 Camm, 'The Queensland Frozen Beef Industry 1890-1914', pp. 31-33.
78 *Pastoral Review*, 14 (1904-05), p. 197.
79 May, *From Bush to Station*, p. 30.
80 *QVP*, 4 (1887), p. 267.
81 H.R.Seddon, *Diseases of Domestic Animals in Australia*, Part 3, Canberra, Department of Health, 1951, pp. 10-2.
82 Land Commissioner to Under Secretary of Lands, Townsville, 7 January 1897, LAN/AF 368, QSA.
83 Bolton, *A Thousand Miles Away*, p. 220.
84 W. Collins to Minister for Lands, 29 May 1897, LAN/AF 368, QSA.
85 Later referred to as a 'bangtail muster', the hair on the animal's tail is cut square to show that it has been counted.
86 W. Forrest, Brisbane, to Under Secretary of Lands, Brisbane, 31 May 1900, LAN AF 891, QSA.
87 *ibid.*
88 G.W. Lilley, *Story of Lansdowne*, Melbourne, Lansdowne Pastoral Company, 1973, p. 77.
89 Balance Sheets, 1903-1904, NAPR.

and dipping, but also of preventative inoculation.[91] The Collins involvement in the industry was evident again in 1897 when Robert was chairman of the government select committee on the branding of cattle in which a new method of earmarking and branding was incorporated into the Brands Act of 1898.[92]

As well as encouraging the meat preservation industry, other profitable activities associated with cattle production were explored by the partners. In 1882 Forrest suggested that with the influx of new settlers through immigration, 'teams are going to be tight for a long time and we might make a lot of money out of bullocks if [we] kept a dozen men at work breaking'.[93] Timber was also sold from the properties, and a sawmill was erected by timber merchant Mather on Inkerman by 1883. Forrest was concerned, however, that Mather sold below Townsville prices and that the wages for his men were too high (£2-£3) where 'equally good men might be got for 25/- and 40/- '.[94] They also had to accommodate licensed timber getters on the properties but the relationship was not an easy one. In 1893, Forrest complained of timber getters disturbing cattle on Callandoon No. 2 block which fronted the Haughton River.[95] Minister Robert Philp was not sympathetic and commented: 'If these men are keeping within the law — which they must do — as to cutting and depasturing we cannot interfere'.[96]

Perhaps the most striking aspect of NAP's involvement in the Burdekin was the sale of its land to the Queensland Government for which the partnership received the enormous sum of £130,142 15s 4d. While similar amounts were paid for Darling Downs freehold estates with viable agricultural and dairy farm land,[97] it cannot be denied that NAP arranged its Burdekin properties so that 'no development could take place in the general Burdekin-Townsville area without traversing some of them'.[98] In 1907, even the Land Commissioner for the Townsville District referred to the 'rich lands now locked up in Inkerman, Woodstock and Salisbury Plains'.[99]

The Burdekin's potential as a sugar growing area was being realised and although 20 hectares of sugarcane was planted for NAP by James MacKenzie in 1882,[100] the partners did not go into large scale sugar production. Very early in the partnership William Forrest had suggested selling the Northern Territory land and going into sugar in order to 'concentrate our business and have it all under one general management'.[101] All the other partners did not agree, however, and when farmers sought support from NAP in the early 1890s to build a cooperatively owned sugar mill the partnership offered little assistance.[102] It saw, however, that money was to be made, in selling the Burdekin land, and from the 1890s the partners were intent on doing so at the highest price.

In 1895 NAP offered to sell Jarvisfield Estate, a 23-square kilometre block on the northern side of the Burdekin River, to the Crown, under the conditions of the Agricultural Lands Purchase Act of 1894.[103] The government decided against the purchase because the Act was designed to provide land for settlers with a small amount of capital, and costly irrigation was required on this block.[104] In 1909 the partners eventually subdivided the land themselves and sold the southern section. The historical interpretation of NAP's ownership of Jarvisfield Estate varies, with the author Connolly suggesting that the partners used a nominee to purchase several subdivisions to dominate the approach to the site of the proposed Jarvisfield Central Mill.[105] On the other hand, Griggs writes that

> this claim could not be verified or refuted, but if it was correct the N.A.P.C.'s plans were squashed in early 1910 when the Kidston Government informed the Lower Burdekin supporters of the Jarvisfield Central Mill Scheme that no further central mills would be built in Queensland for the time being.[106]

90 Perry, *Pioneering*, pp. 286-7; *Pastoral Review*, 6 (1896), 446.
91 Queensland Agricultural Journal, No.18, June 1907, pp. 287-8.
92 *ADB*, Volume 3, 1969, p. 443.
93 W. Forrest, Melbourne, to T. McIlwraith, 25 July 1882, OM64.19, JOL.
94 W. Forrest, Inkerman, to T. McIlwraith, 8 September 1883, OM 64.19, JOL.
95 W. Forrest to Under Secretary for Lands, Brisbane, November 1893, LAN AF 891, QSA.
96 Minister R. Philp, 1893, LAN/AF 891, QSA. Robert Philp later became Premier. A son, Colin John Campbell Philp, married Katharine, daughter of Robert Collins in 1906.
97 Waterson, 'Pastoral capitalism . . . ', p. 405.
98 Roy Connolly, *John Drysdale of the Burdekin*, Sydney, Ure Smith, 1964, p. 37.
99 H.M. Trower, Land Commissioner, Townsville, QPP, 2 (1907), p. 557.
100 T. McIlwraith to J. MacKenzie, 2 September 1882, OM 64.19, JOL.
101 W. Forrest, Adelaide, to T. McIlwraith, 15 February 1882, OM 64.19, JOL.
102 Griggs, Plantation to small farm . . . , p. 188.
103 Register of applications under the Agricultural Lands Purchase Act 1894, LAN AF 154, QSA.
104 Griggs, Plantation to small farm . . . , p. 223.
105 Connolly, *John Drysdale and the Burdekin*, p. 122.
106 *North Queensland Herald*, 15 January 1910, p. 12 in Griggs Thesis, p. 225.

Between 1908 and 1910 the partnership began selling off some of its freehold — 5.2 square kilometres in 1908 and 10.4 square kilometres in 1910.[107] The remaining portions of their holdings of Woodstock, Inkerman and Jarvisfield totalled 325 square kilometres and, as the Inkerman Estate, was sold to the Queensland Government in an agreement dated 8 September 1910. The government had originally offered £1 10s per acre but it finally agreed to £1 12s 6d per acre, with £130,142 15s 4d passing to the partnership over the next three years. Timber and improvements were sold for an additional £1,250 18s 1d.[108] Although the purchase price for the original area is not known, the 1878 mortgage of £47 18s 6d suggests that the partners realised a substantial profit on the sale. The stock from Inkerman was sold to A.J.Cotton of Brunette Downs which adjoined Alexandria.[109]

A ranger's report gives a glowing description of the country:

> The land under offer is well suited for sugar-growing and mixed farming and dairying, as, with the exception of about 1,100 acres in the parishes of Inkerman and Leichhardt Downs, the whole of it can be irrigated, as water can be obtained almost anywhere at from 10 ft. to 22 ft. There is a very large quantity of mill timber, containing ash and gum, along the frontage to the Burdekin River. There is also a big percentage of Moreton Bay ash, gum, etc., suitable for railway sleepers, scattered over the different blocks. In the event of the proposed railway from Ayr to Wangaratta being built, it will enhance the value of this country 30 per cent.[110]

The agreement also incorporated an important arrangement for both NAP and John Drysdale, a prominent sugar-grower and owner of the Pioneer Mill. The Kalamia and Pioneer mills served growers on the northern side of the Burdekin but with no bridge across the river, Drysdale wanted to build a sugar mill on the southern side. NAP held the land and protracted negotiations between John Forrest and Drysdale resulted in the latter's proposal to build a sugar mill (Inkerman) on 518 hectares of NAP's freehold land, three kilometres of which fronted the southern bank of the Burdekin River. The partnership offered 202 hectares (500 acres) to Drysdale at no cost, and the remainder at £5 per acre. The agreement also noted that road and railway access would be built.[111] By ensuring that sugar growers had access to a sugar mill on the southern side of the Burdekin River, the land was more valuable to NAP, and Drysdale's desire to expand his sugar interests was also satisfied. While such an outcome could be interpreted as the result of some political influence, it could also be argued that both Drysdale and the NAP partners were merely astute businessmen.

The NAP sale was a catalyst for the next stage of development in the region. At the first subsequent government auction of 78 square kilometres on 8 December 1911, the land sold for £61,665. The remaining area was not offered immediately, but over a period of a few years the blocks were progressively sold. They were all between forty and eighty hectares and were sold either as agricultural farms, perpetual lease selections or unconditional selections.[112] The opening of the land was very successful, and in the first year there was an increase of 500 in the population of the area. Considering that in 1901 the population of Ayr and district was only 1,838, this was a huge influx.[113]

Improvements were effected and while a few farmers went in for mixed crops, the majority grew only sugarcane. The building of John Drysdale's Inkerman sugar mill progressed satisfactorily, although it was not ready for crushing until after 1913.[114] The railway south of Ayr was opened in 1913 and the town of Home Hill on the southern side of the Burdekin River began to develop. The name was originally to be 'Holme Hill' because of an association with the Crimean War, rather in the same way that military associations were evoked in the naming of Inkerman and other stations. Through an error in making the name for the railway station, the 'l' was omitted, resulting in the name 'Home Hill'.[115]

Although the disposal of the Burdekin properties represented the phasing out of a major pastoral operation in tropical coastal Queensland, by now the partners of the North Australian Pastoral Company had their attentions fully occupied with the extensive run they had been managing for more than three decades in the Northern Territory.

---

107 Balance sheets, 1908-1910, NAPR.

108 Agreement between NAP and the Minister for Lands, Digby Denham, 8 September 1910, NAPR; also reproduced in QPP, 2 (1911-1912), pp. 750-2, 755.

109 P.Forrest, Brisbane, to Major Herbert Ingram, England, 25 January 1922, Herbert Ingram File, NAPR.

110 John J. Byers, Crown Lands Ranger, 24 January 1907, reproduced in 'Report by the Under Secretary for Public Lands under *The Closer Settlement Act of 1906*', 10 August 1911, *QPP*, 2 (1911-1912), p. 747.

111 J. Forrest to J. Drysdale, 6 July 1910, Second schedule of Agreement of 8 September 1910 between NAP and Minister for Lands, NAPR.

112 Inkerman Estate, 1911, LAN AF 163, QSA.

113 Townsville Bulletin, 11 September 1948, p. 3.

114 'Report of H. Trower, Land Commissioner for Townsville, Bowen, Charters Towers and Ravenswood districts', *QPP*, 2 (1913), p. 343.

115 John Henry Peake, *A History of the Burdekin*, (second edition), Ayr, Ayr Shire Council, 1971, p.54.

CHAPTER 4

Alexandria: Early Developments

The first step in the development of Alexandria was to establish a base on the property. By December 1882, there was a 'Top Camp on Ranken River' and a proposed camp on the Playford River, where the homestead was later built.[1] Thomas Harding had been appointed manager, having been manager on Connemara in Queensland. So that stocking could be declared, Harding and John Farrar were appointed Justices of the Peace in February 1883 and sworn in by William Collins when he visited the leases a few months later.[2]

In 1884 the Government Resident reported that Alexandria had 3,500 head of mixed cattle, 70 horses and two stallions, 'all stock being in first-class condition'. At the same time, Avon Downs had 4,000 breeding ewes, Corella (Brunette Downs) had 3,000 head of mixed cattle, Rocklands 7,000 and Lake Nash 6,000.[3] Rocklands township, soon to be Camooweal, had two stores, two public houses, a blacksmith's shop, a butcher and about fifty residents.[4] J.S.Little, of Austral Downs, in a report to the Government Resident in Darwin early 1885, noted that Harding was 'building on the Playford River and has fencing and well sinking going on'.[5] At that time Harding was receiving supplies from Burketown but a newspaper correspondent noted he 'intends sending to the Macarthur next time'.[6] By early 1885, 23,400 square kilometres out of Alexandria's then total of 36,400 square kilometres had been stocked. Excellent seasons followed, although stocking took place erratically with Harding putting the stock 'wherever there was grass'.[7] The grassed areas could not be fully utilised, however, because the cattle could not walk the vast distances which separated the waterholes. With so few cattle over such an enormous area, with very little surface water and no fences, the task was daunting.

In 1885 Lake Nash mustered 500 to 600 fats for Adelaide and 500 bullocks left from Rocklands for southern markets.[8] By 1886 Alexandria had 4,500 cattle with a further 1,500 tracking there from Queensland.[9] Conditions were so good in 1886, that 560 fats and store bullocks, with J. Watson in charge, left Alexandria for southern markets.[10] The 1887 season was 'the best known', and with no

1 T. McIlwraith, Brisbane, to Minister of Education, 20 December 1882, Docket 734, GRS 1, PROSA.

2 W. Collins to T. McIlwraith, 4 January 1883, OM64.19, JOL; T. McIlwraith to Minister for Education, 18 May 1883, Docket 330, GRS 1, PROSA.

3 Government Resident's Report, 1884, quoted in Report of Administrator, 1912, *CPP*, 1913, p.115.

4 J.S.Little, Austral Downs, Quarterly report on Northern Territory, 31 March 1885, *SAPP*, Volume 3, 54 of 1885, p.3.

5 J.S. Little to Parsons, Government Resident, 1 April 1885, Docket A7811, GRS 10, PROSA.

6 Correspondent, 30 November 1885, *N.T. Times and Gazette*, 30 January 1886.

7 Sanders and Packard, Adelaide, to Minister for Justice and Education, 16 January 1885, Docket 58, GRS 1, PROSA.

8 J.S. Little, Austral Downs, to Government Resident, 28 April 1885, Docket A8136, GRS 10, PROSA.

9 J.S. Little, Austral Downs, to Government Resident, 1 September 1886, Docket A8351, GRS 10, PROSA.

10 J.S. Little in Half-yearly Report on Northern Territory, 1886, *SAPP*, Volume 3, 53 of 1887, p.2.

disease among stock, 600 fats and stores were sent from Alexandria to New South Wales.[11]

Stock numbers on Barkly Tableland Stations: 1887[12]

	Sheep	Cattle	Horses
Rocklands	–	13,000	300
Austral Downs	12,000	3,500	250
Lake Nash	–	10,000	150
Avon Downs	14,000	1,200	80
Alexandra*	–	8,500	200
Alroy Downs	–	4,000	50
Herbert Vale	–	1,000	40
Brunette Downs	–	12,000	400
	26,000	53,200	1,470

* Spelling used

Expectations of the Northern Territory pastoral industry declined in the 1890s. Local consumption for the Northern Territory market in 1888 was only 1,344 head, all provided by the North Australian Territory Company which owned properties on the Victoria River.[13] The absence of markets for Tableland cattle, lower prices, poor seasons, lack of water and loss through redwater disease and ticks contributed to the decline. Areas held under lease and stock numbers decreased and although the Government Resident cautioned that Tableland pastoralists should show 'less apathy and more enterprise', the situation worsened.[14]

In order to assist the pastoralists of the northern end of the Territory, the South Australian Government developed the export of cattle to Batavia and Singapore.[15] H.W.H.Stevens, representing Goldsborough, Mort and Company signed a five-year contract for this purpose in 1891. The 600-ton steamer 'Darwin' was built and a subsidy of £5,000 per year was paid to the company. In 1895, in order to assist the Barkly region, the government included Tableland cattle in the shipments but Alexandria did not participate, being too far away in the south-east. Goldsborough Mort took up a two-year contract from 1 April 1897 but because of the tick and redwater problem and small consignments of stock, the service was discontinued from the end of 1897.[16]

In 1894, the Government Resident suggested that freezing works be established on the Roper or McArthur rivers but nothing eventuated.[17] There was renewed interest in this project in 1904 but the most viable option was still to continue overlanding into Queensland, as long as disease was contained.[18] Indeed, the closure of the Vestey meatworks in Darwin in the early 1920s after only a few years of operation had little effect on the Barkly stations. It did, however, serve to illustrate the isolation of the Barkly from the 'Top End' and its closer affinity with Queensland.

For management, the major operational problem confronting Alexandria — and other runs — was lack of water. A newspaper correspondent in 1892 saw Alexandria as being

> magnificent a stretch of pastoral country as I have seen — but not a drop of water on it . . . . The country may be described as slightly undulating downs covered with Mitchell and other grasses, giving the appearance of a cornfield.[19]

Not until much later, in 1920 after bores were put down, was Alexandria independent of natural surface water. Each year managers had the problem of when and where to shift cattle. Harding lamented:

> It is only sheer luck that has pulled us through on Alex so long it being one of the worst watered stations in this part of the Tablelands.[20]

A dry period in 1892 caused great difficulties. Harding was concerned that he would have to make for the Nicholson country with the Soudan cattle. He had already allowed Hutton, the manager of Brunette, to depasture 6,000 there, and realised that

> the only road open to shift the Soudan cattle will be down the Rankin to Austral Downs and then up the Georgina via Rocklands towards the Gulf. Even that way there will be a couple of 40-50 mile dry stages.[21]

Within a month it was impossible to get to the Nicholson, so he started 1,500 store bullocks towards Inkerman. Collins then instructed Harding

11 J.S. Little, Austral Downs, Government Resident's Report for 1887, 53 of 1888, p.4.

12 J.S.Little, Austral Downs, Government Resident's Report, 1887, *SAPP*, 53 of 1888, p.4.

13 J.H. Whitelaw, Stock Inspector, Government Resident's Report for 1888, *SAPP*, Volume 2, 28 of 1889, p.4.

14 Government Resident's Report on the Northern Territory for 1895, *SAPP*, Volume 2, 45 of 1896, p.1.

15 *SAPD*, 3 December 1891, pp. 2301-2.

16 Report of export, 27 December 1897, GRS 9, Volume 5, PROSA.

17 *Register*, 26 July 1894, GRS 9, Volume 4, PROSA.

18 Government Resident's Report on the NT for 1904, *SAPP*, Volume 2, 45 of 1905, p.2.

19 *N.T. Times and Gazette*, 28 October 1892.

20 T. Harding to W. Forrest, 24 August 1892, Docket 481, GRS 1, PROSA.

21 *ibid.*

not to start a further 2,000 head because it 'was cheaper to let them die'.[22]

In 1895 William Forrest noted that more land was being taken up specifically to feed starving stock.[23] But the extra land did not help in 1897 — the driest year on record — when only 119 millimetres (477 points) of rain fell on Alexandria. Despite shifting cattle to the Nicholson country, 5,000 (a third of the herd) perished.[24]

In 1918, Herbert Vale, Lease 1610, totalling 1,071 square kilometres, was purchased by NAP from Kilgour for £8,558 and appropriately named Gallipoli, to remember Australia's sacrifices in World War I. It was not purchased for its grass, however, but rather to prevent it falling into the hands of cattle duffers.[25] Cattle duffing was widespread in the border and Nicholson country. The purchase provided much greater flexibility in the management of Alexandria, as in 1928 when Alexandria received useful coastal rain which did not penetrate further south, thereby allowing management to depasture 15,000 head. Initially there was no stock on Gallipoli but it did have four bores and some fencing.[26] Stocking began in 1920 and soon it became the annual practice to transfer store cattle to the sweet country of Gallipoli for fattening. When the government wanted to resume 268 square kilometres of Gallipoli in 1933, the company argued strongly against it, claiming among other things that the area was not large enough to provide a living.[27]

The 1926-28 drought during Charles Augustin Yorke (CAY) Johnston's time as manager was particularly bad. The annual average rainfall for the three years was only 188 millimetres (750 points),[28] less than half Alexandria's long-term average. Instead of the stock feeding around natural waters until August and then being moved on to the bores as the natural supply gave out, they were at the bores in April and by August had eaten the feed for miles around.[29] Richard Holt, the manager before CAY Johnston, commented in June 1926:

> I never saw the country looking so bad at this time of the year. Most of the herd will be on the bores by the end of July. If no rain falls, . . . we must expect to see breeders start to die.[30]

It was two years before any worthwhile rain fell and it presented Johnston with his most worrying and busy time, shifting cattle and employing pumpers who worked fifteen-hour days at the bores without overtime.[31] Losses were kept to a minimum but cattle could not be moved for sale and returns were badly affected.

Optimum conditions were not always forthcoming and the manager was constantly adjusting the distribution of his stock on the property in relation to the availability of grass and water. In the dry winter months, the Mitchell and Flinders grasses would dry off to form a barely adequate dry feed but the manager hoped it would not rain, lest the pastures blacken and rot in the resulting wet, cold conditions.[32]

But floods could, on occasions, be just as devastating as drought. Cattle often drowned or were swept across to neighbouring properties. An outbreak of redwater usually followed a big wet when ticks flowed down from the Nicholson country. During the last week of 1889, following a protracted dry period when the mails had ceased and some stations were being abandoned, the heaviest floods in memory caused the Tableland rivers to run for the first time in two years.[33] Two years later the flooded Corella Creek claimed the life of Henry Arthur (Harry) Readford,[34] former manager of Brunette Downs and alleged cattle duffer who has been immortalised as Captain Starlight in Rolfe Boldrewood's *Robbery under Arms*. Prior to his

22 W. Forrest, Adelaide Club, to Colonial Treasurer, 3 December 1892, Docket 481, GRS 1, PROSA.

23 W. Forrest to MCNT, 4 April 1895, Docket 109, GRS 1, PROSA.

24 Mounted Constable F.S. Stone, Anthony's Lagoon, 3 February 1898, Government Resident's Report on the Northern Territory for 1897, *SAPP*, Volume 2, No. 45 of 1898-99, p.13.

25 A cattle duffer steals cattle and often alters the brand.

26 P.McI. Forrest, 7 June 1937; P. Forrest to H. Ingram, 3 December 1929, H.Ingram File; Balance Sheet, 30 June 1918, NAPR.

27 F.P.Shepherd, Chief Surveyor, to Administrator, 28 November 1933, PL 189N, NTRS 246, NTAS; P.Forrest, Director's Report, 27 June 1935, NAPR.

28 *Report of the Board of Inquiry appointed to inquire into the Land and Land Industries of the Northern Territory of Australia*, Canberra, Commonwealth Government Printer, 1937, p.99.

29 P.Forrest to Woolley, Beardsleys and Bosworth, 11 November 1926, NAPR.

30 Richard Holt, Soudan, to P.Forrest, 1 June 1926, R.Holt File, NAPR.

31 CAY Johnston Interview, 26 May 1981, NTRS 226, TS 250, NTAS.

32 F.H.Bauer, 'Significant factors in the white settlement of northern Australia', *Australian Geographical Studies*, Vol.1, No.1, April 1963, p.42.

33 Telegram, Government Resident J.Knight to OMCNT, 17 February 1890, Docket 135, GRS 1, PROSA.

34 Telegram, W.Kelly, Powell's Creek to Little, Port Darwin, 10 April 1901, Docket 10442, GRS 10, PROSA.

Brunette days, Readford[35] had been charged with stealing 1,000 head from a station in Queensland and driving them to Adelaide, which led to the use of the Cooper, Diamantina, Birdsville route. Following his acquittal by a sympathetic jury, the judge remarked, 'The verdict, gentlemen is yours, not mine, thank God'.[36]

The continuity of good seasons was unreliable and William Forrest remarked despondently 'If it were not for the water supply — or, rather, the want of it — I should feel more hopeful about the Tableland'.[37] Until bores became commonplace, herd sizes were an indication of the natural surface water of a station, with the Victoria River District well watered in comparison with the Barkly. In 1907, Victoria River Downs carried 69,350 stock on 21,746 square kilometres, Wave Hill had 58,000 on 20,392 square kilometres, while Alexandria carried only 20,000 on 25,732 square kilometres.[38] Brunette Downs, also smaller than Alexandria carried 25,000 head, because it was fortunate to have two permanent lagoons.[39] With bores, however, Alexandria's herd size outgrew Brunette's in 1911.[40]

The NAP partners were anxious to solve the water problem, given the huge size of the run and the knowledge that artesian water did exist on the Barkly. Land usage on Alexandria was confined mainly to the region of the Ranken and Playford rivers, Buchanan and Lorne creeks and country to the north. Thomas Harding, in 1892, said he had also seen every waterhole on these waterways dry.[41] Forrest summed up the situation:

> The rainfall is most uncertain, and experience has shown that there are only three permanent waterholes on the Playford and Buchanan and auction lots included. Dams have been tried and have proved a failure, for when rain does fall it comes down in such tropical torrents that it carries away every obstacle.[42]

There were three wells on the station by 1892 but artesian supplies were favoured. To this end, Forrest consulted with Logan Jack, the Queensland geologist, and others such as the owners of Rocklands.[43]

It was also in the government's interest that artesian supplies be found. Waterless country had been abandoned and poor returns on lowered rents were affecting government revenue. In 1890 Government Resident Knight urgently recommended that the government sink a trial artesian bore on the Tableland. A boring plant was successfully employed on the Queensland side of Rocklands in mid-1890 and Knight suggested it could be used, but no government action was taken.[44]

Despite the Queensland Government's financial aid in artesian boring in that State, with over 200 artesian and sub-artesian watering points by 1896,[45] the South Australian Government seemed loath to help the Territory. A suggestion in late 1890 by the Government Resident to bore on Brunette was rejected in Adelaide; the Minister wrote, '[I] cannot sanction expenditure . . . . Besides the provisions of the new land bill are very liberal in regard to pastoralists'.[46] By the early 1890s, the NAP partners were anxious to undertake the boring themselves. Sir Simon Fraser, who in 1885 had married Anna Bertha Collins, sister of William, was one of the owners of Thurulgoona Station in Queensland where in 1886, the first artesian water in Queensland was successfully tapped. Sir Simon was also a partner in the Collins White Company formed in 1883 and his association with NAP was inevitable and invaluable.[47]

The NAP partners tried to persuade pastoralists on the Tableland to share the cost of putting down a trial bore. Given the financial hardships of the early 1890s which culminated in the 1893 financial crash, Forrest later commented that they 'could not

---

35 The spelling also appears in popular histories as 'Redford'. The signature 'Henry Arthur Readford', Manager of Corella Station (later part of Brunette) appears on a petition from Barkly residents to form a township (Lake Francis, later Camooweal). The petition dated 1 October 1883, is reprinted in L.A. Miller, *The Border and Beyond*, pp.26-7.

36 E.Hill, *The Territory*, p.306.

37 W. Forrest to MCNT, 4 April 1895, Docket 109, GRS 1, PROSA.

38 Ross Duncan, 'South Australia's contribution to the development of the Northern Territory cattle industry', *Historical Studies in Australia and New Zealand*, Volume 11, No.43, October 1964, pp.329; W.T. Hare, *The Early History of Animal Industry in the N.T.*, N.T. Conservation Commission, 1985, p.4.

39 Government Resident's Report, 1907, *SAPP*, Volume 3, 45 of 1908, p.3.

40 Mounted Constable J.J.Kerin, Rankine River, 6 January 1912, Acting Administrator's Report, 1911, *CPP*, 1912, p.35.

41 T. Harding to W. Forrest, 24 August 1892, Docket 481, GRS 1, PROSA.

42 W. Forrest to Minister of Education, 22 September 1891, in SAPD, 11 December 1891, p. 2464.

43 Rocklands Pastoral Company to Minister for Education, 26 March 1889, Docket 251, GRS 1, PROSA.

44 Knight to MCNT, 3 July 1890, GRS 8/5, PROSA.

45 R. Duncan, *The Northern Territory Pastoral Industry 1863-1910*, p.78.

46 Minister for Education and the NT, Adelaide, to Government Resident, 17 September 1890, Docket 1910, GRS 10, PROSA.

47 Harry C. Perry, *Pioneering: the life of R.M. Collins MLC*, p.273.

get sixpence'.[48] It finally took the political influence of McIlwraith and Forrest to persuade the South Australian Government to help financially, so that by November 1891, £4,000 was being voted by the government towards the cost of artesian boring on Alexandria.[49] The aid directed specifically to NAP did not go unnoticed, with comments in the S.A. parliament that the financial help

> would be simply assisting a wealthy pastoral Company to bore on their own land, and they apparently could well afford to do it themselves. The bore could be used only by those who had cattle there.[50]

McIlwraith and Forrest approached the Minister responsible for the Northern Territory but it was Forrest's dogged enthusiasm which saw the project through. His influence as a politician and skill as a negotiator were invaluable in lobbying the government, although it was pointed out in parliament in defence of his actions, that he

> did not ask the Government to undertake all the expense, but was perfectly willing to share part of it and become responsible for the whole outlay if a good supply of water was found.[51]

Certain conditions applied. The site was to be not less than eighty kilometres west of the Queensland border. The government was to pay the actual cost of boring until water was struck, or to a maximum depth of 610 metres (2,000 feet). If there was an insufficient flow the government had to be repaid.

A contract was signed on 21 January 1892 and boring started on 9 August 1892 on block 3107, three kilometres south of Buchanan Creek. Sixty tons of machinery was hauled to the site from Burketown at a cost of £1,200 and when work commenced it was a slow process. The rate of drilling was up to two metres a day but after four months the depth was only fifty-eight metres, with delays due to the 'borer' sticking and the 'sinker' breaking in the hole.[52] A description of the boring equipment was given by a passer-by in 1892:

> The machinery consists of two 8h.p. engines, boiler, and necessary apparatus for pole boring, and an iron derrick 56 feet in height, at the bottom of which the boring is conducted.[53]

Under the supervision of John Henry Bennett, the contractor, four men were employed in twelve-hour shifts carting water from Buchanan's Waterhole for the steam engine and hauling wood six kilometres to the site. Fresh water of excellent quality was struck at seventy-three metres (a sub-artesian depth) and rose about five metres up the casing. While pumps could have been used to draw water from this depth, not until an investigation by the Government Geologist in 1895 was it shown that plans for artesian boring had been misdirected; water supplies from sub-artesian depths were adequate.[54] Prior to this knowledge, however, a flowing artesian bore was preferred from which water could be directed into Buchanan Creek without the use of pumps. By January 1894, after sixteen months, the bore was down 501 metres and a mounted constable from Camooweal reported that the work was proceeding according to plan.[55]

But by March 1895 the water was still not flowing to the surface. Instead, the boring equipment had become jammed and on the recommendation of the Government Geologist it was decided not to recover the lost core-barrel. The bore was abandoned and a new bore further south was planned.[56] At this stage the NAP partners had spent £4,698 on the project[57] while by April 1894 the South Australian Government had spent £3,328.[58] It is not recorded whether the partners repaid any of the money outlaid by the government, but certainly in 1896 they were still in favour when further financial assistance was provided to put down and equip six wells. This time Robert Collins helped with the lobbying, arguing that it would enable the country to be opened up for sheep,[59] although NAP never seriously considered running sheep in view of the prevailing low prices. Perhaps the government's support represented something of a change of

48 W.Forrest, Minutes of Evidence of the Royal Commission into the Northern Territory, *SAPP*, Volume 2, 19 of 1895, p.169.

49 Jenkins, MCNT to W. Forrest, 17 November 1891, GRS 7/6, p. 183, PROSA; *SAPD*, 3 December 1891, pp. 2300-3; 10 December 1891, pp. 2423-25.

50 Mr. Grainger, *SAPD*, 11 December 1891, p. 2464.

51 J.G. Jenkins, Minister of Education, *SAPD*, 3 December 1891, p. 2302.

52 W. Forrest, Adelaide Club, to Colonial Treasurer, 3 December 1892, GRS 1, PROSA.

53 *N.T. Times and Gazette*, 28 October 1892.

54 R. Duncan, *The Northern Territory Pastoral Industry, 1863-1910*, p.79.

55 Newspaper article, 18 January 1894, GRS 9, Volume 4, 465/93, PROSA; Alexandria Bore, A1640, 94/333, Australian Archives, Canberra.

56 Newspaper article, 16 March 1895, GRS 9, Volume 4, PROSA; This first bore was later in use by 1900.

57 Synopsis of Evidence of the Northern Territory Commission, *SAPP*, Volume 2, 16 of 1895, p. xxix.

58 Abandoning Alexandria Bore, 1894, A1640, 94/333, Australian Archives, Canberra.

59 R.M. Collins, Adelaide, to Treasurer, 5 June 1896, Docket 216, GRS 1, PROSA.

No. 18 Bore.

heart; six months earlier the owners of Austral Downs and Milne River stations had asked for, but had been denied, assistance in boring. As the Minister Controlling the Northern Territory endeavoured to explain at the time, there was a 'probability of good water in abundant quantity being obtained by sinking an ordinary well'.[60] Such was the political influence of the NAP partners that the Minister's opinion could change.

Months of indecision by the government as to another site followed the failed attempt on Alexandria. The junction of the James and Rankine rivers was thought to be a more suitable site for a bore, as the Alexandria area was considered too high on the Tableland. The government suggested employing some of Alexandria's men, and financial assistance was discussed. The final decision not to continue deep well-boring was made in July 1896, although the quest for sub-artesian supplies continued.[61] Meanwhile, the government had become generally more accommodating and by late 1896 government assistance for well-boring was available to other Barkly stations, although, with the failure of deep well-boring on Alexandria as well as Rocklands, it was a policy of proceeding 'from the known to the unknown gradually'.[62]

Although it was believed that a sufficient water supply would eventually be obtained from wells since permanent pumping water had been struck at a depth of some forty-six metres,[63] deeper bore drilling was soon under way as a private initiative. In 1900, S. Miller drilled the second bore on Alexandria but he put down casing which was too small for the pump. Although a lawsuit over this failed operation was discussed, nothing eventuated.[64] Work continued, however, and by 1902 NAP had put down four bores and arranged for two more to be sunk.[65] Boring was also proceeding on other stations; in 1904 five bores were put down on Brunette, two on Alroy and two on Rocklands.[66] By 1909 there were forty bores on the Tableland.[67]

As well as deep well-boring, NAP was also first in the Territory — unless Rocklands can be classified as a Northern Territory station — in the use of windmills. William Forrest obtained all the information he could from the South Australian Government and Rocklands Station where 9 metre

60 MCNT to I.S. Little, 2.12.1895, GRS 1, PROSA.

61 Alexandria Bore, 1895, A1640, 95/203, Australian Archives, Canberra.

62 F. Holder, MCNT, to W. Forrest, 18 February 1899, Docket 118, GRS 1, PROSA.

63 W. Forrest, Minutes of Evidence of Royal Commission into the Northern Territory, *SAPP*, Volume 2, 19 of 1895, p. 169.

64 Stock Inspector's Report, 9 January 1899, Docket 8795, GRS 10, PROSA.

65 J. Packard to MCNT, 7 November 1901, Docket 493, GRS 1, PROSA.

66 A.H. Glisson, Manager of Rocklands, Government Resident's Report for 1904, *SAPP*, Volume 2, 45 of 1905, p.2.

67 Anthony Lagoon Dip, 1909, Docket 17901, GRS 10, PROSA.

Windmill repairs by J. Broadbridge. Peg Corbett

Mill and Turkey's Nest at Homestead Bore, Alexandria.

American Halliday Standard Windmills, tanks and troughing had been installed. By early 1899 the partners were ready to use windmills on Alexandria.[68] Mills were erected over the next two decades although by 1922 the number of working windmills on Alexandria bores had fallen to only two, steam engines being preferred to draw the water. Although Rockland bores were later equipped with windmills with huge 11-metre wheels, by the 1930s Barkly windmills were generally 7 to 9-metre wheels on 13 to 17-metre towers made by the Intercolonial Boring Company. Later, Comet and Southern Cross brands were used.[69]

Interestingly, the first Alexandria bore, which had been abandoned in 1895 at 536 metres, was in use by 1900. It is not clear what had happened in the meantime, but certainly it was not until 1921 that a turkey's nest was built on the bore. The idea for these earthen tanks came from Augustus H.Glisson, manager of Rocklands[70] and the first turkey's nest on Alexandria was built in 1910 at No.10 bore. Construction required an army of men with horses, scoops and ploughs to form an earth mound above the ground for storage of approximately 900,000 to 2,200,000 litres of pumped water. Previously, water was held in iron tanks and it was an expensive business to change to earth tanks. As an example, in 1922 W.Cummings, over a period of four months, dismantled the old iron tanks at No.20 bore on

68 W. Forrest, Brisbane, to MCNT, 1 March 1899; Engineer in Chief to W. Forrest, 22 March 1899, Docket 118, GRS 1, PROSA.

69 Ted McFarlane Interview, 19 July 1979, NTRS 226, TS 273, NTAS.

70 F.A.C.Bishop, *Report on an inspection of the pastoral holdings, stock route, bores, and dips on the Barkly Tableland, Northern Territory*, Melbourne, Government Printer, 1923, p.13.

No. 19 Bore: Alexandria. Tony Johnston

Boiler fittings and leather goods. Front: Irene Broadbridge.

Alexandria, built the earth tank and fenced the tank and bore area at a cost of £321 3s 4d. But recycling was in fashion then as now; the roof of the engine house at the bores often came from old curved tank iron.[71]

Subsequent boring never reached the great depths of the 1894-95 attempt. Bores were later drilled to sub-artesian depths of approximately 120 metres. By 1933, Alexandria's bores were increasingly using diesel engines rather than steam engines, with earth tanks often being built years after the bores were put down. In the earlier period the walking beam pumpheads and the distinctive steam engines with their imposing smokestacks

71 Bores and Improvements, Number 20 Bore, 1922, NAPR.

No. 6 Team: Carting firewood for the steam engines at the bores.

were a significant feature of the landscape. Most of them were portable and could be dragged along by bullock teams from one bore to another. As all available timber on the stations was used to fire the steam engines' boilers, the land was being denuded and supplies of timber were scarce. It was at this time that diesel took over.

ALEXANDRIA BORES 1933[72]

Number	Feet	Metres	Year	Engine
1	1,760	536	Bore 1900 Tank 1921	8 h.p. Petter oil engine

Cost over £4,000 — the most expensive; bores drilled later were generally less than £2,000.

Number	Feet	Metres	Year	Engine
2 abandoned			Bore 1900	
3	345	105	Bore 1902 Tank 1918	5.5 hp Davey Paxman steam engine
4	360	110	Bore 1902 Tank 1923	10 h.p. steam engine (1925)
5 (Soudan)	317	97	Bore 1901	8 hp Brittania steam engine
6 abandoned			Bore 1901	
7 (Soudan)	313	95	Bore 1905 Tank 1920	Paxman steam engine
8 abandoned			Bore 1905	
9	306	93	Bore 1905 Tank 1921	Paxman steam engine
10	366	112	Bore 1910 Tank 1910	8 hp Petter oil engine
11	308	94	Bore 1910 Tank 1910	6 hp Paxman steam engine
12	451	138	Bore 1913 Tank 1913	Comet windmill auxiliary 8 hp oil engine
13	384	117	Bore 1912	5 hp Paxman steam engine
14	665	203	Bore 1916 Tank 1922	5 hp Petter oil engine
15	abandoned when tools stuck 1916			
16	278	85	Bore 1917	Steam engine
17	841	256	Bore 1917	8 hp Tange steam engine
18 (Soudan)	487	148	Bore 1918 Tank 1919	Paxman steam engine
19 (Soudan)	263	80	Bore 1917 Tank 1921	7 hp Tange steam engine
20 (Gallipoli)	305	93	Bore 1918 Tank 1922	Comet windmill, 4 hp oil engine
21 (Gallipoli)	358	109	Bore 1922 Tank 1923	6 hp I.B.C. steam engine

GALLIPOLI

Number	Feet	Metres	Year	Engine
1	365	112	Bore 1904 Tank 1929	8 hp Tange steam engine
5	384	117	Bore 1912	5 hp Paxman steam engine
7	383	117	Tank 1929	I.B.C. steam engine

By 1921, with twenty bores down, Alexandria was independent of natural surface water. Cattle tended to stay on the bores during the day and at night would walk up to 12 kilometres to feed. Writing in 1922, managing partner Phil Forrest emphasised the significance of this development. In one season when over 10,000 head were sold, there were between 5,000 and 6,000 cattle

72 Bores, 1933, Holding: Alexandria, A 452/1, 1952/310, Australian Archives, Canberra.

gathered at one of the bores. Fortunately, the bore and engine were good, enabling the stock to drink every second day. On being sold, the bullocks had to go through three dips and yards, numerous gates, trucking yards, spend two days in railway cattle wagons without water on their way to the meatworks, and then had to wait up to a week in a bare paddock before slaughter. Somehow the cattle still averaged 290 kilograms (641 pounds) for bullocks (£4 5s) and 239 kilograms (528 pounds) (£2 2s 6d) for cows.[73]

Herd size comparisons also clearly illustrate the advantages of bores, especially for Alexandria, Lake Nash and Alroy. Brunette did not benefit as much in relative terms because it already had permanent lagoons.

Size of Herds: Barkly Stations 1895, 1908, 1921[74]

	1895	1908	1921
Alexandria Downs	10,136	20,567	45,000
Herbert Vale	1,100		
Brunette Downs	20,050	24,000	30,000
Lake Nash	12,500		25,000
Austral Downs	8,796		6,000
Rocklands	11,000	8,000	12,000
Alroy Downs	3,500	9,000	18,000
Avon Downs	5,758	8,000	5,000
(plus sheep until 1919)			

At the same time, reports on the advantages of bores could be over-optimistic. In 1922 the Chief Inspector of Stock estimated the carrying capacity of Alexandria to be sixteen head to the square mile which would give a total of 160,000. This compared with the herd size at the time of 42,962 [75] and a present optimum capacity of 60,000.

While the scarcity of water was the main problem for management in the early years, disease was also a major cause for concern. Pleuro-pneumonia was noticed in Northern Territory herds in the early 1880s and specifically on the Barkly in 1890 in an Austral Downs mob bound for southern markets.[76] The spread was gradual but in 1902 Alexandria's manager Holt suggested to the Government Resident that inoculation of infected cattle be required at Anthony Lagoon north-west of Alexandria before proceeding further.[77]

Inoculation did occur and as immunisation of cattle became more widespread, herds were treated on the stations prior to travelling. By 1912 a cultivated CPP (contagious pleuro-pneumonia) virus for inoculation was available from Darwin[78] but the managers and stockmen preferred to use a live virus taken from the lungs of a diseased animal. James White of Brunette spoke of the 'mini running battle' with the department over this preference.[79] The live virus was stored in beer bottles which were kept in wet bags in a cool place. The men would laboriously insert a wool seton, which had previously been soaked in the virus, into the lower part of the animal's tail just above the switch. After injection, the cattle were 'bangtailed' to indicate that they had been treated. From the 1920s pleuro-pneumonia treatment was confined to times of outbreak when herds would be inoculated under supervision of the stock inspector who was also the local policeman.[80]

Generally, pleuro-pneumonia was contained but ticks and redwater disease were a more serious problem. Both ticks and the redwater disease had been noticed in the Territory since 1882,[81] and were officially recognised in 1885.[82] Fortunately for Avon and Austral, sheep were not affected and in 1889 record exports of wool were made to London from both stations.[83]

When redwater first appeared, the disease was puzzling. Ticks had fallen off cattle which had previously arrived on the Tableland from Queensland but it was not realised at the time that these cattle had already developed an immunity to the disease in Queensland. While the stocking of Alexandria was unaffected, newspapers reported that vast areas of the Territory were unstocked because of the uncertainty of station owners.[84]

73 P. Forrest to Major H. Ingram, 14 October 1924, H.Ingram File, NAPR.

74 1895: Government Resident's Report on the Northern Territory for 1895, *SAPP*, Volume 2, 45 of 1896, p.15. 1908: Cattle Numbers 1908, Docket 17993, GRS 10, PROSA. 1921: Alfred V.Stretton, Acting Sergeant of Police, Rankine, 19 October 1921, Parliamentary Standing Committee on Public Works, *CPP*, 1921, p.82.

75 F.A.C. Bishop, *Report on an inspection of the pastoral holdings, stock route, bores, and dips on the Barkly Tableland, Northern Territory*, Melbourne, Government Printer, 1923, p.7.

76 Report of Administrator of NT for 1912, *CPP*, 1913, p. 116.

77 R. Holt in Government Resident's Report on NT 1904, *SAPP*, Volume 2, 45 of 1905, p.2.

78 Report of Administrator of NT, 1912, *CPP*, 1913, p.121.

79 James White Interview, NTRS 226, TS 140/1, NTAS.

80 Acting Sergeant A.V.Stretton, Rankine, 19 October 1921, Parliamentary Standing Committee on Public Works, *CPP*, 1921, p.82.

81 H.W.H. Stevens, Government Resident's Report on Northern Territory, 1888, *SAPP*, Volume 2, 28 of 1889, p.2.

82 Report of Veterinary and Stock Department for 1912, *CPP*, 1913, p.114.

83 Report of Administrator of NT for 1912, *CPP*, 1913, p.116.

84 Newspaper article, 2 April 1889, GRS 9, Volume 1, pp. 95-6, PROSA.

It was not until 1895 that the Tableland stations were really affected[85] after Barkly cattle were agisted in northern coastal areas in the drought of the early 1890s. These areas already had a redwater problem and the Barkly cattle were encountering the disease for the first time. Alexandria and Brunette cattle went to the Nicholson region which was infected, and losses were so high that William Forrest said he would rather 'let our cattle die than send them back again to the head of the Nicholson'.[86]

In 1896 Thomas Guthrie, the owner of Avon Downs, expressed concern that diseased cattle with ticks were travelling from the McArthur River to country near the head of the Ranken. Because of the complete absence of fences, he correctly predicted that such a menace would soon spread throughout the Tableland.[87] Queensland Government bacteriologist C.J.Pound was the first to recognise that redwater was caused by ticks. But the public took more convincing[88] because ticks appeared in a new district well before the redwater disease, with many cattle developing a gradual immunity without any signs of redwater.

The Barkly was not affected to the same extent as the northern and coastal areas, and pastoralists tried to play down the extent of the problem, fearing it would affect markets. Queensland had imposed restrictions on the movement of stock within the State in 1894. Two years later the worst fears of Territory pastoralists were realised when South Australia imposed a tick line along latitude 19 degrees 30' south (just below the Alexandria homestead), north of which Northern Territory stock were not allowed into South Australia. At the same time, Queensland prohibited altogether the importation of stock across the border with the Northern Territory.[89] This occurred despite personal representation by Robert Collins to the Minister in Adelaide when Collins gave an assurance that Tableland cattle were clean.[90]

Territory pastoralists had only one market left, the islands to the north of Australia, but most stockowners feared redwater too much to send many cattle to Port Darwin in transit to the northern markets. They were not aware that their cattle would be immune to the disease. An analysis of the situation can be summarised:

> South Australia's restrictions were unnecessary, as ticks could not survive in that colony, and so were the restrictions imposed on Territory cattle by Queensland and Western Australia, for ticky cattle entering infested areas in those colonies could do no harm. These restrictions did not benefit the colonies imposing them. They were cumbersome; they hampered the development of the pastoral industry and they failed to check the spread of the ticks.[91]

Thus the real harm to the Territory came not from the redwater itself, but from the loss of markets resulting from the measures taken to control the disease. With evidence from the stock inspector at Camooweal that the Barkly was clean, a new line was drawn in December 1898 which allowed the tick-free stations of Alexandria, Cresswell, Brunette Downs, Corella, Anthony Lagoon, Eva Downs, Renner Springs, Alroy Downs and Herbert Vale, freedom from quarantine.[92] Under pressure from the pastoralists quarantine restrictions were gradually lightened and eventually removed, but it was a tedious process.

Without fences, the method of exterminating the tick by rotation of pastures, as had been adopted in North America, was impossible.[93] Killing the ticks by dipping cattle in a chemical solution was introduced. Richard Holt was the influence behind the first cattle dip on the Barkly at Anthony Lagoon, some 180 kilometres north-west of Alexandria, in 1903. Holt had reminded the authorities that while ticks had not been a major problem on the Tableland up to that time, a Queensland regulation to be enforced from 1 February 1905 required all cattle to be dipped as a prerequisite for their travel into Queensland.[94] His advice was heeded and although the drovers initially tried to avoid the issue, appropriate regulations were set by the government. The dip mixture consisted of arsenic, caustic soda, Stockholm tar and tallow mixed with water. Fees, collected by the police, were one shilling per head for the first 20,000 and sixpence for every

85 Government Resident's Report on the Northern Territory for 1895, *SAPP*, Volume 2, 45 of 1896, p.2.

86 W.Forrest to MCNT, 4 April 1895, Docket 109, GRS 1, PROSA.

87 T. Guthrie, Geelong, to MCNT, 9 June 1896, Docket 226, GRS 1, PROSA.

88 C.J. Pound, 19 November 1894, GRS 9, Volume 4, PROSA.

89 K.E. McCulloch, Cattle Tick in the Northern Territory in the Nineteenth Century, B.A. (Hons.) Thesis, University of Adelaide, 1962, pp.70-1.

90 William Forrest, Brisbane, to MCNT, 19 June 1896, Docket 245, GRS 1, PROSA.

91 K.E. McCulloch, Cattle Tick in the Northern Territory in the Nineteenth Century, p.73.

92 Stock Inspector's Report, 9 January 1899, Docket 8795, GRS 10, PROSA.

93 Report of Veterinary and Stock Department, NT, 1912, in Report of Administrator of NT, 1912, *CPP*, 1913, p. 121.

94 R.Holt, Government Resident's Report for 1904, *SAPP*, Volume 2, 45 of 1905, p.3.

subsequent head.[95] Later Davies' Dip Mixture was used and supplies sent by steamer through Burketown.[96]

NAP was involved in the construction of the dip at Anthony Lagoon, being paid £250 in 1906 towards its cost.[97] By 1907 the concrete was cracked causing leaking but, with help from Brunette Downs Station, Holt saw to its repair and future upkeep.[98] In 1908, 16,556 head were dipped and Anthony Lagoon became a bustling meeting place with its store, post office and police station.[99]

It seemed the tick problem had been contained — until October 1917, when an outbreak of redwater was reported in an Alexandria mob which had reached Lake Nash Station. The mob was immediately quarantined and as a result the Queensland authorities placed an embargo on all Territory cattle entering Queensland between Camooweal and Lake Nash until they had been dipped. Cattle had to be dipped twice at intervals of seven to ten days prior to crossing the border,[100] and the Queensland Government gazetted the Queensland stock inspectors stationed at Camooweal and Urandangi as Territory stock inspectors. More outbreaks of redwater occurred, and one at the Ranken River in June 1918 prompted NAP to erect a dip there. Of iron construction, it was 12 metres long and had adjoining receiving and drafting yards and crush.[101] Built by NAP staff, Alexandria cattle were now dipped at the Ranken and Lake Nash. The Gulf Cattle Company erected another dip at Brunette bringing the number of dips to five on the Barkly Stock Route so that travelling stock could be dipped every 18 days — at Anthony Lagoon, Brunette, Ranken, Austral Downs and Lake Nash. Dipping charges were standardised at sixpence per head[102] and dips were also erected by NAP and Collins White in Queensland at Noranside Station, 40 kilometres south of Chatsworth Station on the Selwyn-Boulia Road, and, with a syndicate, at Cooridgee, about 30 kilometres south of Boulia.

Ranken Dip, 1920s.

When the ticks disappeared the dips were allowed to fall into disrepair.[103]

In 1920 the Barkly Stock Route, from Newcastle Waters through to Lake Nash, was the first stock route to be gazetted an official government-controlled route. The police constables at Anthony Lagoon and the Ranken played a major part in supervising the dipping along this route, with dipping fees being charged by the cattle station.[104] After a dip was built by the Government at the Ranken in 1940-41, dipping fees of fourpence per head went to the Government.[105] Alexandria was therefore closely associated with the dipping process and in 1922, six hectares of Pastoral Lease 2114 at the Ranken were surrendered to the Crown to enable a separate waterhole to be provided for

95 *South Australian Government Gazette*, 8 January 1903, p.33.

96 John Forrest to MCNT, 18 April 1907, Docket 16285, GRS 10, PROSA.

97 Secretary of MCNT to Inspector of Stock, Camooweal, 3 November 1906, Docket 232, GRS 1, PROSA.

98 Docket 15948, GRS 10, PROSA.

99 Stock Report 1908, Docket 17600, GRS 10, PROSA.

100 C.G.Dickenson, Chief Inspector of Stock, in Administrator's Report, 1918, *CPP*, 1917-18-19, p.39.

101 Alexandria Holding, A452/1, 1952/310, Australian Archives, Canberra.

102 F.H. Bishop, Inspector of Stock, Darwin, 1922, *CPP*, 1923, p.24.

103 D.M. Fraser to Col. A.L.Rose, Director of Animal Industry, Alice Springs, 10 January 1957, Green Copies, Letter File, November 1956 to February 1957, NAPR.

104 C.G.Campbell, Chief Inspector of Stock, Report by Administrator, 1918, *CPP*, 1917-18-19, p. 40.

105 Rankine Dip, 1940-41, File F1, 44/119, Australian Archives, Darwin.

Ear-marking calves at Alexandria. Tony Johnston

Bronchoing at No. 38 Bore: Gallipoli 1958. Left to Right: Jack Isaac, Vivian Chalk, John Ohlsen (at panel) assisted by Aboriginal stockmen. George MacQueen

the use of the travelling public. The area was fenced in order to keep freshly dipped cattle from watering there and contaminating the clean water.[106]

By the 1930s, redwater was largely under control and manager CAY Johnston stated that he dipped only when there was an outbreak.[107]

106 Administrator to Secretary, Home and Territories Department, 30 May 1922, PL2114, F27, NTAS.

107 CAY Johnston Interview, 26 May 1981, NTRS 226, TS 250, NTAS.

Alexandria Stockyards ca. 1934. Australian Archives, Darwin

Compared with the main disease problems of redwater and pleuro-pneumonia, pests were of relatively little importance on the Barkly. With the Queensland Rabbit Board's fence along the eastern border of Queensland and the Northern Territory, rabbits were not a problem, especially to western Barkly stations. In 1905 they were noticed at Lorne Creek and the Ranken with evidence of their presence on Austral Downs and Lake Nash, and although the manager of Rocklands expressed some concern they were of little consequence.[108] Dingoes were another problem through the killing of calves; one estimate of casualties in 1920 was 10 to 15 per cent of all calves dropped in the Territory.[109] The poison 1080 later kept the dingoes under control.

Improvements on Alexandria were gradually effected with new buildings, yards and fences, although fencing was generally for small holding paddocks. By 1933 Alexandria had Surprise Paddock (1905), Home Paddock (1911), Homestead Horse Paddock (1914), Draught Horse Paddock (1914), Weaner Paddock (1922) and 14-mile Bore Paddock. On Soudan, apart from the Horse Paddock (1910) and Bullock Paddock (1917), there were only 56 kilometres of fencing which had been erected in 1915 on the Avon Downs boundary. On Gallipoli there were no fences until a 19-kilometre length of fencing was erected on the Rocklands boundary in 1928.[110]

To enable procedures such as branding, spaying, and treating for diseases such as pleuro-pneumonia, to be carried out efficiently, drafting yards incorporating broncho panels[111] were erected over the years. They included Eastern Creek Yard (pre 1900), Rocky Yard (1902), Baal Yabba (1908), Tobacco Hole Yard (1908), Cigarette Hole Yard (1908), Gidgea Yard (1910), Lorne Creek Yard (1911), Cattle Camp Yard, Ranken (1910), Ibis Yard (1925), Horse-shoe hole Yard (1925), Springs Yard (1927), 8-Mile Yard (1928), and yards at various bores. In 1933 the homestead yard consisted of various receiving yards of post and rail construction, including a horse yard, round breaking yard, calf-branding pen, crush gates, milking yard and killing pen and gallows. There were various yards at the out-stations, including

108 A. Glissan, Rocklands, to MCNT, 5 January 1905, Docket 87, GRS 1, PROSA.

109 Report by Acting Administrator, 1920, *CPP*, 1920-21, p.8.

110 P. Forrest, 6 June 1928, NAPR.

111 A broncho panel is a fixture of post and rails to which a beast is pulled when it is being branded, ear-marked, dehorned, inoculated and in the case of male animals, castrated. To broncho a beast is to put a catching rope on it and pull it up to a broncho panel.

Branding calves at Alexandria, 1920s.

Lenney's Hole Yard (1918) on Soudan and Carrara Creek Yard (1928) north of Gallipoli.[112]

As major fencing was non-existent for many years it was impossible to confidently estimate cattle numbers. The vastness of the run effectively precluded a bangtail[113] muster of any real precision. Indeed, CAY Johnston's herd estimates were based simply on the principle of the number of bullocks he had dispatched. Another check sometimes used was to multiply the number of cattle branded by a factor of four or five. Thus calving rates were difficult to establish with estimates of between 20 per cent and 50 per cent.[114] At mustering time, stockmen from various stations met at tender musters to sort out and claim stock. Referred to as 'tendering', this sorting process was achieved by cattle being cut out from the mob by experienced riders on horses specially trained for the work.

Horse numbers increased as the herd size grew in order to cover the mustering; in 1926 Alexandria had a total of 883.[115] Horse-breeding was taken seriously to the extent that the Government Resident mentioned the purchase in 1907 and 1909 of two thorough-bred stallions for Alexandria.[116] Rocklands Station especially bred horses on a large scale.[117] A problem arose during the 1930s when Alexandria horses branded 676 were sold to or were traded with drovers. Once horses bearing the Alexandria brand were obtained legally, more would be stolen when passing through the Territory where there was always a good market for horses. This difficulty was finally resolved after the [A] symbol brand was registered in 1940 and was used on both cattle (off-rump) and horses (near-thigh), and the rule that 'no horse bearing the station brand could be sold or exchanged' was strongly enforced.[118]

Sidney Kidman, known to CAY Johnston as 'a wonderful fella', was by far the largest buyer of Alexandria cattle although other purchasers appearing in the records include W. Naughton (Durrie), E.Jowett (Palparara) and the Queensland Meat Export and Agency Company.[119] Kidman's son, Walter, used to inspect the store cattle before they left, first for Kidman properties in Queensland for fattening and then on to Marree

112 Holding: Alexandria, 1933, A/452/1, 1952/310, Australian Archives, Canberra; Alexandria Improvements Book, 1892-1943, NAPR.

113 A bangtail muster is a muster which gives an accurate count of cattle. The hair on the tail of a beast is cut square so the animal will not be counted twice.

114 R. Duncan, *The Northern Territory Pastoral Industry 1863-1910*, p.126.

115 Stock Valuation, 31 December 1926, Crouch and Paterson File, NAPR.

116 Government Resident's Reports for 1907, *SAPP*, Volume 3, 45 of 1908, p.4; 1909, Volume 3, 45 of 1909, p.17.

117 L.H. Wilkinson, Manager of Rocklands, Government Resident's Report for 1906, *SAPP*, 45 of 1907, Volume 3, p.5.

118 Harold Chambers to General Manager, Pastoral Inspector for NAP, 12 November 1960, File 15, Correspondence, H.W.Chambers, 1960; R.H.Murray to D.M.Fraser, 21 September 1949, R.H.Murray File, 1948-51, NAPR.

119 Balance Sheets 1913-22, NAPR.

and Adelaide for slaughter. In the early years, some mobs also went to NAP's Burdekin properties and to Burketown to the boiling-down works. Cattle sales varied depending on the season. During the excellent seasons between 1910 and 1919, up to 9,000 head were sold each year. In 1911, at the time of the sale of the Inkerman and Woodstock herds, 11,449 head were sold for £26,736 to A.J.Cotton of Brunette.[120] Because of the drought conditions of 1926-28, none were sold in the 1926 and 1928 financial years but after the reasonable seasons of 1929-30, 11,837 were sold in 1930.[121]

Drovers on contract to the buyers would take stock away from stations in mobs of 1,250 to 1,500 head. Losses depended on the state of the stock route, disease and whether they experienced any rushes[122] in the mob. The Barkly Stock Route was the main outlet for Territory cattle, yet even in the 1930s there were still dry stages of up to 130 kilometres. After dipping at the Ranken and Lake Nash, the mobs would follow the rivers — down the Georgina, Cooper Creek and the Diamantina. Stock also went to fattening properties in the Channel country, or to New South Wales and Victoria, the latter journey taking nine months. By the early 1930s, NAP was sending stock by rail to Townsville for the chilled beef trade from Brisbane.[123]

In 1936, 49,176 head travelled from the Territory to Queensland via Lake Nash and Camooweal, 15,942 travelled to South Australia via Alice Springs, and 10,481 went to Western Australia via Timber Creek.[124] Cattle from the Barkly fetched the highest Territory prices. Estimates in 1937 gave prices at market:

Northern Territory Cattle Prices per head 1937[125]

	£	s	d
Cattle from Barkly Tableland	7	0	0
Cattle from Gulf of Carpentaria	2	10	0
Cattle from Central Australia (to SA)	6	10	0
Cattle from Wyndham meatworks (WA)	2	0	0

Improvements in stock quality were emphasised, as shown in the purchase of good Shorthorn bulls. In 1915, for instance, bulls were purchased from Coorabulka, owned by John Collins and Sons.[126] In 1932 a large consignment of 280 bulls arrived at Alexandria.[127] At one stage, CAY Johnston was making yearly trips to Rockhampton to purchase bulls, and, in turn, other stations purchased stud stock from Alexandria to improve their herds. In a typical example, in 1909 forty stud bulls went to Kilgour's Herbert Vale and thirty cows, thirty bull calves and two stud bulls were sold to Victoria River Downs. As well that year, there was the sale to Kidman of 3,218 bullocks which went to Carandotta Station in Queensland.[128]

Finally, while the establishment and management of Alexandria was well in hand, profits also depended on markets, both domestic and export. With the easing of the economic depression in Australia in the 1890s, there was an increased demand for beef. The marked rise in cattle numbers and prices between 1900 and 1910 increased the value of cattle exports from the Northern Territory by 310 per cent.[129] This was enhanced by the lifting of restrictions on stock movements interstate and the easing of drought conditions particularly in Queensland by 1904.

During these years, the export of Australian beef reflected the country's colonial past and its special relationship with the United Kingdom. Prior to World War I, markets were stable, with the United Kingdom taking the bulk of Australian beef exports as frozen beef. An oversupply of cattle in the 1920s, however, resulted in a slump in cattle prices which affected NAP dividends to the extent that no dividends were paid between 1922 and 1925. In the few years before 1920 prices averaged up to £12 per head but this price decreased dramatically to £4 per head in 1922 and fewer cattle were sold.[130] During this time, the success of Australia's beef trade continued to fluctuate with Britain's prosperity. The General Strike in Britain in 1926 meant that British working families could no longer afford to buy Australian meat and exports to the United Kingdom fell. The Depression of the 1930s hit both the home consumption market and Australia's beef export trade.

By 1935 Alexandria was well-established. Running 45,000 head on its 28,085 square kilometres, its leased area had been stable for nearly two decades; and with bores, yards and fencing, a manageable use of the land had been developed. Disease was largely under control and with its huge size and a reputation to match, Alexandria was showing the way on the Barkly.

---

120 Balance Sheet, Inkerman and Woodstock Stations, 30 June 1911, NAPR.

121 Holding: Alexandria, A452/1, 1952/310, Australian Archives, Canberra.

122 A sudden stampede of cattle through fright.

123 Alexandria Station, 1935, A452, 1952/310, Australian Archives, Canberra.

124 F.A.C. Bishop, Chief Stock Inspector, in Report of the Administration of the NT, 1936, *CPP*, 1937, p. 49.

125 Report of Administration of the NT, 1937, *CPP*, 1937-38, p.7.

126 Balance Sheet, 30 June 1915, NAPR.

127 P.Forrest to Sir Herbert Ingram, 28 June 1932, H.Ingram File, NAPR.

128 Stock Inspector, Rankine River, 7 January 1910, Government Resident's Report for 1909, *SAPP*, Volume 3, 45 of 1910, p.19.

129 P.F.Donovan, *A Land Full of Possibilities*, St.Lucia, University of Queensland Press, 1981, p.196.

130 Balance Sheets, 1910-1925, NAPR.

CHAPTER 5

Partnership to Company: The Forrest Years

When Philip Forrest sold his NAP shares in 1937, it was a watershed in the history of the company. His uncle, William Forrest, had been an original partner and the managing partners of NAP had always included members of the Forrest family. Although William Collins and William Forrest acted jointly in this role, it was Forrest who generally undertook the trips to Adelaide to oversee leases and see to the administration of the partnership. His part in initiating the drilling for artesian water on Alexandria was particularly significant.

His brother John took over as managing partner on 5 March 1902[1] and was instrumental in the

Back: Left to Right: William Tyler Forrest, Philip McIlwraith Forrest. Seated: Edith and John Forrest.

Nancy Grace

Philip McIlwraith Forrest. Managing Partner 1915-1931. Chairman of Directors/Managing Director/ General Manager 1931-1936.

Nancy Grace

1 Deed of Accession, 5 March 1902, NAPR

delicate sale of the Burdekin properties which reaped such financial gains. Following his death in 1911, John's son William Tyler Forrest was the managing partner until 1915 when the position passed to another son, Philip McIlwraith Forrest.

It was Philip who steered the partnership through its dilemma of possible dissolution and subsequent formation into a company; he presided over the decision to purchase the fattening property, Marion Downs; and above all, he kept NAP financially viable in a period of climatic and economic difficulties. Because they were so closely involved in administration, these men crucially shaped NAP's direction over its first six decades.

While there was relative stability in management while the Forrests continued in that position, a period of uncertainty and sadness beset the partnership between 1909 and 1913. In that short time, John Forrest, the three Collins brothers, William, John and Robert, and Captain William Pochin Warner (the brother of John Henry Boyer Warner, the original partner), all died. By late 1913 Sir William Ingram was the only remaining original partner but his isolation restricted his influence. The descendants of the Australian partners were left to either carry on or dissolve the partnership. The original partners had known and trusted each other and had been genuinely interested in the cattle industry. Now it was up to their descendants to determine what the future of NAP might be.

It is significant that the general approach of each of the Forrests was consistent; that is, to acquire land in order to sell for profit. Thomas McIlwraith and the Collins family preferred to concentrate on cattle matters and not be worried with selling, and yet the combination was workable. William Forrest made it his business to be in Adelaide or Brisbane when land became available and to liaise with and lobby government departments and politicians. It was Forrest, for instance, who gave evidence on land issues at the Royal Commission into the Northern Territory in 1895.[2] As a politician, he was in an excellent position to anticipate new land developments and his zest for seeking out possible purchases of adjoining blocks, and for asserting pressure to regain blocks inadvertently forfeited through nonpayment of rent, cannot be denied. Once, in 1880, he was laid up in Melbourne with sciatica at a crucial time when rents which were due went unpaid. Collins was absent; 'he went out west and got out of my reach', said Forrest.[3] He subsequently went to extreme lengths of intervention at ministerial level to make sure the blocks were not forfeited.[4]

Forrest's desire to increase NAP's holdings by the purchase of adjoining blocks or the leasing of more country as it became available meant that the leasing of the Herbert River auction blocks in early 1882 was critical. In Adelaide he successfully leased six of the twenty-nine blocks; always the businessman, he kept a check on strategic blocks and wrote to McIlwraith:

> When looking at the map you might observe how the purchase of 112 auction block improves our position with respect to our country outside of the auction blocks.[5]

Forrest was also prepared to try other approaches. Morehead, a Brisbane politician and pastoral auctioneer well known to Forrest, held 6,760 square kilometres in the Territory, but by 1880 had forfeited his leases. Forrest was sure 'he [Morehead] wd (sic) try to get his back and transfer to us if we only asked him'.[6] The transfer did not eventuate but the attempt illustrates the manoeuvring which the partners would always contemplate.

As early as 1882 William Forrest expressed an interest in selling Alexandria, despite the fact that the Burdekin properties had been purchased specifically in order to stock the former. A profitable sale of the Territory country was clearly more important to him. He summed up his philosophy of obtaining as much land as possible in the Northern Territory and then selling:

> If we could offer say 8000 to 10000 miles of which from 2000 to 3000 was really good even if dry the well watered country on the Nicholson however bad would sell the former part . . . The good country would sell the bad.[7]

John Costello, an early landholder, emphasised the speculative nature of the early land-grab in the Territory.

> Some took up the country by chance on the map, others went out and saw it for themselves; but the speedy result was that all the Crown lands of the

2 William Forrest's evidence was given at Lennon's Hotel, Brisbane on Saturday, 11 May 1895 and is recorded in Minutes of Evidence of Royal Commission into the Northern Territory, 1895, *SAPP*, Volume 2, 19 of 1895, p.170.

3 W.Forrest to T.McIlwraith, 21 May 1880, OM64.19, Sheet 7, JOL.

4 W.Forrest, Sydney, to Morgan, 18 June1880, Docket 309, GRS 1, PROSA.

5 W. Forrest, Melbourne, to T.McIlwraith, 14 December 1882, OM64.19, JOL.

6 W.Forrest to T.McIlwraith, 21 May 1880, OM64.19, Sheet 7, JOL.

7 W. Forrest, Adelaide, to T.McIlwraith, 8.1.1882, OM64.19, Sheet 17, JOL.

Particulars and Conditions of Sale

OF THE

ALEXANDRIA STATION,

TO BE SOLD BY AUCTION BY

MOREHEADS LIMITED,

AT THE

Wool Exchange, Brisbane, on Wednesday, the 15th February, 1911, at 11 a.m.

BY ORDER AND ON ACCOUNT OF

THE NORTH AUSTRALIAN PASTORAL COMPANY.

The ALEXANDRIA STATION situated on Barclays Tableland in the Northern Territory of South Australia.

Comprising the following leases held under "The Northern Territory Land Act, 1899."

CROWN LEASE, Block 2121, containing 1200 square miles: rent, £210 per annum. Tenure, 42 years from 1st July, 1900.

Block 2122, containing 5342 square miles, rent £200 6s. 6d. per annum. Tenure, 42 years from 1st July, 1900.

Block 2123, containing 2610 square miles: rent £195 15s. per annum. Tenure, 42 years from 1st July, 1900.

Note: The above rentals are fixed for the whole term of the leases.

Block 2114, containing 745 square miles, rent £37 5s. for 1st seven years, £74 10s. for 2nd seven years, £149 for 3rd seven years: the rent for the balance of the term to be fixed by the Minister. Tenure, 42 years from 1st January, 1900.

Block 54, containing 300 square miles, is held under annual lease, payable on 15th October. Also £25 per annum is payable annually to the Government as subsidy on the first bore put down.

25,970 mixed cattle more or less, consisting of:—

7,255 males
16,088 females
2,194 do speyed.
433 bulls.

455 horses more or less.

Together with all Station plant, stores, etc., with all improvements erected thereon, and the vendors' interest in and to the registered brand of the said station. The above numbers are believed to be correct but are not guaranteed by the vendors, and are to be taken as more or less, and the words "More or Less" shall include any number in excess or deficient of those mentioned in the particulars irrespective altogether of any legal or conventional interpretation of such words.

Alexandria Auction Notice.

> Territory quickly passed into the hands of lessees, 70 per cent. of whom had no intentions of stocking, but held only with a view to selling before the three years allowed for stocking . . . Most of it all, then, was sheer speculation.[8]

Such speculation in leases at inflated prices was harmful for the industry, because it meant capital for improvements was limited when owners had paid such high prices for the properties.

The NAP partners intended at first to stock within the three-year limit. Forrest, however, realised there were profits to be made with the Territory country and he preferred to sell rather than be bothered with stocking, so the partnership applied to extend stocking by two years from 1882.

He knew that substantial profits had been made by others such as Kilgour who by late 1881 had sold some of his Tableland holdings. But Forrest left his move too late; for, although within months of the auction of the Herbert River blocks in 1882 he tried to attract buyers for Alexandria at three pounds per square mile, no buyers could be found.[9] Forrest was bitterly disappointed that his efforts were unsuccessful.

Even after the two-year extension for stocking was granted (until 31 December 1884), Forrest argued that the period should extend until 1885.[10] But he was beaten on this point — and Alexandria was stocked. Goyder, the South Australian Surveyor General, was against large areas of land being amassed by such partnerships and was opposed to any further accumulation.[11]

The partners, however, had other avenues of protection for their holdings. William Collins, for instance, was a mortgagee with the Australian Joint Stock Bank over an area of the Herbert River auction blocks — later to be Austral Downs.[12] NAP also had access through McIlwraith to sympathetic political hearings on vital issues — in the extension of stocking conditions, in ensuring leases were not forfeited through late payment of rents, and with financial help when it came to artesian bores. McIlwraith no doubt had influence in Queensland and also in South Australia. In 1879 when the partners inadvertently omitted to pay rent, the Chief Secretary and Minister of Education were successfully lobbied by McIlwraith in order that the claims be protected.[13] A year later, Forrest did likewise.[14] They did not succeed in every approach. Despite lobbying by NAP, it was not until 1881 that Adelaide was used as well as Palmerston (Darwin) for payment of rents. Until then rents were payable only in Palmerston, which was very inconvenient.[15]

What the speculators and pastoralists alike did not realise initially was the adverse influences of poor markets, the scarcity of surface water, and the disease which in the 1890s put the industry in a disastrous situation. Not to be beaten, the NAP partners, rather than forfeit the leases, set out to do what they could with Alexandria. The answer lay in improvements, especially in the drilling for sub-artesian water, which made a vast difference to the viability of the property. Yet the same underlying desire to sell continued, especially after the Burdekin properties had been sold.

A public auction was held at the Wool Exchange, Brisbane, on 15 February 1911, with Alexandria advertised as 10,197 square miles (26,512 square kilometres) and carrying 25,970 of mixed cattle — to be sold as either per head for the cattle and horses or on a walk-in walk-out basis.[16] No sale resulted. In fact, no formal bid was made. John Forrest, with his sons William and Philip and some other friends, was prepared to purchase for £70,000 but Robert Collins was not prepared to lower the reserve price of £80,000. John Forrest then proposed that the stock and station be valued for the purposes of the partnership at £70,000 and that any partner be allowed to sell out to the remaining partners. This could have enabled the Alexander, Warner or Ingram interests to sell. No agreement was reached, however, with some consequent disharmony between the Forrest and Collins families. All, however, were in basic agreement to sell and hopes were raised again in late 1911 when an offer was made to purchase Alexandria. By then the asking price had risen to at least £90,000, but still no sale was forthcoming.[17]

---

8 John Costello, Lake Nash, Evidence, Northern Territory Commission, 1895, *SAPP*, No. 19 of 1895, p.181.

9 W. Forrest, Melbourne, to T.McIlwraith, 11 March 1882, 21 March 1882, OM64.19, JOL.

10 Secretary to Minister of Education to W. Forrest, 11 May 1882, GRS 4/7, PROSA; W.Forrest, Inkerman, to T.McIlwraith, 8 September 1883, Sheet 30, 15 September 1883, Sheet 31, OM64.19, JOL.

11 W. Forrest, Adelaide, to T.McIlwraith, 8 January 1882, OM64.19, Sheet 17, JOL.

12 V.T.O'Brien, compiler, 'Kidman Properties in the Northern Territory', Darwin, Department of Lands, n.d., pp.25-6.

13 W. Morgan, Chief Secretary, to Hon. King (Minister of Education), 17 November 1879, Docket 570, GRS 1, PROSA.

14 William Forrest to Thomas McIlwraith, 21 May 1880, OM64.19, Sheet 7, JOL.

15 *SAPP*, Volume 2, 29A of 1881, p.1.

16 Notice of Auction, 15 February 1911, NAPR.

17 R.M. Collins, Tamrookum, to Jane Collins, 21 February 1911, 7 October 1911, NAPR.

There were good seasons between 1910 and 1919 and with the partners enjoying much higher dividends the immediate desire to sell abated. Furthermore at this stage there were major changes in administration, which precluded any major decision making. With the death of John Forrest in 1911, the position of managing partner passed to his son William Tyler Forrest. He was not suited to the task, however, and was not much interested. No significant decisions were made during his tenure from 1911 to 1915 (Alexandria was fortunately enjoying good seasons), and the position of managing partner passed to his younger brother, Philip, before both went to fight in World War I.[18] The only son of Robert Martin Collins, Christopher John Collins, was that family's representative in the partnership but he was never a serious contender for the position of managing partner, being already involved with Collins family business. He died in 1919.

Since the earliest days, the NAP managing partners had been assisted in the secretarial work required in Brisbane, and by agents in Adelaide. F.M.Hannington was employed in Brisbane, from the early 1880s until 1900, although he did not work exclusively for the partnership. The secretarial position did not become full-time until after the three generations of Beardmores. The first, George Oakes Beardmore, had known the partners through being manager of the Queensland Freezing and Food Export Co. Ltd. which had dispatched Queensland's first frozen meat in 1884.[19] During the next decade, he set up an accountancy firm with one of his sons and from 1901 to 1913 he worked for NAP as well as for John Collins and Sons. This began the Beardmore era for NAP, a happy association which lasted forty years.[20] In 1913, George Beardmore's son, Ernest Russell Beardmore, an accountant, took over from his father as secretary and, until Philip Forrest returned from active war service, he attended to NAP business. His role cannot be underestimated as he managed the affairs of the company for some years without assistance from the partners. R.W.(Roy) Beardmore from the third generation of the family, took over as secretary in 1935 following his father's death in 1934. Sanders and Packard had been NAP's agents in South Australia until 1887 when Sanders died and John H.Packard continued alone on behalf of the company. They provided a satisfactory service, although William Forrest liked to be in Adelaide as often as he could to attend to the business himself.

The NAP head office in Brisbane had been in the Dalgety's Building, Elizabeth Street, Brisbane but in 1932 Dalgetys required more floor space and NAP moved to the Commonwealth Bank Buildings in Queen Street. Early in 1935, there was another move, this time to Morehead's Building at 185 Mary Street because, as Phil Forrest explained, the former office was 'very small and the rent high'.[21]

Philip McIlwraith Forrest was born on 3 May 1883 at his father's Avoca sugar plantation at Bundaberg, which had been purchased with financial assistance from McIlwraith three years previously. On learning of the birth, McIlwraith stated he wished to be the child's godfather and henceforth Phil gained McIlwraith as a middle name. During World War I he joined the AIF 11th Light Horse Regiment while his older brother, William Tyler Forrest, served in the Royal Artillery.[22]

Following his war service, Phil Forrest had the responsibility of managing the partnership and, from 1931 until 1936, the company as well. The position was rather thrust on him by force of circumstances after his brother resigned in 1915, and after a decade he remarked that he did not particularly relish the '3 months of the year on station inspection and the balance . . . practically tied to Brisbane'.[23] He found it necessary to keep a constant check on NAP affairs and in 1924 he lamented that although the war intervened he had never taken his wife on an intended European trip; again in 1929 his frustrations were evident when he wanted to take his daughter to England for a year 'to finish as I think it means so much to a girl'.[24]

Forrest could indeed be considered unlucky because of the impact that drought and deteriorating markets had on the fortunes of NAP and subsequently on his role as the company's

18 The 1915-6 Income Tax Return shows William as Managing Partner for three months and Phil for the last nine months, NAPR.

19 OM64.19, Sheet 16, JOL. He was very family oriented although tragically five of his children died from a measles epidemic in 1875. A move to Brisbane to work as manager of the Queensland Meat Export and Agency Company, with which the NAP partners were closely associated, also brought personal loss when the 1893 flood destroyed his home.

20 G.O.Beardmore, 'Glimpses of early Australia', Part V, *Queensland Government Mining Journal*, 65, November 1964, pp.600-1.

21 P.Forrest, Directors' Report, 27 June 1935, NAPR.

22 Edith Nancy Grace, Philip Forrest Typescript, Toowoomba, 1989.

23 P.Forrest to Major H.Ingram, 25 January 1925, H.Ingram File, NAPR.

24 P.Forrest to H.Ingram, 14 October 1924; 3 December 1929, H.Ingram File, NAPR.

manager. The financial statements reveal the story of good profits in the early years of the century followed by a drastic decline after World War I. With the exception of 1912, a dividend was always paid between 1901 and 1919, while between 1919 and 1938 dividends were paid only five times.

Balance Sheets, 1901-1938, NAP.

Dividends paid:	
1901	£ 2,000
1902	£ 4,000
1903	£ 6,000
1904	£10,000
1905	£ 8,000
1906	£14,000
1907	£18,000
1908	£16,000
1909	£14,000
1910	£14,000
1911	£54,000
Sale of Burdekin properties:	
capital repaid 30/6/1911	£28,000
capital repaid 19/9/1911	£30,000
1912	no dividend
1913	£46,000
1914	£58,400
1915	£31,120
1916	£30,592
1917	£29,376
1918	£ 8,640
1919	£25,920
1920	no dividend
1921	£31,968
1922 - 1925	no dividends
1926	£4,320
1927	no dividend
1928	£3,456
1929	no dividend
1930	no dividend
1931	£4,320
1932 - 1938	no dividends

Every year seemed to bring new anxieties; in 1926, at the beginning of a protracted drought Forrest commented:

> My policy in looking after the North Australian Pastoral Company's affairs has been to hope for the best but be prepared for a bad time . . . .[25]

Unfortunately the 1926-28 drought was particularly bad, one of the worst Alexandria had experienced, and it put the partnership into debt. Very few of the cattle could be sold because it was too dry for them to travel and the partnership was overdrawn on its running expenses.[26] Forrest was aware of the financial difficulties of the company, with no dividends being paid. Like his uncle and father, however, he pushed on with 'dogged perseverance'. With no outlet in the way of fattening properties, NAP was at the mercy of store stock buyers such as Sir Sidney Kidman. In 1926 in discussions with Kidman over the sale price of Alexandria cattle, Forrest took two hours to get an extra 1s 3d per head which pushed the price up to £5 1s 3d each for a mob of 8,000 bullocks.[27]

He travelled to Alexandria in October 1928 to see the effects of the drought for himself and agreed with manager Johnston that 'Alexandria could hardly look worse. Nothing but bare ground as far as the eye could see'. The cattle were walking out from bores up to 13 kilometres (eight miles) to feed. But Forrest also commented that it was necessary to keep purchasing bulls. 'It would be false economy to let the herd deteriorate'.[28] The position on Alexandria could change so quickly. In 1932, the position was 'very grave', with stock so low in condition that it was useless to attempt to make sales, but an unexpected fall of fifty to seventy-five millimetres (two to three inches) of rain late in May saved what appeared to be a desperate situation.[29]

The difficult times of dry seasons and economic depression created new pressures in the partnership and an unfortunate but significant series of events occurred between 1921 and 1936 which led to Philip Forrest's resignation and departure after twenty-five years service with the company. The basis for these developments was a re-awakening of interest in NAP from the English partners, particularly the Warners. Until the 1920s they had concurred with whatever the Australian partners had wanted, but now complacency had given way to renewed interest — and influence.

The English-Australian connection had always been successful while NAP was financially buoyant, but in times of downturn, perhaps inevitably, people began to express concern. Not only did distance create communication difficulties, but the English partners found it difficult to imagine what it was like in the vast waterless plains of Alexandria, with drought, disease and difficulties in marketing. The situation was particularly remote in the Warner case since

25 P.Forrest to Woolley, Beardsleys and Bosworth, 11 November 1926, NAPR.

26 *ibid.*

27 P. Forrest to Richard Holt, 7 May 1926, Richard Holt File, NAPR.

28 P.Forrest to D.M.Fraser, 1 October 1928, NAPR.

29 P.Forrest, Directors' Report, 30 September 1932, NAPR.

communication was through their solicitor. By contrast, Sir Herbert Ingram often sent letters to Phil Forrest. In turn, Phil kept him informed, a situation upon which Ingram was later to comment: 'Forrest used to send to me reports from time to time which I much appreciated'.[30] Indeed, following a meeting between them in England, their letters became less businesslike and were interspersed with some references to their families, and to cricket.

The first problem arose in 1922 when, with a slump in cattle prices and no dividend paid, the sale of Alexandria was again being contemplated. Woolley, Beardsleys and Bosworth, the solicitors in England for the William Pochin Warner Estate,[31] wrote to Phil Forrest in 1922 expressing their dissatisfaction at the 'unbusinesslike way' their clients had been treated. Having received no communication for three years, their approval was suddenly and hurriedly being sought to sell Alexandria to a prospective buyer for £250,000. They questioned how a considered opinion could be given when in three years the only communication from their Brisbane legal representatives had been the arrival without explanation of £3,600, which they presumed to be dividends.[32] Forrest was so disappointed that the Warner interests had obviously lost confidence in the company, a situation for which he felt personally responsible, that he tendered his resignation. A meeting of the partners on 6 June 1923, however, persuaded him to withdraw it and stay on as managing partner.[33]

With solicitors Thynne and Macartney as the new agents for the Warners in Brisbane, Forrest went out of his way to keep the Warner interests informed and wrote numerous letters to England, in addition to sending annual reports. There was still an undercurrent from England, however, especially from the Warner interests, which were indeed keen to sell Alexandria. Disappointed that the £250,000 sale to A.J.Cotton of Brunette did not eventuate — the purchaser could not raise the capital[34] — they were also critical that Alexandria was doing so poorly. Furthermore, they correctly pointed out that being a partnership, rather than a limited liability company, the partners were susceptible to serious liabilities. The partnership was limited to twenty members which also made dealings difficult when shares were left to several beneficiaries. Finally, the Warner interests were particularly anxious because the estate of Captain Warner could not be wound up; the beneficiaries much preferred to have access to their money rather than to leave it in a trust.[35]

Clearly, the Warner interests wanted NAP to be sold and for the money to be divided. They were not alone in their opinion. Phil Forrest commented in 1926: 'All partners here realise how complicated the partnership is becoming and are anxious to realise . . . '.[36] CAY Johnston later summed up the situation,

> There was that many [shareholders] in it that the accountant . . . told me one time the only way he could balance it up was to draw a big tree with a lot of branches on — he'd start at the top and run that one down — . . . then he'd run another line down.[37]

But it was difficult to find a buyer for such a large property and as with Brunette which was also for sale, no buyers were forthcoming.

As early as 1927 Phil Forrest expressed a desire to form a limited liability company. Disagreement with the company's solicitors over their handling of this matter and a lingering discontent over the manner in which the same firm had earlier neglected to represent the Warners adequately led him to transfer NAP's legal business. The decision to break with the solicitors was not taken lightly, Phil Forrest and his father having been personal friends with partners in the firm for many years. But Forrest was not one to stand for inefficiency and made the decision 'both in the company's interests and also my own'.[38]

In February 1930 Crouch and Paterson and in particular Edward Robert Crouch, formally became the solicitors to NAP. Crouch had acted on behalf of the executors of the will of R.M.Collins and more recently, in 1927, on behalf of the estate of Jane Collins. After NAP was formed into a limited liability company in 1931, both English interests appointed him as their alternative[39] director, and he

30 H.Ingram to E.R.Crouch, 17 January 1947, Crouch and Paterson File, NAPR.

31 This estate held 7/10 of the Warner shares; the C.B.Warner Estate held 3/10 of the Warner shares.

32 Woolley, Beardsleys and Bosworth to P.Forrest, 5 December, 1922, 11 December 1922, 18 January 1923, NAPR.

33 H. Ingram to E.R.Crouch, 17 January 1947, Crouch and Paterson File, NAPR.

34 A.J.Cotton, *With the big herds in Australia*, Hidden Vale, A.J.Cotton, 1931, p.189; P.Forrest to Maj.H.Ingram, 25 January 1922, H.Ingram File, NAPR.

35 Woolley, Beardsleys and Bosworth to P.Forrest, 23 September 1926; P. Forrest to Sir Herbert Ingram, 8 January 1930, Crouch and Paterson File, NAPR.

36 P.Forrest to Woolley, Beardsleys and Bosworth, 23 July 1926, NAPR.

37 CAY Johnston Interview, 26 May 1981, NTRS 226, TS 250, NTAS.

38 P.Forrest, 12 October 1927, NAPR.

39 Generally referred to as 'alternate director'.

Edward Robert Crouch. E.M. Crouch
NAP's Solicitor 1930-1955
Alternate Director for English Shareholders 1932-1955.

also had a significant role in advising on legal matters as a member of the board.

Edward Robert Crouch was born in England on 11 January 1873. Educated at Brisbane Grammar School, he represented Queensland at tennis and cricket and was admitted as a solicitor on 4 December 1894. He formed several partnerships throughout his legal career — Crouch and Darvall (1896-1909), Crouch and Eden (1909-22), Crouch and Paterson (1930-52); Crouch and Crouch (1952-88). The firm presently continues as Crouch and Lyndon. E.R.Crouch died in Brisbane on 8 August 1962.

The proceedings for the formation of the company were relatively quiet until trustees of the will of Captain William Pochin Warner, solicitor Walter Ernest Puckering and an accountant John Robert Humphrey, arrived from England in November 1930 to make enquiries about the best means of winding up the partnership. Realising the impossibility of a sale of Alexandria they stayed for two months and were very helpful in the conversion of the partnership to a limited liability company. They had planned a trip to Alexandria but were advised against this because of the heat and the 'primitive conditions'[40] on the property. At a meeting of partners on 12 December 1930, a subcommittee, Phil Forrest, W.E.Puckering, J.R.Humphrey, D.M.Fraser and E.R.Crouch, was set up to decide on the best form of company for the sake of tax advantages. At the same meeting it was decided that the partnership should buy out William T.Forrest's interest (5/32) which had to be sold to satisfy the Queensland Trustees holding the W.T. Forrest shares. Subsequently, this was done for £10,000.

While the advice from the English visitors was helpful and contributed to improved relationships between the English and Australian shareholders, NAP also sought advice from William Kelly, a taxation expert from a Brisbane firm. He advised having the registered office at Alexandria because if it were, leases would have to be registered only in Darwin, and double registration fees would be avoided. The company would also be exempt from income tax because under Section 5A (1) of the Income Taxation Act, income derived from primary production in the Northern Territory was exempt. Under the existing partnership, members could not claim this exemption because they were not themselves resident in the Northern Territory.

The North Australian Pastoral Company Limited was incorporated under the South Australian Companies Act of 1892 and became a limited liability company on 30 March 1931. The new company commenced business on 1 July 1931, and the first annual meeting was held on 22 September 1931 at the company's branch office in Dalgety's Buildings, Elizabeth Street, Brisbane. At the meeting of directors that morning Phil Forrest was elected Chairman and Managing Director, while other directors present were Edward M.Shaw, trustee of the R.M.Collins Estate; John W.F.Collins, son of William Collins, representing the William Collins Estate; Thomas McIlwraith Taylor, representing the Alexander shareholders (the position being transferred in 1934 to William Forrester Alexander);[41] and E.R.Crouch, company solicitor and later alternate director representing the English interests — Walter Puckering (1932-34)[42]

40 W.F.Alexander, 'Ninety years of the N.A.P.Co. 1878-1967', Unpublished Manuscript, Brisbane, ca.1968, p.22.

41 P.Forrest to Sir Herbert Ingram, 4 September 1931; 5 July 1932, H.Ingram File; Minutes of Directors' meeting, 22 September, 1931, NAPR.

42 Minutes of Directors' Meeting, 21 January 1932, NAPR. Puckering asked E.R. Crouch (letter dated 27 November 1931) to be his alternate director. The board subsequently approved this at a Directors' Meeting on 21 Janaury 1932.

and Sir Herbert Ingram (1934-55).[43] Included in the next meeting of directors four months later were Bill Frith, the company's pastoral inspector, and Douglas M.Fraser, representing the holding of his mother Lady Anna Bertha Fraser. The registered office was at Alexandria with CAY Johnston, Alexandria's manager, as secretary.

The shareholders' meeting in the afternoon of 22 September 1931 was a relatively small and short affair, with only five shareholders present. Mrs Mary Adelaide Gwendoline Collins, the widow of William, was the only woman present. Forrest read a report from the directors and the meeting voted on formal matters concerning the powers of the directors. The death of Richard Holt, the long-time manager of Alexandria,[44] was noted with regret.

Under the partnership, meetings had been quite informal; even if partners were absent, generally their views were known. Under the new arrangement, business had to be carried out formally. At a general meeting in October 1932, therefore, proxies from Mrs William Collins and Mr John Collins were ruled out of order having been received within forty-eight hours of the meeting.[45]

DISTRIBUTION OF SHARES 1932

Shareholder	Number	Nominees	Number
A.C. Collins	1,975	A.C. Collins	1,975
Est late W.P.Warner	22,222	Edward G. Warner	8,889
		Dorothy Kate Eddowes	2,222
		Grace Warner	2,222
		Amy Wilkins	2,222
		W.E.Puckering M.F.Warner J.R.Humphrey } in joint names	6,667
P.McI. Forrest	22,222	P.McI. Forrest	22,222
Trustees Est. R.M.Collins	6,585	Henry Bruxner) Edward M. Shaw) Edward R. Crouch) as trustees in estate R.M.Collins	6,585
M.A.G.Collins	6,913	M.A.G.Collins	6,913
Blanche Margaret Alexander	8,889	Harriette Margaret Forrester Alexander	500
		Janet McIlwraith) William Forrester Alexander)	1,000
		Blanche M.Alexander	7,389
Trustees est. J.G.Collins	9,877	D.M.Fraser(trustee)	9,877
H.Bruxner as Exec. est.C.J.Collins	13,169	Henry Bruxner(as exec. will J.G.Collins)	2,303
		Henry Bruxner	10,866
Lady Fraser	5,926	John Simon Fraser	100
		Peter Fraser	100
		John Neville Fraser	2,863
		Douglas Martin Fraser	2,863
Sir H. Ingram		Sir H.Ingram	22,222
		TOTAL	120,000

With the company now in place, the directors turned to the problem of low returns from Alexandria. Phil Forrest commented in 1932 that current prices 'would only return enough to pay off our overdraft and provide expenses for the ensuing year'.[46] Two years later he summed up the situation:'We have had practically no return on our capital for more than ten years and small prospect of any dividends for some years to come'.[47] Even though bullocks fetched up to £14 on the Melbourne and Adelaide markets while making only £3 if sold on the station, the directors discounted the possibility of selling them in the south as fats because they would lose too much condition from travelling.

The possibility of selling Alexandria was discussed again in 1934 but it was accepted that obtaining a reasonable sum for it seemed impossible. The alternative was to secure a fattening depot and thereby secure the difference between store and fattened cattle prices which was presently going to people like Kidman. The decision was made to purchase Marion Downs, a station of 10,956 square kilometres on the Georgina River in western Queensland, which at the time had 7,500 cattle. As Phil Forrest explained:

> At present values, Alexandria alone, could only sink into debt, but with Marion, granted a few decent seasons, we may not only pay our way, but have a balance to divide.[48]

NAP took a month's option to purchase Marion Downs for £40,000; and Phil Forrest and Bill Frith, the pastoral inspector, set out to inspect the property. They were held up by the flooded Burke River for eight days, and so had to extend the

43 Minutes of Second Half-Yearly General Meeting of Shareholders, 1 November 1934, NAPR.

44 General meeting, Brisbane, 22 September 1931, NAPR.

45 Minutes of general meeting, 20 October 1932, Brisbane, NAPR.

46 P.Forrest to Solicitors for Warner Estate, 22 October 1932, Crouch and Paterson File, NAPR.

47 P. Forrest to E.R.Crouch, 1 February 1934, Crouch and Paterson File, NAPR.

48 P.Forrest to E.R.Crouch, 24 March 1934, Crouch and Paterson File, NAPR.

option by two weeks in order to inspect the property. Rather than wire their offer of £30,000 which was to 'court a refusal and end negotiations', two directors, Douglas Fraser and Bill Frith, went to Melbourne to meet the vendors, the Mackinnon brothers, in person. They were pleasantly surprised when the Mackinnons accepted the £30,000 without a counter offer. They had held Marion Downs for forty years but of late had not been seriously attempting to use it. Forrest added: ' . . . they are wealthy people and held it more from sentiment than business reasons'.[49] A deposit of £15,000 was paid in cash and the balance of £15,000 was due in twelve months at 4 per cent interest. The bank would have provided the total required but its interest of 5.25 per cent was considered too high.[50]

The subsidiary company of Marion Downs Pty. Limited was formed to save separate income taxes having to be paid in either the Northern Territory or Queensland. The directors were the same for both companies. Phil Forrest, however, had some doubts about the acquisition, claiming that since the Georgina and Diamantina country was affected at the same time as Alexandria, no benefit could result. Nevertheless he yielded to the majority board decision. D.M.Fraser was very much in favour, however, and this decision to purchase was the first obvious example of Fraser's influence in the company.[51]

The problem of distance was always an inconvenience, and the trust and confidence placed in the Australian shareholders by the English interests continued to be of utmost importance. While Herbert Ingram and Phil Forrest were the two largest individual shareholders, Forrest had the added responsibility of managing the company. Ingram commented to him about the Marion Downs purchase: ' . . . you have more to lose than any of us . . . . I am content to abide by your advice in this matter'.[52] But there were again misgivings by the Warner interests who could not understand expanding the company when NAP was in such a poor financial state. Forrest tried to explain:

> . . . this new departure is not our wish to expand but an endeavour to make dividends and thereby make the North Australian Pastoral Coy's assets saleable.[53]

It is interesting that the intention to sell was still in Forrest's mind.

While in the first year no fats were turned off Marion Downs, it soon proved to be a profitable asset and well worth the risk taken in purchasing it. For the first time, Phil Forrest's opinion (not to purchase Marion Downs) was under question and an uneasiness was developing between him and D.M.Fraser. In this context, the events to unfold in 1935-36, although they seemed trivial, betrayed deeper feelings. History has shown that D.M.Fraser was such a strong personality that perhaps there was not room for them both. Forrest, described as a 'shy, touchy man [who] did a difficult task well',[54] decided to leave while he was still in control, the best possible exit after such a long association with NAP.

In 1935 the problem arose once again, mostly because of the great distance which separated the English and Australian partners. Following the death in June 1934 of Director Walter Puckering, Sir Herbert Ingram was elected a director on behalf of the English shareholders. Solicitors in England, Woolley, Beardsleys and Bosworth, also attended to NAP matters on behalf of the Warner family.[55] They wrote from England in December 1935 requesting more copies of the balance sheets and reports of June 1935. They went on to comment how much they regretted the state of the company since the *Times* and *Financial Times* in England had reported

> that Pastoral Companies in Australia were greatly improving their position . . . . [Yet from NAP] for the last five or ten years we have received nothing but pessimistic reports.[56]

They were also critical of both the amount of wages paid (which represented 66 per cent of the value of the stock sold), and also of the cost of pumping.

The newspaper reports were inaccurate. Certainly the company, in particular Phil Forrest, was not to blame for the situation. He replied, giving an honest assessment of the situation. From January to December 1935, neither Alexandria nor

49 P.Forrest to E.R.Crouch, 24 March 1934, Crouch and Paterson File; P.Forrest to Woolley, Beardsleys and Bosworth, 6 April 1934, NAPR.

50 P.Forrest to Sir Herbert Ingram, 26 March 1934, H.Ingram File, NAPR.

51 P.Forrest, Brisbane, to D.M.Fraser, Mundoolun, 5 July 1928, Amalgamation File, NAPR.

52 Sir Herbert Ingram, Cirencester, to P.Forrest, 30 April 1934, Crouch and Paterson File, NAPR.

53 P. Forrest to Woolley, Beardsleys and Bosworth, 6 April 1934, NAPR.

54 W.F.Alexander, 'The N.A.P.Co., 1878-1967', Unpublished Manuscript, ca.1968, p.1.

55 Minutes of Second Half-Yearly General Meeting of Shareholders, 1 November 1934, NAPR.

56 Woolley, Beardsleys and Bosworth, Loughborough to NAP, 9 December 1935, NAPR.

Marion Downs could make any sales, and 'if they could have, it would not have been possible to get the stock away as the rains having failed, the roads were closed'. This was the reason for the high wages to sales ratio. He explained the necessity to pump for water, some bores being situated 60 kilometres from any track and visited only once a fortnight. While present values for stock were low, it did not pay to pump for water; 'But what are we to do? Stop pumping, and in a few days our main asset would disappear'. He highlighted the instance of low prices for cattle. For example, 5,000 head netted just 13s 8d each, and this was the only offer. The buyer gambled on early rains but the drought persisted and he nearly lost the lot. As Forrest explained: ' . . . it was better to have over £3,000 in the bank and ease the position for the remainder of the herd'. He gave examples of other stations in the area, such as Austral Downs which made a loss of £62,000 in the previous year, and of the losses being experienced on Lake Nash, Barclay Downs and Brunette. In conclusion he commented: 'I have done my best over a long trying period, but I cannot make it rain'.[57]

Unfortunately he took the Warner correspondence as a personal criticism and resigned from 29 February 1936, despite pleas by the board and shareholders not to do so. The Warner interests were astonished, claiming that the comments were healthy criticism and nothing 'to which a reasonably minded person could object'.[58] Nothing would change Forrest's mind, and he left NAP and Moreheads, severing the family interest in NAP which had lasted fifty-nine years. He purchased a dairy farm at Seventeen Mile Rocks near Brisbane and commuted between the farm and his St Lucia home. He died in 1964, aged 81. His joint position of managing director and chairman was filled by Douglas Martin Fraser and his shares were bought by Francis Foster, of Hobart. Henry Bruxner, husband of Annie, the eldest daughter of Robert Collins, filled the vacancy created on the board.

Although still essentially a family company, NAP was distanced from its original configuration by the departure of Forrest. While the English interests remained as isolated as they always had been, the McIlwraith interests had decreased to a minority and the Collins brothers, with all their enterprise for exploration and stock matters, had little control over the fact that in the next generation they produced mainly daughters; women still had no real place in the male-dominated world of the pastoral industry.

One person to emerge from the Collins family who would, as some would argue, more than make up for this lack of numbers, was Douglas Martin Fraser or DM as he was known. He was joined in the next phase of the company's development by Francis Henry Foster from Hobart, a pastoralist, businessman and, for a time, politician. The two brought about change and expansion reminiscent of the early days of NAP, but with one purpose: NAP was going to survive and prosper.

57 P. Forrest, Brisbane, to Woolley, Beardsleys and Bosworth, 21 January 1936, NAPR.

58 Woolley, Beardsleys and Bosworth to Secretary, NAP, 17 March 1936, NAPR.

CHAPTER 6

Station Life on Alexandria: 1877-1938

Everybody spoke well of it. Everybody talked about working there. The men that worked there never left . . . they stayed till they died . . . I'd have died there.[1]

Alexandria Homestead from Stockyard 1920. Peg Corbett

The sentiments of CAY Johnston[2], manager from 1924 to 1938, spoken two years before he died in 1983, sum up the effect that Alexandria had on the people who lived and worked there. Despite what some might see as a life of isolation, desolation and inadequacies, for others it was all they needed or ever wanted. Was it the impressive size which kept them there? Perhaps the managers were particularly good or, with the stability of ownership, NAP was seen as a fair employer.

The issues can be debated and in the fullness of time one could romanticise. The bush ethos is compelling and can colour the interpretation of the past. But for Alexandria, sufficient evidence exists to show that the place developed in its people a sense of belonging, that the managers in this period were fitted to their task and that NAP proved to be as sympathetic and understanding as

1 CAY Johnston Interview, 26 May 1981, NTRS 226, TS 250, NTAS.

2 Charles Augustin Yorke (CAY) Johnston.

it could, given the distance of the property from its Brisbane headquarters.

There were problems, however, and while the basis of any major concern in the pastoral industry is environmental, such influences in the early period were compounded by other factors. With no fences, very little surface water, disease and distant markets, the problems often seemed insurmountable. When selling-out was no longer an option for the NAP partners, they set about solving their problems and they did it with enthusiasm. Bores were put down; markets were encouraged; and Marion Downs, and later other fattening properties, were acquired to enhance profit margins. Two groups made all this possible; the partners making the business decisions and the station employees who implemented their policies. The role of both parties must be recognised; and this chapter concentrates on the latter and their life in the outback.

The first manager of Alexandria, Thomas Harding, had taken up and managed Connemara[3] for De Burgh Persse in 1875; Hardings Ranges on the property were named after him. He took up runs of his own in the same area in 1878 but sold these in 1881 prior to his departure for Alexandria. He also knew the Collins family, having taken up Toolah run for Robert Collins in 1875.[4] In 1882, he established NAP's presence on Alexandria at 'Top Camp' on the Ranken River and by early 1885 had moved to the present site of Alexandria's homestead. He was there at a time when patterns of contact were first being established with the Aborigines. Nat Buchanan found the Aborigines on the Tablelands showed little hostility to whites and only on one occasion did he have any trouble — 'a few shots' soon scattered them.[5] Initial encounters on Alexandria, however, were not so easy. While the Wakaja[6] tribe of the Soudan, Avon Downs and Camooweal area seemed to be the original tribe of most of Alexandria's station Aborigines, other tribes in the vicinity included the Wanji (Nicholson),[7] Indjilandji (Ranken),[8] Wambaia (Brunette),[9] Kunindiri (Cresswell)[10] and Jaroinga (Lake Nash).[11]

While there were only 500 Europeans[12] in the Territory in 1882, white settlement brought inevitable disagreements with the blacks. In 1885, just as Alexandria was being established, two whites, John Ross and a man named Ben, both well-sinkers, were reported murdered at the station.[13] J.S.Little, who was manager of Austral Downs at the time, spoke of 'a large tribe about there, they evidently meant to clear the whites out'. He went on, 'I do not know what step Mr. Harding took to punish the murderers, but in my opinion there is only one remedy, which Northern Queenslanders have never found to fail'.[14] This of course was to 'disperse' the blacks, which in reality meant 'shoot them'.

By the 1890s, cattle were stirring up waterholes which subsequently caused the holes to silt up. The Aborigines' eating grounds were also being threatened, so they were allowed to come into the stations. Tensions between black and white, however, still ran high, especially after the murder of a well-sinker at Alexandria in 1892. The man had walked a little distance from camp one night and was attacked, his skull being split in two with a tomahawk.[15] Tom Law, a stockman from Alexandria during the 1920s, continues the story:

> A punitive party was organised, and at Pika Pika, N.W. of the Homestead it engaged an aboriginal tribe who was thought to be sheltering the murderer. From what old Victor told me, it must have been just plain slaughter. When passing the area with him one time on patrol with the police, he said the place just stank with dead blackfellows.[16]

The real murderer had made his way to Herbert Vale Station still carrying the axe, but before the

3 In Queensland's Channel country, Connemara was purchased by NAP in 1986.

4 Gregory North Pastoral Holdings, LAN N42, N44, N45, QSA.

5 G.Buchanan, *Packhorse and Waterhole*, p.39.

6 Also spelt Wogaia, Wagaja, Waggaia, Wagai, Waagai, Wagaiau, Waagi, Warkya, Worgaia, Workaia, Warkaia, Workia, Workii. Norman B.Tinsdale, *Aboriginal Tribes of Australia*, Berkeley, University of California Press, 1974, p.236.

7 Also spelt Wanyi, Wanyee, Wanee, Waangyee, Wonyee. Tindale, *op.cit.*, p.237.

8 Also spelt Indilandji, Indjilindji, Injilinji, Intjilatja, Indjurandji. Tindale, *op.cit.*, p.226.

9 Also spelt Wombaia, Wambaja, Wampaja, Wonbaia, Wombya, Yumpia, Umbaia, Umbia. Tindale, *op.cit.*, p.237.

10 Also spelt Goonanderry, Leecundundeerie, Cundundeerie, Kunandra. Tindale, *op.cit.*, p.229.

11 Also spelt Yaroinga, Yarroinga, Yaringa, Yorrawinga, Yarrowin, Jurangka. Tindale, *op.cit.*, p.227.

12 Government Resident, Quarterly Report on the Northern Territory, *SAPP*, Volume 4, 147 of 1882, p.2.

13 Report of Administrator, 1912, *CPP*, 1913, p.102.

14 The man, Ben, is also referred to as Benigan or Donovan by J.S.Little; J.S. Little, Austral Downs Station, 22 April 1885, in Quarterly Report on the Northern Territory, 30 June 1885, *SAPP*, Volume 3, 55 of 1885, p.3; J.S. Little to Government Resident, 28 April,1885, Docket A8136, GRS 10, PROSA.

15 W. Forrest, Brisbane, to Treasurer, 23 March 1892, Docket 171, GRS 1, PROSA.

16 T.Law letter in W.F. Alexander, 'Ninety years of the N.A.P.Co. 1878-1967'.

Richard Holt (manager of Alexandria 1900-1924) leaving for Brisbane in 1916. Grant Dearden

police arrived the natives there had killed him — in the same way he had killed the white man.

During the 1890s, Harding reported that the blacks were troublesome in the Nicholson country at a time when Brunette and Alexandria cattle were being depastured there, because of drought on those properties. Harding and the Brunette manager, James Hutton, commented how difficult it was to get men to go there.[17] Tensions subsequently eased and the tribes in the area acquired the reputation of being more peaceful than those in other areas, such as the Roper River region where they were reputed to be 'very treacherous',[18] or in the Gulf country where Landsborough had 'found them anything but friendly . . . and actually deemed it dangerous to camp near the waterholes'.[19] By the late 1890s, Aborigines were employed on Alexandria; indeed most stations by then were being worked by blacks under a manager and a few white stockmen.[20] Aborigines were not mentioned, however, in the staff list of 1895.

ALEXANDRIA STAFF 1895[21]

Thomas Harding	Manager
Thomas D. Seymour	Bookkeeper
Moritz Helwig	Station hand
J. O'Brien Hobson	Pastoral student
William Jackson	Horserider
John McNamara	Well sinker
John Bennett	Artesian well borer
William Aspland	Labourer
George Cant	Station hand
Frank Strahan	Labourer

Harding managed Alexandria, staying on until 1898 when, as the evidence suggests, he left to take up pastoral leases of his own.[22] The staff list also shows that artesian boring was under way and a pastoral student, probably out from Britain to gain colonial experience, was in residence.

Tennison H.Robbins, who had been a drover in the Territory, followed Thomas Harding as manager in 1898; but there is also reference to a Mr Martin during this time.[23] By May 1900, however,

17 T. Harding to W. Forrest, 17 June 1892, Docket 481, GRS 1, PROSA.

18 Mr.Crawford of Hodgson Downs quoted in Telegram, Government Resident to Minister of Education, 1 June 1888, Docket 541, GRS 1, PROSA.

19 G.Sutherland, *Pioneering Days; In the early Sixties*, Brisbane, Wendt, 1913, p.12.

20 Mounted Constable Stone, 9 March 1898, Docket 333, GRS 1, PROSA.

21 *N.T. Times and Gazette*, 21 June 1895.

22 Docket 8369, 11 July 1898, GRS 10, PROSA.

23 Stated in letter by Thomas Law, 30 May 1968, in W.F.Alexander, 'Ninety years of the N.A.P.Co. 1878-1967'.

Richard Holt had started his long period of management. The story goes that Martin mistimed the move to shift cattle from Soudan to the Nicholson, and 4,000 head perished. He made a hasty exit from the station and recommended Holt for the job.[24]

Richard Tweedy (Dickie) Holt was born in Melbourne in 1859; seven years later his family travelled north to Rockhampton. At 15 he was employed by surveyors on Westland Station in the Longreach district, and having a liking for station life, he became a stockman there. Following an unsuccessful trip to the Kimberley goldfields in 1887, Holt came to Alexandria and was soon head stockman.[25] In time, as manager, he became a well-respected and influential identity on the Barkly, often consulted by the Government Resident in Darwin to report on the Tableland. The dip at Anthony Lagoon was the result of such advice. Alexandria was rather a showplace and in 1922 a report towards the end of Holt's management stated that,

> much valuable information may be obtained by an inspection of Alexandria Station by pastoralists desiring to take up land on dry areas of the Northern Territory . . . The improvements on this station are good, and maintained in excellent condition.[26]

Holt was further praised for his 'efficiency, combined with economy, in running the station plant'.[27] He was a Justice of the Peace and signed death certificates for those who died in the area. He never married but had a part-Aboriginal son, Bruce Holt, who worked on Alexandria and was a very good mechanic. Although drink sometimes proved his weakness, Dickie was an excellent cattleman and at home in the harsh conditions.

He was a rough bushman — it was certainly a difficult place for a white woman — and a stockman in Holt's time tells of the time the bookkeeper managed to employ a Mrs Walker as housekeeper.

> Poor Mrs Walker. Holt would always address her as 'Woman'. At the table for instance it was 'What will you have woman'. Not satisfied to have her keep her position as Housekeeper, he found out that she hated dogs . . . He promptly sent out for the biggest and ugliest bulldog he could find, and installed it in the house . . . We lost our housekeeper.[28]

Victor, King of Alexandria, his wife Maggie and their son Fulche. Tony Johnston

Hospitality, however, was generally very much part of station life and when a long-standing resident left, the district always turned out to the farewell. One such event occurred on 8-9 May 1901, when a two-day race meeting was held at Corella Station for James Hutton who had managed Corella, later incorporated into Brunette Downs. There was success in the Flying Handicap for Richard Holt's horse Whalebone, while he also won on Susanna in the one mile Hack Race.[29] Holt was particularly fond of horses and paid nearly £200 to purchase and transport a stallion from Brisbane, a huge amount considering his annual salary at the time was £350.[30]

---

24 *ibid.*

25 Newspaper cutting, n.d. E.M.Crouch Papers, Brisbane.

26 Chief Inspector of Stock, F.A.C.Bishop, *Report on an inspection of the pastoral holdings, stock route, bores, and dips on the Barkly Tableland, Northern Territory*, Melbourne, Government Printer, 1923, p.9.

27 *ibid.*

28 T. Law, in Alexander, *op. cit.*

29 'Hutton Farewell Meeting', *N.T. Times and Gazette*, 14 June 1901.

30 Richard Holt Typescript, E.M.Crouch Papers.; Income Tax Return, 30 June 1915, NAPR.

CAY (Charles Augustin Yorke) Johnston: Manager of Alexandria 1924-1938. Tony Johnston

After the next manager, CAY Johnston, took over in 1924, Holt stayed on briefly in an advisory capacity before retiring to Brisbane, where he died seven years later. His service to the company had been notable, not the least in the station's dealing with the Aborigines. He had a great affection for Victor, 'King' of the Wakaja, the tribe in the area, and on more than one occasion Victor, an excellent tracker, had saved Holt's life when the blacks were 'troublesome'. A friendly fight he once had with Victor was recalled with amusement by the revered Aborigine many years later:

> on account of we two fella being mates, we been all about roll alonga ground locked together, and finished in a water hole.[31]

CAY Johnston also had great respect for Victor[32] and his wife Maggie. But CAY retired him even though he still wanted to work. As CAY's wife Ellen recalls:

> He'd come and sit on the steps and talk to Dad and tell him all the news about all the boys . . . He was a King — they looked up to him.[33]

31 Victor in T.Law's letter, in W.F.Alexander, 'Ninety Years of the N.A.P.Co. 1878-1967'.

32 Victor died in 1954 and is buried at Alexandria.

33 Ellen Johnston Interview, 26 May 1981, NTRS 226, TS 250, NTAS.

Ellen (Nell) Johnston; the first white woman to live permanently on Alexandria. Tony Johnston

Charles Augustin Yorke Johnston, or CAY as he was known, was born in 1892[34] and arrived at Alexandria in 1919 as head stockman. He had worked on Queensland pastoral properties (Arabella and Alpha stations) before serving with the Second Light Horse in World War I. On Alexandria, he used a compass from his army experience to find his way around; with one of his early problems being the intrusion of brumby runners, he was constantly on the move, riding the 137 kilometres to Gallipoli and 121 kilometres to Soudan on horseback. He took over from Holt in 1924 and remained as manager until 1938.[35]

An early task for CAY was to reform some of the stockmen who had taken to drink. One of the stockmen involved recalls:

> He told the men that he wasn't born a b . . . . . d, but they could very easily make him one if they refused to cooperate.[36]

He willingly shared in the hard work with his men and was justly rewarded by their loyalty; he had fond memories of his staff.

34 Born 6 August 1892; died 12 November 1983.

35 After 1938, Johnston managed the Post Office Hotel at Cloncurry for the Staunton family, and later had a job at Cloncurry with Burns Philp. He managed the sheep property Cammeray outside Julia Creek, and relieved at Victoria River Downs Station before retiring to Townsville.

36 T.Law in W.F.Alexander, 'Ninety years of the N.A.P.Co. 1878-1967'.

> They never left you . . . The old saddler, the old bullock team driver, the blacksmith . . . they were there for thirty years. Even before we went there.[37]

He married Ellen (Nell) Staunton in 1925;[38] her family had a hotel at Duchess and later at Cloncurry, and her brothers Dinny and John were carriers for stations on the Barkly. Nell became the first white woman to live permanently on Alexandria, and she was the only one until CAY relented and allowed her to have a woman as cook. 'There'd be too much ruption on the place if you bring a white woman here', he had said. Chinese and some Japanese men were also employed to cook, although this did cause one incident of racial tension at Soudan in 1925-26. George Fong, a Chinese, shot at and wounded a part-Aborigine, Jack Green, and as a result was sentenced to two years in prison.[39]

Although kept busy raising their three children, Nell missed contact with the outside world and eagerly awaited the mail. It came to Alexandria fortnightly by packhorse, but in the wet no mail came for months on end. Christmas presents for the children had to be bought in September so they would arrive before the wet season, which was always anticipated in November.

Nell felt the loneliness, CAY being often away. Unlike sheep stations, where at least during the shearing there would be more people around, cattle properties had comparatively few hands. 'He'd only get home to sign the cheques and pay the men and go again', she said. Her daily contact was with the bookkeeper, Jack Owen (who had arrived in 1927), the cook, her children and the Aboriginal people. With no electricity, entertainment at night consisted of either playing the gramophone or reading by the kerosene light. The ritual for dinner remained the same. The bookkeeper, the engineer and the mechanic came to dinner every night; when CAY was away, the bookkeeper would do the carving. Because of the loneliness, Nell had a young black girl to sleep at the homestead with her.

In the Johnstons' time, Brunette, Alexandria and Alroy used to have monthly social gatherings where the managers, their wives and usually the bookkeepers would meet. On Alexandria, there was usually a game of tennis, with a game of bridge at night followed by a sing-song around the piano or the gramophone. They would dress for dinner, the men in ties, and coffee would be served in the lounge. 'You had to keep up the morale', said Nell. For something different, she would take her sewing over to Alroy for the day. Nell felt, however, a certain sense of separation at these gatherings, because unlike the Whites at Brunette and the Schmidts at Alroy who owned their stations, the Johnstons did not own Alexandria.

Hospitality was important and visitors were always welcome. All ministers of religion were received, with services held in the dining room. A highlight was the visit of the Administrator from Darwin once a year. This was reciprocated in 1926 when Nell and other managers' wives drove to Darwin in the Johnston's Oldsmobile and stayed at the Administrator's residence. They were invited on board a naval ship for dinner at which they

> had a wonderful time. Imagine going onto a naval ship and everything being lovely, and us never seeing anything like that — really a red letter day.[40]

And friendliness with one's neighbours was essential:

> You'd never let anybody go by and you wouldn't *dream* of going by anybody's house without calling in and saying you went through their property . . . If any of our boys went into Camooweal they'd pull into Avon Downs, tell the manager they were going through and could they bring anything out of town for them.[41]

Personalities abound, particularly in remote pastoral communities. During Holt's time a cunning scoundrel named Paddy Lenny became a thorough nuisance on Alexandria and Avon Downs by running a mob of horses there. He had been a labourer on Alexandria in the early 1890s and had begun breeding horses without consent. With a stockyard camp on Avon where he branded, and with food given to him by friendly drovers, he was entrenched. He had hundreds of horses which roamed in search of feed in the vicinity of Soudan and Avon Downs and disturbed the cattle. When the backwaters dried up he put his horses on Lorne Creek and when this was dry, he used the station's bores. But he would leave the engine 'snorting under a 95lb pressure of steam' and it was feared that others would emulate him by helping themselves to the bores.[42] He persisted despite pleas from John Forrest to desist. In 1907 Forrest even asked his old friend Parry Okeden, former Queensland Commissioner of Police, for advice but he was told it was the lessees who had to take action. In 1911 Lenny was still on Alexandria and

37 *ibid.*

38 Born 17 January 1904; died 11 June 1990.

39 G.V.Dudley, Commissioner of Police, Report of Acting Administrator, 1926, *CPP*, 1926-27, p.25.

40 Ellen Johnston Interview.

41 *ibid.*

42 John Forrest, July 1907, Docket 16593, GRS 10, PROSA.

the story goes that he died in about 1923 on his way back from Darwin, where he had ridden to obtain 'justice for himself and his horses'. A permanent reminder of him is Lenny's Hole Yard on Alexandria and Lenney[43] Waterhole on Avon Downs.[44]

Count Mario Carlos Pereferrio (Charlie) Biondi was another local identity. At one time a stock rider on Alexandria, Biondi was said to hold the title of a once rich and powerful Italian family; he had heard of the Territory as a place to restore the Biondi fortunes and to build a new principality. Instead he married Queenie Hennessy (of Aboriginal-European descent), raised eleven children, and died in the 1940s after spending a lifetime in the Territory. He was found to be excellent company on droving trips, being able to

> converse fluently on European classics and history, and would enliven the long rides and night camps with operatic excerpts rendered in a fine tenor.[45]

After working on Alexandria he packed the mail from Camooweal to Borroloola and later took possession of the store at Anthony Lagoon where his

> hospitality [was] a legend and his exquisitely plaited goods — stockwhips, belts and hatbands — prized as a local art form.[46]

It is reported that his daughter from a previous marriage in Italy, while on tour with an opera company, tried to visit him. Troubled by the news, he dressed as smartly as possible and met her in Camooweal with the news that it was impossible to travel further because of the wet.[47]

From the 1910s to the 1930s, James Broadbridge made a considerable contribution to Alexandria. The earliest record of his presence was in 1913 when he built the Alexandria homestead. After starting with accommodation in a small partitioned portion of Alexandria's store which had only an antbed floor, he proved to be a jack of all trades and to NAP's advantage he seemed to be master of them all. His skills as a carpenter were legendary and as well as his work as Alexandria's bookkeeper, itself time consuming, he managed to erect various buildings at Alexandria, the police stations at Ranken River and Anthony Lagoon, and the Ranken Dance Hall.[48] He made significant alterations and improvements to other buildings, and NAP's balance sheets in the 1920s make numerous references to these, including the bachelors', gardener's, house gins', and blacks' quarters. In 1920 Broadbridge built the men's quarters at Soudan without any assistance. On two-metre stumps with four bedrooms, this house took him six months to complete.[49] A noted

James Broadbridge was one of NAP's most capable employees. On Alexandria from 1913 to the 1930s, he was bookkeeper, carpenter, sawmiller, photographer and gardener. He was also secretary of the ABC Races, the Camooweal Hospital and the Camooweal Jockey Club.

43 The spelling on maps is Lenney as well as Lenny.

44 Report by Richard Holt, 24 March 1907, John Forrest to Government Resident, 3 May 1907, 31 January 1908, Docket 16285, GRS 10, PROSA; Letter by T. Law, 30 May 1968 in W.F.Alexander, 'Ninety Years of the N.A.P.Co. 1878-1967'.

45 Mary Durack, *Sons in the Saddle*, London, Corgi, 1985, p.38.

46 *ibid.*

47 *ibid*, p. 373; L.A. Miller, *The Border and Beyond*, p. 51.

48 *Northern Standard*, 13 October 1933, PL2114, F27, NTAS.

49 J. Broadbridge, Cost sheet of Soudan Men's Quarters, 1920, NAPR.

Alexandria Homestead in the late 1920s. John Oxley Library

photographer and gardener, and an able bush poet, he also busied himself as secretary of the local Ranken (or ABC) Races, the Camooweal Cottage Hospital and the Camooweal Jockey Club. As he wrote to Beardmore at the Brisbane office:

> I cannot sit and talk with the other employees hour after hour, all about stock, each other (generally pulling each other to pieces and telling dam[sic] lies) and Gins in particular and really all about nothing at all.[50]

The new Alexandria homestead with a veranda all round was of ripple iron construction. It had an office and an overseer's room which Broadbridge added in 1921. The homestead was ceiled and lined with tongue and groove boards but the office was not. French doors opened on to the veranda from each room. Lattice work was erected on the veranda and blinds fitted in 1926.[51] CAY and Nell Johnston can therefore be pictured as they looked out for storms:

> We'd walk to the front door, we'd walk to the back door, we'd walk to the right door, we'd walk to the left door, we'd say, oh there's a storm falling over there.[52]

They each continued to do this in their homes long after they had left Alexandria.

By the 1920s the station resembled a small township with buildings for the butcher, the carpenter, the mechanic, the blacksmith, and the saddler (Harry Lanson, a long-term employee). The Aboriginal quarters consisted of nine galvanised huts. The house kitchen, a separate four-metre square building, was erected in 1926. The large station kitchen was erected in 1908 and renewed in 1917, but it did not have a floor until 1924. The overseer's house at Soudan was on two-metre blocks and consisted of three rooms with verandas front and back. On Soudan there also were men's quarters, a blacksmith's shop and a kitchen with its frame of bore casing. The accommodation at Gallipoli was more primitive; in 1933 it consisted simply of a men's hut, five metres by 11 metres, of galvanised iron on sawn timber and bore casing, with two rooms, and a veranda in front, all having an earth floor. Bush timber brought in from the top end of Wild Horse Creek above Ibis Hole was milled in the sawmill erected by James Broadbridge. The frames of buildings were often made of bore casing, a matter of using whatever was available because of the enormous cost of transporting it.[53]

Staff on the stations of the Barkly also included pumpers (or boremen) who were colourful characters and considered somewhat eccentric. They were a mixture of alcoholics, loners, bush

50 J.Broadbridge (signed 'Broady') to Mr Beardmore, n.d., Peg Corbett Papers, Brisbane.

51 Holding: Alexandria, 1933, 452/1, 1952/310, Australian Archives, Canberra.

52 Ellen Johnston Interview.

53 Holding: Alexandria, 1933, A/452/1, 1952/310, Australian Archives, Canberra; T.Law, 30 May 1968, in W.F.Alexander, 'Ninety years of the N.A.P.Co. 1878-1967'.

Kitchen, Alexandria, 1920s.

Blacks' Quarters, Alexandria, 1920s.

lawyers and renegades from justice. In the terrible year of 1928, pumping went on all year; but in a good season it would occur for just three or four months. Up to three pumpers were employed to tend each bore, to chop wood and to fire up the engines, while Aborigines were also employed to chop wood. The individual pumpers often kept to themselves.

Men's Dining Room, Alexandria, 1920s.

Soudan out-station ca. 1934.

Australian Archives, Darwin

That inevitably meant three separate camps and three separate camp fires as these were difficult, independent, interesting individuals. If you chanced by, you could be invited to partake of three separate cups of tea or three feeds of corned beef by three very different characters who probably were not on very good terms with one another. Diplomacy was a prerequisite for a manager's job.[54]

54 L.A. Miller, *The Border and Beyond*, p. 60.

In Queensland a third class ticket (or licence) was required to drive a steam engine used for pumping, but in the Territory a permit was easily obtained. The introduction of diesel engines in the 1930s reduced the workload, and greater use was made of Aborigines as pumpers even though they generally knew nothing about mechanics. Ted McFarlane, a windmill erector and repairer, describes a situation that often arose:

Circular saw operator J. Broadbridge.

> They knew how to crank it and how to switch it on ... Well, if they broke down, they couldn't tell you what was wrong. They'd just go back — 'bore broke down.' 'Well, what's the matter? Pump head bin break or engine bin break down?' 'Everything bin break!' Well, you'd go out there and find that the nozzle in the injector was blocked up or something like that.[55]

To check on the bores, CAY Johnston would drive around in a buggy with a horse tailer to mind the spare horses. The horses would be changed every 30 kilometres. The engineer proceeded in the same manner if there was a breakdown.[56] In 1936 there were twenty-two effective bores on Alexandria of which sixteen were equipped with steam or oil engines, and four with nine-metre windmills, together with pumping gear and large earth tanks. With each beast consuming eight gallons per day, and allowing for evaporation, the herd of 46,000 required 360,000 gallons (1,620,000 litres) per day.[57]

Drovers were an obvious and essential part of the pastoral industry, and since the stock route passed through the Ranken, they were a common sight on Alexandria. They dreaded crossing the Ranken Plain, 50 kilometres of treeless monotony, where firewood had to be carried on packhorses. Cattle dung was sometimes used in the fires; it was also an excellent mosquito and fly repellent but it required continual stoking.[58]

Droving was fraught with disappointments and delays, as drover Walter J. Rose revealed in a letter written by firelight at his camp at Marion Downs in 1909, while on a droving trip down the Georgina with a mob from Newcastle Waters:

> From Roxborough there is not a bite to eat. The road is fearfull (sic) . . . I have struck a patch of old stuff today and am making the best of it. From here down there is very little I am told. My cattle are becoming to get a bit of a tail to them . . . . Hoping the Georgina is better as I go down. Anybody that tells you the Georgina is good send them to me.[59]

By watering on Alexandria bores, cattle could make it as far as Lake Nash, but a very difficult dry stage followed, of 166 kilometres to Carandotta in Queensland. In 1905, Holt notified the Government Resident that after the creeks stopped running, waterholes would remain on the stock route through Alexandria at the Buchanan

---

55 Ted McFarlane Interview, 19 July 1979, NTRS 226, TS 273, NTAS.

56 CAY Johnson Interview.

57 P.Forrest to Woolley, Beardsleys and Bosworth, 21 January 1936, NAPR.

58 L.A. Miller, *The Border and Beyond*, p. 80; Peter Muir, *Pituri Pete: a saga of cattle land and desert sand*, Leonara, Muir, 1985, p.268.

59 W.J. Rose, Marion Downs, to J. Lewis, 12 June 1909, John Lewis Papers, PRG 247, Mortlock Library, Adelaide.

Alexandria and Avon Downs Stockcamp at Lorne Creek ca. 1920. John Oxley Library

A typical stock-woman.

Creek for nine months, at the Ranken River for ten months and at Soudan for twelve months.[60]

The problem of the lack of permanent water on stock routes gradually eased as government bores were put down, but the routes were often closed in especially dry times. In 1917 the first government bore was put down on the mile-wide Barkly stock route which stretched from Newcastle Waters to Lake Nash.[61] The route was formally gazetted by government in May 1920, and four more stock routes were gazetted in late 1920, including the infamous Murranji Track which was dangerous because of the prevalence of poisonous plants and the absence of permanent water.[62] Cattle were walked from the Victoria River District down the Murranji, across the Barkly to the Georgina Stock Route. After the railway reached Dajarra in 1917, during good seasons, cattle would be moved from Soudan to Dajarra and thence by rail to the meatworks at Townsville, although they lost condition on the month-long trip.

Early droving was by packhorse and dray. Conditions were tough, with long dry stages of 'the never ending sameness of those everlasting plains',[63] no medical help, no telegraph and only basic food of salt beef, damper and jam.

60 Government Resident's Report for 1905, *SAPP*, Volume 2, 45 of 1906, p.7.

61 Report of Administrator, 1915-16 and 1916-17, *CPP*, 1917-8, p.9.

62 Report by Acting Administrator, 1920, *CPP*, 1920-21, p.7.

63 Quoted in L.A. Miller, *The Border and Beyond*, p.81.

Alexandria Station Aborigines ca. 1920s. Tony Johnston

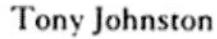

Avon Willie. The Rainmaker at Alexandria. William Foster

James Broadbridge wrote of drovers in the 1920s:[64]

Some deliver at Goondiwindi
Others drove to New South Wales
To meet them on their long trek back
They tell some stirring tales.

Of how the cattle galloped
When the nights were inky black
And they felt the icy water
Softly trickling down their back.

Sometimes wives accompanied their husbands but it was a tough life. As one drover, Jack Charlton, later recalled:

> I admire the men who put up with those hardships, but more so the women. They stuck with their men and the problems. The women of the west were as tough as the men of the Murranji.[65]

Drovers lived in fear of a cattle rush; and they often had no sleep for days on end because, as CAY Johnston recalled, 'if a bullock jumps or goes, they'd go with him'.[66]

Drovers through the Ranken until the 1930s included W.Williams, D.Cameron, A.Nelson, J.Fowler, E.Larkins, Ted and Tom Starr, Tom Laffin (Snr.), John Miller, Walter Rose, Walter Green, Walter Hooker and Stan Fowler. The Crouch brothers, Bert, Jack and Jim, were also prominent as drovers; Jim Crouch became head stockman at Gallipoli in 1930 before becoming

64 James Broadbridge quoted in L.A. Miller, *The Border and Beyond*, p.79.

65 Jack Charlton, 'Moleskin-moulded in the Murranji', *Stockman's Hall of Fame Newspaper*, Volume 42, March 1992, p.7.

66 Cay Johnston Interview.

manager at Alroy.[67] Long before their brands and earmarks could be read, experienced stockmen could tell the mobs by 'their size, their horns, their colour and their conformation'.[68] CAY Johnston could always pick Wave Hill cattle:

> terrible cattle to walk . . . When they were settled down, they'd still walk — walk over your fences. They were that used to it. But they were noted for that.[69]

Stock camps on Alexandria consisted of twelve to fifteen stockmen, including four or five Aboriginals who would muster, brand, earmark, dehorn, spay, castrate and inoculate as required. Cattle would be drafted from the mob for turnoff, a process which usually demanded three changes of horses — a total of fifty or sixty horses. The 'bronchoing'[70] method was used and the Aboriginal stockmen were experts on horses and with ropes. As James White of Brunette commented:

> There was a lot of daylight between saddle and man many a time, but they stayed there. They loved their gallops, they loved chasing something.[71]

The Aboriginal women were also excellent with stock and while clothing would be standard issue, managers had them dressed the same as the men so they would not be noticeable, especially when mustering near civilisation.

Aboriginal women played a major part in the domestic life of the station carrying out tasks of cleaning, washing, cooking and waiting on tables. Not accustomed to the white toil of scrubbing floors — the Aborigines lived in nine galvanised huts with dirt floors, each hut barely 2.5 metres square[72] — their method was to sit on the floor of the homestead and wash all round them, then move on to the next place and wash that. 'You can imagine how terrible it looked', said Nell Johnston years later. They washed clothes in tubs while all the time chewing tobacco. Said Nell:

> It would come all round their mouth, dirty green. Well, I wouldn't allow them to come up to the house . . . but they'd stick it behind their ears. And I'd go over to the laundry sometimes when they didn't know, and I'd hear them say, Missus coming! And . . . they'd hide it.[73]

In the homestead another of their tasks was to keep the punkah[74] moving for coolness; pulled by string tied around the big toe. The blacks would eat away from the whites, even when they were out in the camps. They never came to the kitchens. CAY Johnston's explanation was 'if you give a blackfella a bit of leniency he'd take a [mile] you know. Always let them think you don't want them'.[75] Rowand Murray, a later manager of Alexandria from 1946 to 1955, gave his summation:

> An Aborigine is a natural psychologist, whose mind is ever on the alert to assess the exact amount of latitude he can take.[76]

Cultural differences abounded. They were 'honest thieves', openly displaying ribbons for instance that they had taken from the nightgown belonging to the station manager's wife as it hung on the line; or they would take cutlery from the kitchen. To retrieve it Mrs Johnston would say 'whoever brings me the most teaspoons today I'll give them a tin of jam'. The teaspoons would be returned.

Part of station life included walkabout time during the wet, between November and March. Originally Aborigines would leave *en masse* for ceremonies, often at crucial times of the working year. The compromise of the off-season break was reached, and a station dray stocked with food supplies would take the Aborigines and their dogs away from the station, usually towards the Nicholson country. (The Alexandria Aborigines spoke Wanyi which could be understood in this area.[77]) At the same time, some white employees took holidays; others would stay and see to repairing machinery and saddles, horseshoeing, and making ropes and hobbles. 'You were occupied . . . .You didn't sit down and do nothing'[78] said CAY Johnston.

Aborigines seemed happy with their lot; to Europeans they were always laughing and giggling. These were the days before liquor was allowed and when the extended Aboriginal kinship network would be 'kept' by the station. Until 1910 there was little effective administrative control over relations between settlers and Aborigines. In 1904, legislation restricted the sale of opium, and in 1908

67 Rankine River Police Station Journal, 1931, F300, NTAS.
68 L.A. Miller, *The Border and Beyond*, p. 78.
69 CAY Johnston Interview.
70 To broncho a beast is to put a catching rope on it and pull it up to a broncho panel using a strong horse.
71 James White Interview, NTRS 226, TS 140, NTAS.
72 Holding: Alexandria, ca. 1933, A452/1, 1952/310, Australian Archives, Canberra.
73 Ellen Johnston Interview.
74 Large swinging, screen like fan hung from the ceiling; of Indian origin.
75 CAY Johnston Interview.
76 R.H. Murray, 'Management of Northern Stations', in Frank O'Loghlen (ed.) *Beef Cattle in Australia 1958*, Sydney, Johnston, 1958, p.61.
77 J.Dymock, *Nicholson River (Waanyi-Garawa) Land Claim*, Darwin, Northern Land Council, 1982, p.121.
78 Cay Johnston Interview.

a register of 'half-castes' was begun and a full-time Protector appointed. Between 1910 and 1953 the protective legislation was restrictive, in effect eroding the position of Aborigines. Employers of Aborigines had to be licensed, the Protector became the legal guardian of every Aboriginal and part-Aboriginal child, reserves were declared and crown land leased to missions. Marriage between Aborigines and non-Aborigines was subject to ministerial approval, and it became an offence for a non-Aborigine to be found near an Aboriginal camp.[79] The position in 1933 was clarified by the Chief Protector of Aboriginals:

> In the Territory the mating of an aboriginal with any person other than an aboriginal is prohibited. The mating of coloured aliens with any female of part aboriginal blood is also forbidden. Every endeavour is being made to breed out the colour by elevating female half-castes to the white standard with a view to their absorption by mating into the white population.[80]

But while it was illegal for a white man to cohabit with an Aboriginal woman, there were the inevitable clandestine affairs on the stations. As Ellen Johnston said, 'we've taught them anything bad they know. Their morals were very high'. The situation has been summarised:

> The shortage of white women in the Territory left many white men, especially the working class, with little choice but to associate with Aboriginal women. And some of those belonging to the upper echelons surreptitiously indulged out of preference, temporarily forgetting their 'higher ideals'.[81]

Special permission was required to take Aborigines on droving trips, and work permits were required from whites wanting to employ Aborigines. So, in 1933 for instance, permission was sought from Darwin for Aboriginal Sambo to be employed by Jim Fowler on a droving trip.[82] Police appointed as protectors of Aborigines arranged work contracts, negotiated suitable wages and generally saw to their welfare. Of the Aborigines' wages, part was remitted monthly to a trust account which the police supervised.

The Aborigines were not only invaluable with the stock but also in finding the way. Where a white man would generally panic and gallop his horse if he were lost, the Aborigines never got lost. CAY Johnston always made sure that an Aborigine accompanied a white stockman — although, as he explained:

> the only thing about the black fella, he'd always tell you something that would please you. If you said, how many cattle do you see over there, if he was riding and this waterhole got dry, he'd tell you anything to please you.[83]

In 1936 the new chairman of directors, D.M.Fraser, wrote to the Chief Protector of Aborigines at Darwin asking about the Federal Government's policy regarding the maintenance of Aborigines. NAP was claiming that the cost of feeding and maintaining the aged and infirm blacks was heavy.[84] Both government and station-owners had a responsibility to station blacks and in the decades to follow this was to become a greater challenge.

The company's financial and other records for the early 1930s give some indication of how the ordinary pastoral worker was coping during the Depression. The company's income tax return for 1930 listed eight Alexandria employees as liable for assessment.

ALEXANDRIA STAFF 1930[85]

		Wage plus board
CAY Johnston	Manager	£500
W.C. Frith	Pastoral Inspector (Brisbane)	£333
D. McInnes	Engineer	£313
J.H. Crouch (Gallipoli)	Head Stockman	£299
J. Ormiston	Blacksmith	£260
E.D. Lemon (Soudan)	Head Stockman	£260
A. Malpass	Carpenter	£125
E. Larkins	Drover	£ 95

One result of the worsening economy of the Depression was NAP's successful case in the Arbitration Court in 1932, against the Northern Territory Workers Union, securing a 20 per cent reduction in wages.[86] It is interesting that at a shareholders' meeting in September 1931, Phil Forrest as chairman of directors, was asked to consider a reduction of 15 per cent in salary for himself and his staff. He agreed to give the matter

79 Ian Keen, 'The Alligator Rivers Aborigines - Retrospect and Prospect', R. Jones (ed.) *Northern Australia: Options and Implications*, Canberra, A.N.U., 1980, p.177.

80 C.Cook, Chief Protector of Aboriginals, in Report on the Administration of the NT, 1933, *CPP*, 1932-33-34, p.7

81 Ann McGrath, *Born in the Cattle*, Sydney, Allen and Unwin, 1987, p.70.

82 Rankine River Police Station Journal, 2 May 1933, F300, NTAS.

83 CAY Johnston Interview.

84 D.M.Fraser to Chief Protector of Aboriginals, 28 May 1936, NAPR.

85 Income Tax Return 1930, NAPR.

86 P. Forrest to Sir Herbert Ingram, 22 October 1932, H. Ingram File, NAPR.

Ranken Store, 1920s.

consideration but there is no record of his decision.[87]

In 1933 there were thirty white men employed on Alexandria excluding the manager, and six white casual workers. Fifteen Aborigines employed as stockmen were paid eight shillings per week with food and clothing for themselves and dependants.[88] This compared favourably with the minimum wage under the terms of the Aboriginals' Ordinance of 1918 of 5s per week.[89] White stockmen meanwhile, were receiving £2-£3 per week. The basic wage for European male employees at the time was £2 8s per week.[90]

During the Depression unemployed men walked from station to station, were given 'a feed' and had to move on. Sometimes there were over a dozen men at the store getting 'tucker'. For those permanently on the station, life went on as before, with plenty of beef, much of it salted, and vegetables from the station gardens. The rib bones cooked in a camp oven were a favourite on killing night. Food arrived a few times a year from Burketown. The dried fruit had to be washed and the Aboriginal kitchen hands would sit and shake it through gauze before the fruit was rebagged and hung in the store. Ti-tree bark was put in with the flour and fruit to keep out the weevils.[91]

Station workers were paid only once a year, although CAY Johnston said they were never formally dismissed after the cattle season each year as occurred on some stations. They could draw a cheque whenever they wanted, but they seldom did so since there was generally nothing to buy. Stores in Burketown, at the Ranken and in Boulia permitted a form of currency — calabashes — which were effectively IOU's. They were pieces of confetti-like paper which generally fell to pieces in the pockets of jackaroos before they could be used.

The station workers really wanted money only at one time of the year, for an event they looked forward to — the Ranken Races. There is some evidence that these races, referred to as the ABC Races (the initials being taken from Alexandria, Brunette and Cresswell), began as early as the 1900s although officially they began in 1910 at Brunette.[92] The driving force behind them was Ted Lowe, who was manager of Brunette 1911-1913, and after the war 1918-1923. In 1914 the races shifted to the Ranken, presumably when Lowe left for the war. George Watson, the owner of the Ranken store, was a racing enthusiast and took over as secretary, followed in 1915 by James Broadbridge. The races continued at the Ranken

87 General Meeting of Shareholders, 22 September 1931, NAPR.

88 Holding: Alexandria, A452/1, 1952/310, Australian Archives, Canberra.

89 Frank Stevens, *Equal Wages for Aborigines*, Sydney, Aura Press, 1968, p.11.

90 *ibid.*

91 Ellen Johnston Interview.

92 M. Mahood, *The Australian Stockman*, Sydney, Landsowne, 1988, p.94; J. Macdonald Holmes, *Australia's Open North*, Sydney, Angus and Robertson, 1963, p.225.

The ABC Race Club Executive July 1922. Left to Right: George Watson (Owner of Ranken Store); Herbert Lloyd (Manager of Avon Downs); Pat Synnott (Managing Director of Synnott, Murray and Scholes, General stores Burketown and Camooweal); Richard Holt (Manager of Alexandria); Bill Reilly (Owner of Landsborough Hotel in Camooweal). In Front: James Broadbridge (Bookkeeper at Alexandria and Secretary of the Race Club).

until 1949 when disapproval by the new store manager caused them to return to Brunette Downs The races shifted to a new site on Brunette from 1951.[93] Alexandria was always heavily involved; for example, in 1951 Alexandria people had successes in both the 'blackboy's race' and the bullock riding championship, while jackaroo Rod Bellette won the ride against all-comers.[94]

Much of the social life for the stockmen centred on these meetings, and as CAY Johnston said 'they never looked for anything else'.[95] Twelve months' entertainment was packed into one glorious week. The race committee would send a buggy into Camooweal to pick up all the eligible girls and then chaperone them. As historian Ada Miller explained, 'For the dancing they had rubberoid on a sand floor surrounded by hessian and they'd dance all night'.[96] James Broadbridge built a small hall which was later used as the Ranken Store. Branding in the stock camps was staggered to fit in with the races, which were held in June each year. Competing horses were restricted to local grass-fed stock. Committee meetings were held to organise such events — such as at the Ranken in August 1930, where following the meeting there was a cricket match between Camooweal and the Rankine Club.[97] In the 1920s, the races had become so popular that more room was needed in the horse paddock. George Watson, who ran the Ranken store, at that stage successfully applied to increase the area by five square kilometres.[98]

At the time Alexandria was being established, the complete absence of South Australian police in the district was a matter of concern to the pastoralists. In the 1880s a sly grog shanty on Austral Downs was selling liquor without intervention.[99] Dutiable goods were being smuggled across the border into the Territory[100] and overlanders to the Kimberley goldfields were

93 J. Macdonald Holmes, *Australia's Open North*, p.228.

94 R.H.Murray to D.M.Fraser, 21 July 1951, R.H.Murray File, 1951-55, NAPR.

95 CAY Johnston Interview.

96 L.A. Miller, *The Border and Beyond*, p. 83.

97 Rankine River Police Station Journal 2 August, 1930, F300, NTAS. In earlier times the spelling Rankine was commonly used.

98 Director of Lands, Darwin, to NAP, 11 September 1923, PL 2114, F27, NTAS.

99 John Farrar to Government Resident, 11 August 1885, Docket A8501, GRS 10, PROSA.

100 Sub Collector Customs, Port Darwin, to Collector of Customs, Adelaide, 3 August 1886, Docket 637, GRS 1, PROSA.

Ranken 1948: Hugh Deviney, Clarrie and Ella Hudson, Jean Deviney at the marriage of Clarrie and Ella.

June Blaschek, nee Hudson

helping themselves to horses on stations en route.[101] In 1893, the site for a proposed store on Happy Creek on Rocklands was strongly opposed by pastoralists who saw it as an excuse for a sly grog shop.[102]

Some explanation is needed, therefore, how George Watson who established the Ranken store in 1904 after hawking goods from Cloncurry for a time,[103] could obtain a liquor licence under the South Australian Licensing Act of 1908.[104] There was nothing grand about the store's appearance. After Herbert Lloyd (Manager of Avon Downs 1913-1929)[105] and his wife Gwen took over from Watson in 1929, it was described as

> a humble little galvanised iron structure . . . The rafters are of rustic, unsawn timber . . . It is pleasantly unconventional in plan, giving the effect that its architecture grew and was not preconceived.[106]

It was a hub of activity, selling alcohol, food and requisites to the drovers but it was also an obstacle for boss or head drovers to overcome. Perhaps the police station established there in 1911 was meant to solve possible drinking problems. The arrival of a police presence, however, was said to have followed complaints of horse thieves and cattle duffers.[107] At the same time, police were established at Newcastle Waters and Lake Nash. Previously, the nearest police had been at Borroloola, Anthony Lagoon and Camooweal; at the latter, Queensland police were sworn in as South Australian constables.[108]

Hughie Deviney was a popular and respected policeman on the Barkly. He was posted to Ranken River from 1929-31, to Anthony Lagoon from 1931-34, and after the war served again at the Ranken until his retirement in 1959.[109] As well as carrying out his police duties, he checked stock movements, supervised dipping, counted stock, and patrolled for cattleduffers. His trackers were often on duty at Soudan and he left horses at various places on Alexandria for his use. A perusal of the police station journal shows that in 1930 eight horses were at the Ranken police station, one at the Buchanan, two at Soudan, three at Alexandria homestead, two 'bush', and four 'alleged to be dead'.[110]

He lived with his sister Jean (another sister, Mary, lived with them at Anthony Lagoon) who was also remembered with fondness by people such as James White, whose family owned Brunette for many years.

> She [Jean] never, for one minute, of that thirty years let herself slide in any way at all. When they went to the ABC races, she had a special tent, she had a special table, she had beautiful linen on that table and served afternoon tea in Doulton china. She never did a single thing wrong.[111]

Social gatherings were more frequent once white women settled on the Barkly, and with them came 'proper' standards. 'I wouldn't be seen in a pair of trousers on the property . . . . You have to remain a

101 Government Resident, Half-yearly report on Northern Territory, 30 June 1886, *SAPP*, 54 of 1886, p.2.

102 1893 Map, Docket 406, GRS 1, PROSA.

103 J.Macdonald Holmes, *Australia's Open North*, p.225.

104 Mounted Constable R. Stott, Borroloola, 29 December 1905, Government Resident's Report, 1904, *SAPP*, Volume 2, 45 of 1905, p.29; V.T.O'Brien, 'Miscellaneous Lease No.44 and later No.115 — George Watson Ranken Store — 1904', 19 November 1987, Department of Lands, Darwin.

105 C. Wagstaff, *Avon Downs N.T. 1882-1982*, p.8.

106 A.P.Noonan, 'The Rankine', *Walkabout*, June 1938, p.49.

107 L.A.Miller, *The Border and Beyond*, p.63.

108 Government Resident, 8 November 1886, Docket 1064, GRS 1, PROSA.

109 Born Helidon, 1900; joined NT Mounted Police 1928; died, Toowoomba, 29 February 1984.

110 Rankine River Police Station Journal, 31 December 1930, F300, NTAS.

111 James White, Oral Interview, 8 January 1983, NTRS 226, TS 140/1, NTAS.

Captain Wrigley at Alexandria, 8 December 1919. Tony Johnston

woman . . . always stockings on and shoes', said Nell Johnston.[112]

There were few health problems on the station apart from sandy blight which was widespread in the inland. Four men, one a miner on his way to Pine Creek, died from beri beri in 1909,[113] and men on the government bores suffered from beri beri because of the lack of vegetables. Death also sometimes occurred by misadventure. John Corbett in 1885 was unfortunately killed while well-sinking on Alexandria; through a mishap with the winding gear he was not able to get up the shaft in time to escape a dynamite charge.[114]

The few women in the Territory were automatically expected to be nurses. Nell coped:

> I was never frightened of sickness or that I wouldn't be able to cope to do anything. But I did feel much better when we got the flying doctor.[115]

The Flying Doctor Service began in 1928 with Dr.K.St.Vincent Welch as the first doctor. A large inaugural meeting was held at Alexandria at the time, with the first base being established at Cloncurry. Nell remembered John Flynn with plans spread out as he designed buildings for the Inland Mission Hospitals.

Alexandria has played a significant part in the history of aviation in Australia. The first landing of an aircraft in the Territory was at the station, when, on 8 December 1919, Captain Wrigley landed on his way to meet Ross and Keith Smith who flew into Darwin from Great Britain two days later.[116] Nine years later, on 24 February 1928, Bert Hinkler landed at the homestead for breakfast on his record-breaking solo flight between England and Australia. He had landed at a bore and windmill the day before to quench his thirst, slept there overnight and took off the next morning to see the homestead only a few miles away. He spent the day at the homestead and gave CAY several London newspapers before flying to the Ranken where he stayed the night.[117]

Two years later, Amy Johnson landed at Alexandria during her historic flight between England and Australia, the first woman to complete such a trip. Flyers had the wind with them coming down from Darwin but in the vicinity of Alexandria, found the wind against them, which often caused them to lose their way.[118] Air traffic increased and by 1930 commercial planes left Brisbane every week, arriving at Alexandria forty-eight hours later.[119]

CAY Johnston had been on the inaugural Qantas flight in February 1925, which extended the airmail service from Cloncurry to Camooweal.

---

112 Ellen Johnston Interview.

113 *NT Times*, 4 June 1909, Docket 18267, GRS 10, PROSA.

114 J.S.Little, Austral Downs, 22 April 1885, in Quarterly Report on the Northern Territory, 30 June 1885, *SAPP*, Volume 3, 55 of 1885, p.3.

115 Ellen Johnston Interview.

116 Report of Acting Administrator of NT, 1920, *CPP*, 1920-21, p.1.

117 A record of Bert Hinkler's flight is graphically recorded at Bundaberg in what had been his home in England from 1926 until his death in 1933. The home was demolished and reconstructed in his hometown of Bundaberg.

118 *North Queensland Register*, 3 March 1979.

119 P. Forrest to Thynne Macartney, 28 April 1930, C.B.Warner Trust File, NAPR.

Broken down at Gum Flat – McArthur River N.T. May 1920. James Broadbridge seated in front. Tony Johnston

Mount Isa was added as a stopover in March 1925, and after 1934 Qantas was flying once a week from Cloncurry to Daly Waters via Camooweal, Alexandria, Brunette, Anthony Lagoon and Newcastle Waters. Not only was mail delivered to Alexandria by air, the pilots used to bring lollies and fruit for the children and library books for Mrs Johnston.

Transport had come a long way since the days when, in order to collect the mail, Thomas Harding had to ride to Rocklands. In the early 1880s mail came from Queensland via Boulia, Lake Nash and Rocklands every two weeks, and despite McIlwraith's lobbying to have it extended to Alexandria, it was not until 1886 that a service ran between Camooweal and McArthur River via Top Camp (Alexandria).[120] Queensland stamps were always used as there was no South Australian post office in the area. Queensland also collected all duty on goods coming over the border or via their ports in the Gulf, so strong links were established with Queensland.[121]

Although there was a break in the service, by the early 1890's the mail was again being run from Camooweal (surveyed in 1887) to McArthur River (by then Borroloola established in 1885) via Top Camp. The service was often closed, however, or delayed by drought or flood. Mailmen such as Alex Grant, Charlie Biondi, Jack Fuller, Jim Grace, Bert O'Keefe, Clarrie Hudson and the Booth brothers became vital links with the outside world.[122]

So too were the teamsters who carted goods to the stations from Burketown and Borroloola. Burketown had been an established township in the 1860s but was considered 'very rough and unhealthy . . . a place only fit for demons'.[123] Supplies had to last twelve months and the story goes that Nat Buchanan was met in a camp by another searcher for country who said, as Buchanan was sieving out the weevils and grubs in the flour before making a damper,

> 'Why, . . . that's rotten flour you have there, Buchanan.'
> 'Oh, it's not too bad while these things can live on it,' was the drawled out reply.[124]

With resettlement of the Barkly in the 1880s, the Carpentaria Road Board was anxious to improve

120 T.McIlwraith to Minister of Education, 20 December 1882, Docket 734, GRS 1, PROSA; T.McIlwraith, 5 March 1886, Docket 265, GRS 1, PROSA; Spelling of McArthur is often Macarthur.

121 J.S. Little to Parsons, Government Resident, 1 April 1885, Docket A7811, GRS 10, PROSA.

122 L.A. Miller, *The Border and Beyond*, p. 51.

123 G.Sutherland, *Pioneering Days: in the early sixties*, p.17.

124 G.Buchanan, *Packhorse and Waterhole*, p.28.

Teamster's Offsiders – Alexandria.

the road to Burketown from the Barkly, and set about undertaking repairs in late 1884. The Territory administration was also anxious for trade and the road from McArthur River to the Tableland was pushed through in 1884-85.[125]

The government had grand ideas for Borroloola. It had been seen as a place for 'roughs from North Queensland. Horse and cattle stealers made it their head-quarters'.[126] The author Ernestine Hill commented that

> whenever a trooper rode down to arrange a census for a packhorse mail, three-quarters of the population, horse thieves and dead-beats, went bush.[127]

Services were provided, including those of a resident magistrate, police and an impressive library. By the 1920s the latter was languishing:

> As the white ants, they say, will eat the inside out of a billiard ball in this part of the country, the books require considerable attention.[128]

Tableland stations tended to favour Burketown because of economics, despite the saving of 320 kilometres which the McArthur River road to Borroloola afforded.[129] Consignments were landed at Burketown from Sydney for £4 per ton while from Sydney at Borroloola via Port Darwin it was £8 per ton.[130] In 1905 R. Farrar had one table valued at £1 10s consigned from Palmerston via Borroloola; the freight was a staggering £4 8s.[131]

Residents of the McArthur complained that trade from that district was going to Queensland because of South Australian Government neglect of the roads.[132] There were also complaints that Tableland stations were receiving goods over the border free of duty, either directly or through hawkers (who sold principally tobacco and drapery). The Northern Territory administration had hoped that the customs paid at Camooweal would drive the trade back to Borroloola. Although Queensland police were sworn in as South Australian constables to ensure that duty was paid, the trouble and complaints persisted.[133]

Teamsters were spoken of with great respect on the Tableland. Yet some could not resist the

125 Managers of stations to Government Resident, 10 September 1885, in Half-Yearly Report on the Northern Territory, 31 December 1885, *SAPP*, Volume 3, 53 of 1886, pp.3-4.

126 G.R.McMinn, 1887, GRS 9, Volume 1, PROSA.

127 E. Hill, *The Territory*, p.208.

128 A.J.Cotton, *With the big herds in Australia*, p.170.

129 Managers of stations to Government Resident, 10 September 1885, in Half-Yearly Report on the Northern Territory, 31 December 1885, *SAPP*, Volume 3, 53 of 1886, pp. 3-4.

130 *Northern Territory Gazette*, 12 February 1892, Docket 155, GRS 1, PROSA.

131 Mounted Constable R. Stott, Borroloola, 29 December 1905, Government Resident's Report, 1904, SAPP, Volume 2, 45 of 1905, p.29.

132 *South Australian Register*, 3 March 1892, GRS 1, PROSA.

133 W.G. Stretton, Borroloola, Government Resident's Report, 1888, *SAPP*, Volume 2, 28 of 1889, p.11.

demon drink — such as the team in the 1910s which left Brunette for Borroloola

> with the best intention . . . but alas . . . there was a pub at Borroloola and, on arriving there, they got on the spree. They lost the horses, which instinctively came home, but not along the track, and so there was no loading for the [next] twelve months.[134]

Teamster stories abound, like the one of the teamster who died en route to a station. By the time the team had reached the Ranken store, the teamster's widow who was accompanying the team decided to marry one of the employees. When George Rankine J.P., who performed the ceremony, asked the groom 'Wilt though take . . . ', his reply was 'She's not the one I wanted, but she'll bloody well have to do'.[135]

In the early days the purchase and transport of Alexandria's supplies was arranged by William Collins. The goods travelled by sea to Burketown and thence by teamster to the station.[136] Later the station dealt directly with the store of Synnott, Murray and Scholes at Burketown.[137] The wagons would take six to eight weeks and their arrival was always a big event. The Aborigines would know when they were coming:

> Team! Team! They'd sing out, and they'd all run along the road — my kids too, and the teamster used to have a bed underneath the back where he used to sleep, well, they'd get on that bed and drive back to the homestead.[138]

Although camels were a common sight on Alexandria in the early days, they were used only once after 1925, when they carted supplies and bore casing from the boat at Burketown. Horse teams were much quicker and with fewer side effects as CAY Johnston explained:

> You'd have to put the tucker out in the sun to get the smell of the camel off it. Especially the flour. The horses hated them. They only had to smell a camel coming, they'd go mad.[139]

The Barkly was very dependent on teamsters until the advent of motor transport in the 1920s; at the time the teamsters complained that 'the car fumes would poison the grass and kill the horses'.[140] CAY Johnston bought his own Oldsmobile in the 1920s while the company bought a number of T Model Fords for working cars. One was purchased in 1917 and there was another on Alexandria by July 1919.[141]

Telegraphic communication was limited, with the nearest station for Alexandria being Powell's Creek on the overland telegraph line. In 1891, Barkly residents unsuccessfully petitioned the Minister responsible for the Northern Territory to connect Camooweal, Anthony Lagoon and Powell's Creek by telegraph.[142] Not until 1912 when Avon Downs was linked by telephone to Camooweal did Barkly managers have access to closer telegraphic contact. As well, an internal phone system, quite sophisticated for the time, connected the major sheep camps on Avon with the homestead.[143] The isolation of the inland districts of the Territory was also gradually overcome by the advent of wireless. By 1932, the police stations at Borroloola and Anthony Lagoon were connected by wireless,[144] and in 1934 radio and transmitting sets were installed at the police stations at Daly River, Roper River, Ranken River and Timber Creek, which not only facilitated police administration but also aided communication for those isolated communities.[145]

Alexandria had a pedal wireless in the mid-1930s although CAY Johnston still had to rely on letters and telegrams for communication with head office. He would go over to Avon to ring through a telegram, giving, for example the number of bullocks he had to send, then wait for a reply. This was obviously time consuming unlike the situation for the owner-managers who could make decisions for themselves. But relations with NAP in Brisbane were good, with Phil Forrest or the pastoral inspector Bill Frith or both coming up once a year to 'have a look around'.[146] But distance and communication difficulties left the detailed running of the station pretty well up to the managers, a situation which suited them.

At various times, there was proposed a transcontinental railway which would have enhanced the development of the Tableland. As early as 1879, Messrs Briggs, Headley and Favenc

134 A.J. Cotton, *With the big herds in Australia*, p.175.
135 L.A. Miller, *The Border and Beyond*, p.75.
136 William Collins, Mundoolun, to Thomas Harding, Alexandria, 11 July 1887, Letterbook, J.Collins and Sons, Fraser Family Papers, Mundoolun.
137 Patrick Synnott, Camooweal, 21 October 1921, Parliamentary Standing Committee on Public Works, *CPP*, 1921, p.85.
138 Ellen Johnston Interview.
139 CAY Johnston Interview.
140 L.A. Miller, *The Border and Beyond*, p.75.
141 Income Tax Return, 1918, NAPR.
142 Petition by pastoralists and residents of the Tablelands, 1891, Docket 34, GRS 1, PROSA.
143 C. Wagstaff, *Avon Downs NT 1882-1982*, p.18.
144 A.V. Stretton, Superintendent of Police, Darwin, 1932, in Report of the Administration of the NT, 1932, *CPP*, 1933, p.33.
145 A.V. Stretton, Superintendent of Police, Darwin, in Report of the Administration of the NT, 1934, *CPP*, 1934-35, p.45.
146 CAY Johnston interview.

James Broadbridge's pocket garden and bush-house.

crossed the northern section of Alexandria between the 17th and 19th parallels to report on a probable railway route.[147] The Pine Creek to Darwin railway was begun in 1887 as the first section of the Transcontinental Railway. Only this section was built although the NAP partners supported the suggestion of a link between the Tablelands and Pine Creek.[148] The Transcontinental Railway Act was passed in 1902 with the intention that the line be built on the land grant system, whereby the railway line would be laid progressively as land along the route was settled. The railway was never built.[149] Another scheme proposed in 1902 was to link Camooweal with Cloncurry, Winton, Longreach, Charleville and Cunnamulla.[150] Then in 1909 it was proposed to link Port Darwin with Camooweal, Bourke and Sydney.[151] In 1920, those responsible for administering the Northern Territory were still pleading for a railway, saying that the two great natural industries of the Territory, pastoralism and mining, were dependent to a large extent on a railway. Railway services in the Territory then were limited to those between Darwin and Katherine,[152] but in 1928 the line was extended to Mataranka and in 1929 to Birdum 483 kilometres (300 miles) south of Darwin. A southern railway from Alice Spings to Oodnadatta was completed in 1929 but the Barkly never did receive its much-promised railway despite continuing calls over ensuing decades for a railway to be built from Birdum through the Barkly to the railhead at Dajarra in Queensland[153].

By the onset of World War II, Alexandria had an outlet for store cattle to its Channel properties and cattle for slaughter went to coastal meatworks in Queensland. In 1934, an ambitious government scheme had been investigated to build a port and meatworks on Vanderlin Island, 45 kilometres from the mouth of the McArthur River. Cattle were to be shipped to the island for slaughter and export, following transport by diesel tractors on an all-weather road from the Barkly to Borroloola. There were, however, too many difficulties for the scheme to be viable. The jetty needed to withstand cyclones; it was realised that an all-weather road was impossible; the shallow river required dredging, and most important of all, the cost was prohibitive.[154]

There was a sense of the continuity of life and life-style on Alexandria from the 1880s to the 1930s. After World War II, however, the pace and nature of living there were to undergo many changes.

147 E. Favenc, 'Report on Country in the Northern Territory', *SAPP*, Volume 4, 181 of 1883-84, p.1.

148 W. Forrest, Minutes of Evidence of the Royal Commission into the NT, *SAPP*, Volume 2, 19 of 1895, p.169.

149 Government Resident's Report, 1905, *SAPP*, Volume 2, 45 of 1906, p.9.

150 Transcontinental Railway, 3 March 1902, Docket 11227, GRS 10, PROSA.

151 *Register*, Adelaide, 20 March 1909, GRS 9, Volume 6, PROSA.

152 Report of Acting Administrator, 1920, *CPP*, 1920-21, p.16.

153 J.H. Kelly, *Report on the Beef Cattle Industry in Northern Australia*, Canberra, Bureau of Agricultural Economics, 1952, pp.213-8.

154 'Report of the Board of Inquiry; Land and Land Industries of the N.T. of Australia', CPP, 4/1937, Volume 3, 1937, pp.858-62.

CHAPTER 7

A Foundation For the Future: Douglas Martin Fraser and Francis Henry Foster

Ken McGuire, a former NAP station manager who knew Douglas Martin Fraser for nearly twenty years, has described him as:

> a dynamic person who could tell you a lie with a twinkle in his eye; always asked everyone their opinion on a matter and then did exactly as he intended to do in the first place. Never committed himself to anything until he was sure it was the correct decision. [He was] always a good touch for broke ringers and drovers in the big smoke after a Christmas binge . . . A drover once called at the old Morehead's Building in Mary Street half under the weather waiting to see Mr Fraser. The receptionist said 'Which Mr Fraser do you want to see sir, Mr D.M. Fraser or Mr W.M. [son Bill]'. The drover replied 'I'm not sure, only that little b . . . who cracks the big whip'.[1]

Such a statement shows Fraser's varied qualities; he was one of the most influential men in the company's history. Though only 170 centimetres (5 ft 7 ins) tall, he was forceful, arrogant and pig-headed, yet he could also be warmhearted and generous. As a member of the board from 1932-1968, he served as chairman of directors from 1936 until 1967, and as managing director from 1936 until 1956. His enormous influence was felt throughout the company. Bob Kirk, a ringer at Marion Downs in the early 1950s, remembers that:

> He'd always get on a horse . . . , he wore these jodhpurs . . . and leggings . . . boots all polished, little tie and a whip . . . my God, he [was] like a general . . . We couldn't work him out at all . . . , [he] never spoke much.[2]

Douglas Martin Fraser: Director 1932-1968; Managing Director/General Manager 1936-1956; Chairman of Directors 1936-1967. Bill Fraser

Douglas Fraser was born in Melbourne in 1888, the second son of Sir Simon Fraser and Lady Anna Bertha Fraser (called Bertha). His mother was the younger sister of William Collins, one of the founding partners of NAP, while his father was a wealthy pastoralist and politician. Apart from the

1 Ken McGuire Interview, Bell, 5 February 1990. McGuire was manager on Soudan, Marion Downs and Alexandria.

2 Bob Kirk Interview, Herbert Downs, 14 November 1990.

fact that Bertha was a shareholder, Sir Simon also had various indirect connections with NAP. With William Forrest and others, he formed the Squatting Investment Company which purchased properties on the Dawson River. With Sir Malcolm McEachern of Melbourne and Albert W.D. White of Bluff Downs in north Queensland, he was a partner in Collins White and Company which had been formed in 1883. This company owned Glenormiston, Eulolo, Strathfield and Beaudesert stations. Bertha was Sir Simon's second wife, and after their marriage in 1885 they returned to Melbourne, where he continued his career in politics. He was first elected to the Legislative Assembly in 1874 and to the Legislative Council in 1886. As a staunch liberal and ardent imperialist, he was a delegate to the Federal Convention of 1897-98 and helped frame the Constitution for the new Commonwealth of Australia.

The Frasers lived at Toorak, where Bertha was one of the leading figures in Melbourne's social life. Douglas, with his brothers Simon and John Neville (father of future Prime Minister Malcolm Fraser), was educated at Melbourne Grammar School. While studying law at Melbourne University, he stayed at Trinity College where he met Francis Foster from Tasmania, who was studying Civil Engineering. A firm friendship developed between them, which was to be of great importance to the future of NAP.

In 1911, while Douglas was still a law student, Bertha gave him her Power of Attorney which gave him access to the North Australian Pastoral Company. He also went to Queensland about this time, visiting the Collins family's pastoral interests. In 1913 he was admitted as a Barrister to the Victorian Supreme Court but he never practised, moving instead to the 4,858 hectare (12,000 acre) Collins family property, Mundoolun, following his marriage to a cousin, Marion Dorothea Jane Collins (called Jane or, affectionately, Bub or Baby), who was the youngest daughter of Robert Collins. At that time, their aunt, Jane Collins, was the only surviving Collins family member living at Mundoolun, and the family needed male leadership. Douglas was well suited to the task. With a new house — New Mundoolun — erected on the property following the wedding,[3] he occupied himself energetically with managing both the property and the interests of the two companies, John Collins and Sons, and the Collins White Company.

From early 1918 Douglas Fraser served in the Middle East with the 5th Light Horse First AIF. On his return he continued to work at Mundoolun. His office was a small shed at the back of the house; here he was ably assisted by Agnes MacGregor. When he became managing director of NAP in 1936, Miss MacGregor moved to Brisbane to work at the NAP office in Morehead's Building. Thereafter Fraser worked in Brisbane three days a week, from Tuesday to Thursday. His son Bill recalls: 'He used to work us like hell all weekend ... and we used to sigh with relief when he left'.[4]

His capacity for work was enormous. As well as managing NAP and Collins White, he was a director, and also, for a period, Chairman of Directors, of the Queensland Meat Export Company. He was Chairman of Moreheads from 1951 and in 1956 supervised the merger of Moreheads with the Elder Smith Company. He was managing partner of the family company, John Collins and Sons (which was dissolved in 1941) until 1936, when John W.F.Collins took over; he was a councillor on the Beaudesert Shire Council; he was on the advisory council of General Accident, Fire and Life Assurance Corpn. Ltd; he was President of the Queensland Club in 1947; he was Chairman of the Society of St Andrew, and for thirty-seven years he was on the board of Emmanuel College at The University of Queensland.

Fraser's approach to the management of NAP was businesslike; he worked tirelessly for the company. It must be remembered that as well as being chairman of the NAP board, he was also managing director until 1956, when Ken Moore was appointed general manager. Although station managers recruited some staff locally, Fraser always made the final decision concerning senior appointments. Even with appointees at lower levels, managers were aware of his preference for Protestants. In 1944, when NAP's pastoral inspector Harold Chambers cautiously employed a Catholic, Malcolm O'Flannagan, he explained to Fraser that O'Flannagan was a 'good horseman, cattleman and worker' and a bad Catholic since he referred to priests as 'b.bs and other very uncomplimentary things'.[5] Even after 1956, when a general manager was appointed, Fraser took a keen practical interest in the properties. For instance, he added his own words of wisdom about the transfer of Rod Bellette from manager of Gallipoli to manager of Soudan in 1958:

3 This house was destroyed by fire in 1939 and a new one completed in 1941.

4 Bill Fraser Interview, Mundoolun, 15 September 1990.

5 H.W.Chambers to D.M.Fraser, 27 April 1946, Alexandria File, 1944-1947, NAPR.

> So far as Bellette is concerned, history relates that nobody really does very good work when deeply in love and the sooner he marries his girl and settles down the better for everybody.[6]

Fraser was responsible for negotiating the sale of cattle with the meatworks as well as the overall management of the properties. He did not trust machinery, and was not a good judge of cattle and horses, but his strength lay with his legal training which he drew upon to contribute to the company's sound economic position. He was instrumental in obtaining very favourable lease conditions for Alexandria in 1941, and in 1965 he presented a strong case to contain Government resumptions. His assistance in proving the company's residency in the Northern Territory, which had major tax implications for NAP in the 1946 High Court case, was considerable. Over a decade later, during NAP's road train era he and Ken Moore instigated the transportation of goods from Tweed Heads thus saving permit fees.

He was respected by his peers and feared by his opponents. While he did not like 'yes men' among his employees, actually preferring those who stood up to him, he could be severe with any who were wrong. While he observed the courtesies, he did not care what people thought. He particularly disliked smoking, and he did all he could to discourage it, a difficult task in an era when smoking was quite acceptable. Harry Bruxner, a long-serving director, had to sit next to the window at board meetings if he wanted to smoke. In an incident at the Queensland Club, Fraser, as the senior member present, was once asked by another member for permission to smoke in the dining room. Although it happened on a day when only a handful were present in the huge room, Fraser bellowed for all to hear, 'I'll be damned if I'll have anyone smoking in this room'.[7] While he enthusiastically took up the challenge of managing NAP, it was his association with Francis Henry Foster which played a crucial part in strengthening the company's potential for expansion.

Foster was born in 1888 into an established family of land-holders who for two generations had owned properties in Tasmania and Victoria. A good sportsman, he was educated at Launceston Grammar School and later graduated in Civil Engineering from the University of Melbourne. He then worked briefly as an engineer with the Victorian Railways, the Shire of Heidelberg, and Monier Pipe Construction under Major John Monash. He spent 1912 in England with his parents and sisters and on his return in 1913 toured central and western Queensland with Douglas Fraser. This was an exciting time for them both and it cemented their mutual interest in the pastoral industry.

Francis Henry Foster, Director 1937-1971. Henry Foster

They journeyed by rail to Longreach and by car to various properties including Kynuna, Eulolo, Chatsworth, Coorabulka, Springvale, Marion Downs and Glenormiston.[8] They were to visit the Northern Territory but the death of Robert Collins in August 1913 made it necessary for Douglas to return home. Although the Northern Territory section of the tour was abandoned, Francis continued with his trip to western Queensland and spent a total of five months in the State.[9] He was so impressed with the country that in 1914 he returned north with his father, travelling up the Birdsville track in an early model Studebaker car.

---

6 D.M. Fraser to Bill Young, 24 July 1958, Green Copies, Letter File, June-September 1958, NAPR.

7 Michael Fraser Interview, Mundoolun, 15 September 1990.

8 Kynuna, Coorabulka, Marion Downs and Glenormiston were all later owned by NAP.

9 F.H.Foster, Record of Trip to Queensland, 1913, Foster Family Papers, Tasmania.

Francis Foster and Douglas Fraser: 1913. Henry Foster

Foster's interest in the region was already well developed, and he had inquired, in early 1914, about the possibility of purchasing property in western Queensland,[10] although he did not pursue the matter then. When war broke out they were camped at Lake Nash Station on the Barkly Tableland. His father returned immediately to Tasmania in order to join the Expeditionary Forces, after which he spent several months at Gallipoli. With his brother John also overseas at war, Francis was left to see to the family's business affairs while also serving with the Australian Imperial Forces in various capacities in Tasmania, mainly training men for overseas units. Although keen to serve overseas himself, he was not sent until early 1918 when he served in the Middle East and Europe. He returned to Tasmania in 1919 to resume his business career, after surviving both the Western Front and the devastating influenza epidemic that followed the war.[11]

Foster, like his friend D.M. Fraser, had an immense capacity for work, with a broad range of business and community interests. Apart from his involvement with NAP, he became a director and often chairman of numerous businesses including Perpetual Trustees Executors and Agency Co. Ltd., Industrial Engineering Ltd., Industrial and Pastoral Holdings Ltd., Murex (Australasia) Pty. Ltd., Prestige Ltd., Perpetual Insurance and Securities Ltd., Colonial Mutual Life Assurance Co. Ltd. and Perpetual Trustees and National Executors in Tasmania. His prominence within the Tasmanian Farmers, Stock Owners and Orchardists Association led to significant positions in other rural-based organisations — for example, as Tasmanian representative on the Australian Meat Board, as Chairman of the Tasmanian Meat Board and as a Member of the Graziers Federal Council. He was Chairman of the Tasmanian State Committee of the CSIRO,[12] and Chairman of the Board of Management of the Hutchins School in Hobart. He was elected to the Tasmanian Parliament in 1937, representing the predominantly rural electorate of Wilmot, but he found this frustrating because of the slowness of government to initiate change. In 1941, he happily returned to the more active pursuit of his many business and other interests, including the management of the family's sheep, cattle and agricultural properties in Tasmania (Armitstead, Mineral Banks, Fosterville, Cape Portland, Windfalls and Merton Vale) and in Victoria (Langi-Kal-Kal).

Beneath a quiet, unassuming exterior was a man of keen intellect and authority, whose strength of character, although manifested differently, quite matched that of Douglas Fraser. Foster came increasingly to take charge of his family's business interests from the start of World War I. The original family plan had been for him to concentrate on engineering, scientific and business matters while his brother John would involve himself in the family's farming interests. The war, however, substantially broadened his involvement in family affairs, particularly after John was killed overseas. His role grew through the 1920s and 1930s, until, following his father's death in 1944, he assumed total responsibility.

Through his years of increasing involvement in rural affairs he responded effectively to challenge, applying the systematic approach of his engineering and business training to the pastoral industry. As a result, his interest and knowledge in general business, financial affairs and pastoral

10 F.A.Brodie and Co., Stock and Station Agents, Sydney to F.H. Foster, Mundoolun, 4 May 1914, Foster Family Papers.

11 'Francis Henry Foster — A Biography', Typescript, n.d., Foster Family Papers.

12 This was initially CSIR.

Douglas Fraser in the Minerva car in which he and Francis Foster toured Queensland in 1913. Henry Foster

matters was respected and sought after by many organisations, including the NAP board. He was particularly receptive to new ideas and, while always giving careful consideration to investment decisions, he was not afraid to invest in a wide variety of areas. For example, he gave strong support to NAP's purchase of the road trains, and it was at Foster's instigation that NAP diversified into shares in 1946. The resulting portfolio consisted mainly of industrial shares, and by 1967 it was worth $1 million. The shares were a vital source of revenue, especially when new properties were being purchased, but the portfolio was gradually sold down over the next two decades as NAP concentrated more on its major activity, cattle production.

Although he was involved in a wide range of business and agricultural pursuits, and despite living in Tasmania, Foster almost certainly considered NAP his greatest business love. During his thirty-four years on the NAP board, he was very much an 'active player' and was always kept well informed on all matters by the Brisbane office. His opinions and advice were sought as a matter of course before any major decisions were made. For example, Michael Crouch explains the purchase of the property for the company's head office in Brisbane in 1968:

> The decision to buy this property was chiefly made by Mr. Foster who flew into Brisbane, inspected and persuaded the Board all within a matter of hours. Mr. Foster works like this . . . [13]

Regular visits to Melbourne over a long period of time enabled Foster to establish and maintain valuable contacts in the legal, financial and political world, often to the benefit of NAP. His interest in and support for the company continued after the death of D.M.Fraser in 1968 and after Foster's own resignation from the board in 1971. Even at the age of 85 he made a trip to Alexandria and the Channel properties. 'There was no stopping him; he wanted to be in everything,' recalled Michael Crouch, who with Ken Moore, accompanied him on this trip in 1973. Foster's marriage to Patricia Ainslie Wood produced six children, with the two sons, Henry and William, presently on the NAP board. He died in 1979 at the age of 91.[14]

The long-term friendship between Fraser and Foster was of great significance to the company's stability and development. Their respect for each other was unquestioned and although Fraser was by nature resistant to change, he would listen to Foster. The introduction of the road trains was initiated essentially by Francis Foster and Bill Fraser, DM's son. Fraser and Foster took pains to see the company improve financially and in this

13 E.M.Crouch, Chairman's Address, Annual General Meeting, 14 November 1968, E.M. Crouch Papers.

14 Foster Family Papers.

they succeeded. Although only two properties, Monkira and Coorabulka, were added during Fraser's thirty-one years as chairman, the established base enabled the purchase of a further seven over the following twenty-two years.[15] The personal contrasts between Fraser and Foster were complementary just as their engineering/law backgrounds had been, and only towards the very end of Fraser's life did they have marked differences of opinion. But both men shared a common goal — productive investment in the pastoral industry.

The earliest example of their shared pastoral interests came in 1917, when Fraser became very interested in purchasing Brunette Downs. He envisaged a partnership to cover the cost of £200,000, consisting of his brother Simon, their cousin Christopher Collins (only son of Robert Collins), and Francis Foster. The trust between the Foster, Fraser and Collins families can be seen in the following extract from a letter written by Francis Foster to his father Henry in December 1917:

> The Collins and Frasers won't buy unless they are quite satisfied there is something pretty good in it and as they know the country and have access to all the Books and as they know the profits from "Alexandria" they will make no mistakes. So you need not be worried about a bad purchase being made.[16]

Richard Holt, manager of Alexandria, inspected and reported on Brunette on their behalf. All plans were quashed, however, when both DM and Francis Foster left for the Australian Imperial Forces overseas, Foster in March 1918 and Fraser soon after. Furthermore, redwater disease was becoming a problem on the Barkly and the Queensland Government had closed the border to all stock movements until more dips had been installed.[17] The deaths of both Simon Fraser and Christopher Collins in 1919 finally put paid to any thoughts of a partnership.

The next more successful attempt to invest jointly came in July 1925 when the Foster family (76.2 per cent) and Douglas Fraser (23.8 per cent) formed the Monkira Pastoral Company in order to purchase Monkira Station from Sidney Kidman.

Share Holdings: Monkira Pastoral Company[18]

Henry Foster	13,450
Blanche Laura Foster	1,000
Jeanne Blanche Foster	3,000
Edith Florence Foster	3,000
Marion Dorothea Jane Fraser	1
Douglas Martin Fraser	9,999
Francis Henry Foster	11,550
TOTAL	42,000

The price was £42,000 on a walk-in, walk-out basis, upon the assumption that there were 10,000 to 11,000 Monkira cattle on the property or on adjoining stations. Kidman had expressed a desire to sell Monkira as early as 1921. He was losing money but not specifically on Monkira, and he needed to reduce his holdings.[19] Morney Plains was also on offer from Kidman and an option to purchase was taken on both. Foster and Fraser inspected the properties and Francis wrote to his father of the strenuous 1,514-mile journey (2,438 kilometres) they made by car:

> We got lost for a day & a half on the way from Morney to Monkira and the last 350 miles back to Charleville we had a broken back spring which however we managed to build up with fencing wire.[20]

They found Morney far inferior to Monkira of which they spoke in glowing terms: the plentiful water supply, the splendid condition of the cattle, the excellent fattening country. Kidman knew it was one of his best properties and afterwards he regretted the decision to sell, offering some thousands of pounds to call the deal off. Walter, his son, and Ted Pratt, his travelling manager, were both against selling,[21] but it was too late. Furthermore, one thousand bullocks had been sold on the station for £5 2s 6d per head, the proceeds of which went to the new owners.

The lease had fourteen years to run but, unfortunately, by 1928 Monkira was in the grip of a devastating drought. Henry Foster explained to the Land Administration Board:

> With the exception of approximately 2,000 mixed cattle — which are travelling East and are at present

15 Glenormiston, Islay Plains, Connemara, Kynuna, Wainui, Boomarra, Vergemont.

16 F.H.Foster, Sydney, to H. Foster, 16 December 1917, Foster Family Papers.

17 D.M.Fraser, Mundoolun, to H. Foster, 18 March 1918, Foster Family Papers.

18 Share Holdings, Monkira Pastoral Company, Foster Family Papers.

19 Sidney Kidman to Under Secretary of Lands, 23 December 1921, Monkira, LAN AF 290, QSA.

20 F.H.Foster, Mundoolun, to Henry Foster, 2 August 1925, Foster Family Papers.

21 'Monkira', Ken Moore Typescript, Brisbane, 1990; F.H.Foster, Mundoolun, to Henry Foster, 2 August 1925, Foster Family Papers.

> very weak, and too poor to sell — the Coy has nothing left but around 30 horses away on agistment, . . . the run is empty.[22]

The partners, although very concerned, had no intention of being beaten. Fraser and Foster met with the Land Commissioner in 1929 and secured a reduction in rent from twelve shillings to eight shillings per square mile. They also thought of raising additional capital to restock, but this was not necessary thanks to reasonable seasons in the early 1930s.

Although the first meeting of the Monkira Pastoral Company was held in Brisbane on 9 July 1926, thereafter meetings were held in Tasmania. Henry Foster, usually referred to as Colonel Foster, was the majority shareholder and his son Francis Foster was managing director. Douglas Fraser also had considerable involvement, attending to the sale of Monkira cattle. The two friends were continually thinking of ways in which they could expand, and in 1928, they seriously proposed the amalgamation of Alexandria (NAP), Chatsworth and Coorabulka (John Collins and Sons), Glenormiston (Collins, White and Co.), and Monkira (Foster/Fraser).[23] The managing partner of NAP, Philip Forrest, was not keen and so the merger did not take place. The merger of all the properties except Alexandria was then proposed and by 1933 Fraser and Foster were considering amalgamating Monkira with Coorabulka. 'I don't mind about Chatsworth' said Fraser.[24] History has shown that, while it took another forty years, they did achieve their wish; all except Chatsworth were finally owned by the one company, with NAP providing the avenue through which the amalgamation could take place.

At this stage Foster was not directly involved in NAP, but in 1937 the 22,222 Forrest shares (some 19 per cent of the company) were offered to him for fifteen shillings each. Foster offered eleven shillings and they were sold for thirteen shillings and sixpence. He joined the board the same year and from the late 1930s set about increasing the Foster family's NAP holdings whenever shares became available. By 1967, once the family were the largest shareholders, they halted this policy of purchase, considering that a non-controlling interest of around 40 per cent made for the most harmonious situation among the various shareholder groups. The family was later to regret taking this position, when in 1985 they were suddenly obliged to increase their holdings beyond 50 per cent to deter other parties from attempting to take control of the company.

Francis Foster strongly advocated a merger between the Monkira Pastoral Company and NAP, which could be achieved by an exchange of shares. NAP had been looking for another property to take the turnoff from Alexandria, as Marion Downs was unable to cope with the huge numbers. In 1938, for instance, there were 15,000 steers and 4,000 spayed cows to send away from Alexandria. The purchase of any other property comparable to Monkira would involve a cash outlay and for this reason Monkira was favoured. Although Brighton Downs came up for sale in 1938, it was not purchased because it was non-flooded country, the purchase would bite heavily into the cash position of the company, and it would make future amalgamation of Monkira less likely. Furthermore, DM wanted to pay a dividend as soon as possible in order 'to justify the acquisition of Marion and my management'.[25]

The amalgamation of NAP with the Monkira Pastoral Company in 1939 was conditional on two factors. The Monkira Pastoral Company was to purchase Coorabulka from John Collins and Sons, which was being wound up due to dissension among its members; and Francis Foster was to purchase the NAP shares belonging to Mary Adelaide Gwendoline Collins, wife of the late William Collins, and those of her son John.[26] Relations between Douglas Fraser, a director of John Collins and Sons, and John Collins, the chairman of John Collins and Sons, had been strained for some time. At the Directors' meeting of John Collins and Sons on 2 March 1939, for instance, Collins stated that in his opinion 'Mr Fraser had been antagonistic to him over a long period'. It was at this time that John Collins left the NAP board. Coorabulka was purchased by the Monkira Pastoral Company from John Collins and Sons for £40,000 walk-in, walk-out, and then Monkira Pastoral Company amalgamated with NAP, becoming a subsidiary company. A total of 42,000 fully paid shares of £1 each were issued by NAP, increasing the company's paid up capital from £120,000 to £162,000.[27]

---

22 H. Foster to Chairman, Land Administration Board, 8 August 1928, Monkira, LAN AF 290, QSA.

23 Francis Foster to Douglas Fraser, 12 July 1928, Foster Family Papers.

24 D.M. Fraser, Mundoolun, to F.Foster, 1 January 1933, Foster Family Papers.

25 Directors' opinions on purchase of Brighton Downs, Minutes of Directors' Meeting, 19 May 1938, NAPR.

26 For biographical details of John W.F. Collins, see *ADB*, Volume 8, 1981, pp. 78-9.

27 Minutes of Directors' Meeting, 6 June 1939.

Douglas Fraser's strategic and tactical skills cannot be illustrated better than by his efforts on behalf of NAP in negotiations with government on the Alexandria leases. Although these were not to expire until 1941 and 1942, he was at pains, for example, not to antagonise the Federal Government, in the aftermath of the 1937 Payne Fletcher Report.[28] Revisiting much earlier views from the last century, the Report advocated sheep for the Barkly, estimating it could carry 1,250,000. It stated, however, that unless a railway was built to the region — it was suggested from Dajarra to the Ranken River — there was no prospect of sheep being run successfully. Fraser felt that the Commonwealth was unlikely to build a railway because of the small area involved.[29] No opposition was shown by NAP at the mention of sheep, even though they did not agree. At any rate, Fraser felt that the Federal Government would not act on any of the recommendations in the Report, and he was correct.[30] He could focus more directly on the real issues of importance to NAP, the Alexandria leases and their renewal.

Under the Crown Lands Ordinance 1924 and the Crown Lands Ordinance 1927 of North Australia, provision was made for the surrender of existing leases in exchange for leases with an extended term until 30 June 1965. Unlike neighbouring properties Brunette Downs and Alroy Downs, NAP chose not to take up the extended lease, preferring the security of tenure and lower rentals of their South Australian leases.[31] From 1924 until 1940 there was a saving of £9,000 in rents.[32]

In 1934, Philip Forrest began to communicate with the government on the issue of the 1941-42 renewal.[33] While the minister hinted at resumption, he would not be drawn on a decision, merely saying he acknowledged that

> the holders of Alexandria Station are exceptionally good tenants, and every endeavour is being made to try and meet their wishes to obtain an extended lease of at least a substantial portion of their holding.[34]

This indecision meant uncertainty for NAP. There was a huge amount of capital invested and if the government were to resume part of the land or fail to renew leases, the company wanted to be ready for the consequences.

The subsequent negotiations were long and drawn out, and the favourable outcome[35] in 1942 was due in no small way to the persistence of Douglas Fraser and the support of Francis Foster. They were not afraid to put NAP's case to politicians, and they sought legal opinion from Brisbane counsel, L.G.Lukin.[36] As a result, the lease numbers 2122, 2123, 2114, 2148, 2488 and 189, a total of 28,119 square kilometres, were replaced by Lots 1, 2, 3 and 4, a total of 28,078 square kilometres, 50 per cent of which was resumable on 30 June 1965,[37] the rest being held under lease until 30 June 1984.

Lot No.	square miles	square klms.	Region
Lot 1	8,276	21,518	Homestead and northern
Lot 2	1,106	2,876	South-western
Lot 3	554	1,440	Ranken
Lot 4	863	2,244	Soudan

Land had been excised for a military road, a stock route, an aerodrome and, at the Ranken River, a police station.[38] Priority was given to NAP for Lots 1 and 3 and while there was open

28 'Report of the Board of Inquiry; Land and Land Industries of the Northern Territory of Australia', 10 October 1937, *CPP*, 4/1937, Volume 3, 1937, pp. 813-926. In March 1937 the Commonwealth Government appointed a Committee of Investigation to inquire into and report upon the resources of the Northern Territory, particularly the future of the pastoral industry. Messrs. W.L. Payne, Chairman of the Land Administration Board in Queensland, and J.W. Fletcher, a Queensland pastoralist, were prominent on the Committee. The Committee took evidence in most of the capital cities as well as the Northern Territory. Alexandria was visited.

29 D.Fraser to F.Foster, 11 February 1938, Foster Family Papers.

30 For discussion see W.Wynne Williams, 'The Payne-Fletcher Report on the Northern Territory', *Australian Quarterly*, March 1938, pp.53-60.

31 D.Fraser to F.Foster, 26 January 1939, Foster Family Papers.

32 Hicks, Memo, Department of the Interior, 29 October 1940, Alexandria Leases, A 452, 1952/310, Australian Archives, Canberra.

33 P.Forrest to Minister, Department of Interior, 23 March 1934, Alexandria Leases, A 452, 1952/310, Australian Archives, Canberra.

34 T. Patterson, Minister for the Interior to Sir Donald Cameron, 10 January 1936, Foster Family Papers; Alexandria Leases, F27, PL 2114, NTAS.

35 *Commonwealth Government Gazette*, No. 157, 4 June 1942, p. 1354.

36 Opinion re Northern Territory Pastoral Leases, L.G.Lukin, 25 September 1941, Foster Family Papers.

37 C.L.A.Abbott, Administrator, 11 May 1942, Alexandria Subdivision, *Northern Territory Government Gazette*, No. 14, 15 May 1942, p.102.

38 Alexandria Leases, F27, PL 2114, NTAS; E.G.W.Wood to Administrator, 19 June 1940, A452/1, 1952/310, Australian Archives, Canberra.

competition for Lots 2 and 4, these too went to NAP. Ultimately it was a question of who would use the land best. It was not a place for small businessmen and in the final surrender and application for new leases, NAP's only rival was the Vestey family, owners of Wave Hill.[39] Rents were reasonable; there was an increase of only £253 per year, the rate varying from nine pence to four shillings per square mile.[40] Even DM was satisfied with the conditions of the leases, commenting to Foster: 'I feel we should accept this offer and get the matter settled while we have a sympathetic Administrator [Abbott] in charge'.[41] The Adder Block of 1,204 square kilometres had been resumed from Rocklands in 1935 and was taken up by NAP in 1943 [42] to bring the total area of Alexandria to 29,281 square kilometres (11,262 square miles).[43]

In the early 1960s, with the possibility of 50 per cent being resumed in 1965 under the N.T. Crown Lands Act of 1946,[44] Fraser and Ken Moore, the general manager, entered into discussions with the Federal Department of Internal Territories and the Northern Territory Administrator. They realised that some country would be resumed. The question was, how much? The Administrator favoured consolidation, after which part of each of the five leases would be resumed. But this was impractical. The fact that there were five leases and only half could be taken from each meant that it was virtually impossible for anyone to make workable living areas from the resumed leases. At the same time, the remaining non-contiguous areas would be insufficient for NAP, since they would not allow a practical and viable operation. In mid-1961, a meeting was held in Darwin between Fraser, Moore, Hugh Barclay, head of the N.T. Lands Department, and the Administrator, Roger Nott. Nott insisted on consolidation and began to read the N.T. Ordinances to make his point. Fraser interrupted, 'Your Excellency, would you desist reading ordinances or quoting them to us. Moore and I would be better informed than you are in relation to this matter.' In Ken Moore's words:

> the Administrator would not discuss any other matter until we would agree to consolidate. D.M.Fraser, with full dignity said 'Your Excellency, we have reached an impasse.' Mr. Nott said, 'I thank you for your courtesy but my title is not Your Excellency but Your Honour.'
> 'Oh!' said Fraser, 'You were not appointed by the Queen — you are a political appointee. Come Moore we will depart we are wasting our time,' said DMF. With which we left — much to everyone's embarrassment.[45]

Negotiations continued with a meeting in Canberra arranged by Fraser with Prime Minister Menzies, Sir Magnus Cormack and Paul Hasluck, the Minister for Internal Territories. Bill Young, manager of Alexandria, was called to Brisbane to give his advice. A subsequent meeting in Brisbane at NAP's office settled the issue. Fraser, Moore, Barclay as Acting Administrator, and Lambert, Head of Internal Territories, were present. A map of Alexandria was produced and Lambert drew a horizontal line dividing the property in half, the bottom half for NAP and the top half to be resumed. He invited Fraser to convince him and Barclay that the line should be moved further north. It was soon well above the Alexandria homestead and serious argument back and forth moved the line to its final position. The line was inked in on the map and initialled by Fraser and Barclay.[46] With the changes effective from 1 July 1965 with leases expiring on 30 June 1984, Alexandria was now down to 16,359 square kilometres (6,292 square miles) which is very close to its size today of 16,116 square kilometres (6,198 square miles).

The improvements on the resumed area for which compensation of $26,390 was paid to NAP, consisted of four bores fully equipped, and 87 kilometres of fencing.[47] Bill Young had advised that such an excision would have virtually no effect on Alexandria's management because the resumed country ran very few stock. It did, however, provide three blocks for selection, plus a fauna and flora reserve. Benmarra was taken up by Luke Wise, and Mittiebah and Mount Drummond by Bruce Walker and Dave Joseland. Joseland later bought Walker out and took in partners including Lee McNicholl, at that time NAP's veterinarian. The partnership later sold to Norland Pastoral Pty.

39 Renewal of Alexandria Leases, 1932-42, A 452, 1952/310, Australian Archives, Canberra.

40 C.L.Abbott, Administrator, Darwin, to Managing Director NAP, 17 April 1941, Foster Family Papers.

41 D.Fraser, Mundoolun, to F.Foster, 21 January 1942, Foster Family Papers.

42 G.Pigott, Chief Clerk, Northern Territory Administration, Alice Springs, to W.Kelly, 20 February 1943, Renewal and Resumption of Alexandria Leases 1940-45, NAPR.

43 Alexandria Leases, NTRS 246, PL 431, NTAS.

44 H. Barclay, 'Land Tenure in Relation to Agricultural and Pastoral Development', Occasional Paper, No. 6, CSIRO, n.d., pp.6-10, Vern O'Brien Personal Papers, Darwin.

45 Ken Moore Typescript, Brisbane, 1990.

46 The map is still in the NAP office.

47 Directors' Annual Report, NAP, 30 June 1965, 30 June 1966.

Limited owned by W.F. and A.F. Munro of Weebollabolla, Moree.

Fraser and Foster were also very conscious of the tax implications of NAP's activities. In the early 1940s, the Taxation Department began to ask whether the company could legitimately claim residency in the Northern Territory and thus avoid paying income tax under Section 23(m) of the Income Tax Assessment Act 1936. Fraser acted swiftly and from July 1941 until 1954, when counsel advised that under the Income Tax and Social Services Act No.3 of 1953 the company would not lose the benefit,[48] all directors and shareholders' meetings were held at Alexandria. As well as the distance involved, war-time restrictions on travel into the Northern Territory often precluded the attendance of directors and shareholders, so meetings were not held regularly. The 1941 shareholders' meeting even had to be adjourned because of the lack of a quorum. To prove further the point of residency, in 1943 Harry Barnes, the manager of Alexandria, was appointed managing director, although in reality Fraser, as chairman of the board, maintained control. Management was cumbersome, however, with two sets of books being kept, one at Alexandria and the other in Brisbane. Mrs Weston, the Alexandria housekeeper, was even made an alternate director and Jack Owen, the bookkeeper, was appointed secretary of NAP. Usually, at least two out of Fraser, Bruxner and Alexander were present at the meeting.

In order to prove the case, NAP in 1946 went to the High Court of Australia — and won.[49] The fact that it came into existence as a partnership in the Northern Territory where it subsequently registered as a company was in NAP's favour. It was shown by NAP's Melbourne solicitors, Arthur Robinson and Company, that the directors met in Brisbane only because of convenience. The High Court decision meant that £122,587 12s in tax was saved between 1941 and 1946, which was obviously to NAP's great advantage.[50] The case set a precedent with a similar successful appeal in 1946 by the Vestey-owned Waterloo Pastoral Company Limited.[51]

48 Minutes of Directors' Meeting, NAP, 185 Mary Street, 4 March 1954.

49 North Australian Pastoral Company Limited (Appellant) versus Federal Commissioner of Taxation (Respondent), *Commonwealth Law Reports*, 1945-1946, Volume 71, Part 5, Sydney, Law Book Company, 1946, pp. 623-36.

50 Minutes of Directors' Meeting, NAP, 27 November 1946.

51 *Commonwealth Law Reports*, 1945-46, Volume 72, p.267.

Charles Bruxner, Director 1957-1976. Mary Bruxner

In 1951 another precedent was set, this time for NAP shareholders, when Francis Foster was successful in his appeal to the High Court of Australia about a rebate on dividends which had been allowed to individual shareholders under section 107 of the Income Tax Assessment Act of 1936. Some time after the 1946 residency case, the Commissioner of Taxation reassessed this latter situation and sought to recover the rebate already granted to one of the shareholders, Bertha Clotilda Harris. Francis Foster took the matter up in relation to his own assessment, in a Melbourne sitting of the High Court. The Melbourne solicitors Arthur Robinson and Company were engaged again and the case is still quoted in law texts, with the argument that under section 170 (3) of the Act the assessment could only be amended if there had been an error in calculation or a mistake in fact. It was found that neither had been the case. In particular, it was found that failure by the taxpayer to inform the Commissioner of a fact already known by the Commissioner (the appeal by NAP) did not constitute a failure to disclose a fact. Foster's appeal was upheld and tax rebates paid before 1946 were retained by all NAP shareholders.[52]

52 Foster V Federal Commissioner of Taxation, Full High Court (1951) 82 C.L.R. 604 quoted in N.E. Challoner and J.M. Greenwood, *Income Tax Law and Practice (Commonwealth)*, second edition, Sydney, Law Book Company, 1962, pp.1018-24; Reginald Barrett, *Principles of Income Taxation*, Sydney, Butterworths, 1975, pp.19-20; H.H. Mason and L.G. Priddle, *Case Companion to Ryan's Income Tax Manual*, Sydney, Law Book Company, 1976, pp.230-1; K.W.Ryan, *Manual of the Law of Income Tax in Australia*, third edition, Sydney, Law Book Company, 1972, p.247.

Other taxation advantages resulted when the company, Alexandria Station Pty. Limited, was formed in 1946 for Alexandria's dealings in livestock, plant and stores. In addition, on advice from legal counsel and their tax adviser, W.T. Kelly, NAP chose to base their tax assessments on the market value of cattle rather than average cost values.[53] This meant thousands of pounds in savings to the company.

In April 1937, it was resolved to increase the board from seven to eight members in order to allow Francis Foster to join.

NAP Board 1937	Origin
Douglas Martin Fraser	Collins
William Forrester Alexander	McIlwraith
Edward Robert Crouch	Ingram
Henry Bruxner	Collins
Edward Morton Shaw	Collins
John William Fitzclarence Collins	Collins
Francis Henry Foster	New shareholder
William Chard Frith	Pastoral Inspector

William Forrester Alexander, Director 1934-1976.

David Alexander

With the exception of Francis Foster and Bill Frith, the composition of the board continued to be either direct descendants of the original partners or their representatives. John Collins, son of William Collins, resigned from the board in 1939 and was killed two years later in an aircraft accident at Archerfield. Edward Shaw, a grazier from Telemon near Beaudesert, who had been a director representing the estate of Robert Collins, resigned in 1943 and was replaced temporarily by Harry Barnes, manager of Alexandria, to allow the holding of directors' and shareholders' meetings at Alexandria. There was a close association between the Collins and Bruxner families. Henry Bruxner, who was married to Annie, daughter of Robert Collins, was appointed chairman of directors during DM's absence from Australia in 1951. Bill Fraser, DM's son, was appointed on 3 October 1951 and is the longest-serving director on the present NAP board.[54]

Although Francis Foster was kept well-informed through correspondence, by telephone and through periodic meetings in Melbourne with Douglas Fraser, business commitments in the south prevented his attending board meetings or visiting the properties as often as he would have liked. He always played a key role in major company decisions, but because of his absences he appointed alternate directors: Harry Barnes in 1943, Harold Wentworth Chambers (Pastoral Inspector) in 1945,[55] Bill Fraser in 1946-47, Charles Bruxner in 1957[56] and his elder son, Henry Francis Foster, in 1968. In 1971, Foster retired from the board after thirty-four years as a director but he remained Henry's alternate and was retained by the company as a consultant, until his death eight years later.[57]

Edward Robert Crouch retired from his position on the board in 1955 and was followed by his son, Edward Michael Crouch, also a solicitor. When Sir Herbert Ingram died in 1958, his son Herbert requested that Michael Crouch continue as the family's alternate director. Henry Bruxner died in 1957, and his son Charles gained a place on the board in his own right. By 1968, the number of board members had fallen to five.

53 Ken Moore Typescript, Brisbane, 1990; Memo from D.M.Fraser, 4 September 1952, Foster Family Papers.

54 The longest serving director in the company's history to date has been Bill Alexander, a McIlwraith descendant, who served for 42 years from 1934-1976.

55 Minutes of Directors' Meeting, NAP, Alexandria, 8 September 1945.

56 Meeting of Directors, NAP, 21 February 1957.

57 Ken Moore, General Manager to Station Managers, 2 August 1971, File 1, Correspondence, Alexandria, 1971, NAPR.

NAP Board 1968
Edward Michael Crouch (Chairman)
William Martin Fraser
Charles Robert Bruxner
Francis Henry Foster
William Forrester Alexander

Shareholders' meetings were held half-yearly until 1942 when it was decided to hold them annually. But between 1941 and 1954 with such meetings at Alexandria, it was virtually impossible for shareholders to attend. In 1954, a further Article was added to the Articles of Association restricting the number of members to fifty.[58] The Warner interests were registered as a trust, releasing seven places; the Fosters subsequently transferred their shares to a single company, Wivenhoe Pty. Ltd., and the Bruxner family used two family companies, Jelbyn and Mallapunya. This flexibility allowed the families to pass shares on to other family members in the event of death.

After various family deaths and transfers, in 1967 the shareholders of The North Australian Pastoral Company Pty. Limited (the 'proprietary' having been added in 1964 to comply with new company laws) were as follows:

SHAREHOLDERS 1967

ALEXANDER	David Seymour Forrester (as Trustee for Miss Janet McIlwraith Alexander)
ALEXANDER	Harriet Margaret Forrester
ALEXANDER	Estate B.M. (Dec'd)
ALEXANDER	William Forrester
BRUXNER	Charles Robert
CAINE	Jean Collins
CLARKE	Anne Dorothea
CONNOLLY	Mary Blair de Burgh
CROUCH	Edward Michael
(FOSTER	Francis Henry
("	" " (Ann Patricia Foster Trust)
("	" " (Jane Marian Foster Trust)
FOSTER	Patricia Ainslie
FOSTER	William Francis
FOSTER	Henry Francis
FOSTER	Misses Jean Blanche and Edith Florence
FRASER	Douglas Martin
FRASER	William Martin
FRASER	William Martin and CLARKE Alexander H.B.
FRASER	John Malcolm
FRASER	Peter
HOOPER	Arabella Georgina Collins
HOWE	Betty Bannatyne de Burgh
INGRAM	Estate Herbert (dec'd)
INGRAM	Herbert and TOMKINSON, William Robert
INGRAM	Herbert and INGRAM, Michael Warren
INGRAM	Sir Herbert
INGRAM	Michael Warren
JOSHUA	Mary
MOORE	Clothilde Rawson
PERSSE	Desmond de Burgh
PERSSE	Beryl Margaret de Burgh
PERSSE	Beryl Mary
PERSSE	Margaret Anne Janette
PERSSE	Beryl Mary and RODEN, Olof Clarence
PERSSE	Burton William de Burgh
PHILP	Katharine Mary
PURVIS-SMITH	Nan Gertrude
RALSTON	Rosamond Anne Collins
SCOTT	Dorothea Gwendoline Martin
WARNER	Christopher F., CARTER, W.O., & PERKS, C.G.H.
WHILEY	Alison Collins
WIVENHOE PROPRIETARY LIMITED	

The family group proportions had changed to some extent by this time:

	1937 %	1966 %
Foster	18.5	43.2
Fraser	13.2	16.4
Collins	23.9	17.1
Ingram	18.5	13.5
Alexander	7.4	5.5
Warner	18.5	4.1
Crouch	0.2	
	100.0	100.0

The notable changes were the substantial increase in the Foster holding (mainly from the absorption of the Monkira Pastoral Company by NAP in 1939) and the marked decrease in the Warner proportion. There was some reduction in the proportions held by the Collins and Ingram groups, and some increase in the Fraser holding if considered separately from the other Collins shareholders. The Alexander distribution remained about the same, while in 1962 director Michael Crouch was given the opportunity, with the board's approval, to purchase shares from the Ingram family.

Following the low returns of the 1920s, the second half of the 1930s showed a considerable improvement for Australian beef producers. One

58 Minutes of Directors' Meeting, NAP, 185 Mary Street, 7 October 1954.

catalyst for change after 1934 was the shift towards the export of chilled rather than frozen meat, which was not only popular with consumers because of its tenderness and palatability, but also attracted higher export prices. At about the same time, Australian exporters benefited from the much reduced competition from Argentina in the British market.

World War II increased demand for Australian beef exports to Britain, with supplies of both high and lower quality frozen meat (the latter for bully-beef) required in the war effort. Chilled meat shipments commenced again in 1956, and prices for better quality beef rose steadily in the post-war years.

A Fifteen Year Meat Agreement from 1952 until 1967 between Australia and the United Kingdom guaranteed a market which enabled Australia to increase production, although the floor price did not necessarily translate to more profitable returns. The agreement also denied Australia taking full advantage from the late 1950s of the burgeoning United States market which required third-quality meat for hamburgers.[59]

The European Economic Community put a virtual end to Australia's exports to the United Kingdom in the early 1970s but this was offset by the excellent United States trade. Coinciding with slumps in both American and Argentinian production, the volume of Australian meat exports between 1957-58 and 1958-59 to the United States increased by a factor of ten.[60] By 1969, Britain's share of Australia's meat and veal exports had fallen from nearly 80 per cent of the total between 1953 and 1958 to around 20 per cent between 1963 and 1968.[61]

Other factors enhanced the market prospects of Australian beef producers. Australia's relative isolation and stringent quarantine precautions meant that it was free of most of the dangerous livestock diseases such as foot-and-mouth disease, rinderpest and bluetongue.[62] Producers were now able to move cattle quickly, through the development of beef roads and road transport, which provided much greater opportunities to resist the devastating effects of drought. As living standards rose in Asia, new markets were opened up for Australian beef.

Large cattle concerns such as NAP were well-placed to benefit from these generally positive developments. In 1939, the company began to pay dividends again for the first time since 1931. Seasons and markets were good and profits increased. This improved situation continued after Ken Moore became general manager in 1956, Douglas Fraser having relinquished this role to remain as chairman of directors. With reasonable seasons and markets, the 1960s were especially good for dividends, which peaked at $324,000 in 1969.[63] In 1970, assets were revalued to bring them up to current standards, when a nine for one bonus share issue was made and the capital increased from $324,000 to $3,240,000.[64]

The NAP's Brisbane office moved three times between 1935 and 1968. In 1935 it had moved to Morehead's Buildings, 185 Mary Street. From September 1958 until October 1968 it rented office space at 39 Creek Street in a building owned by the insurance company, General Accident, Fire and Life Assurance Corporation Limited. In the 1960s, with the company in a healthy financial position, the board set about acquiring the company's own premises. A building was purchased in 1968 at 46 Edward Street for $135,000. On the corner of Edward and Margaret Streets,[65] it became known as NAPCO House.[66]

Meanwhile the company's improved financial position brought increased benefits to employees. A superannuation scheme was introduced in 1961, and Ken Moore, the general manager, was the first to take part. A policy of financing senior permanent employees for the purchase of a home was implemented in the 1960s, and yearly bonuses were given to station managers.

Just as the company has changed premises, so it has changed banks. NAP had always dealt with the Queensland National Bank, Thomas McIlwraith being one of its strongest supporters. In mid-1929, however, the Monkira Pastoral Company transferred its business from the Queensland National Bank to the Union Bank of Australia Limited because of a tightening in credit policies of the former.[67] In November 1937 NAP changed to the Union Bank of Australia for similar

59 Dawn May, 'The Impact of the 15 Year Meat Agreement on the Development of Northern Australia', Unpublished Paper, Australian Historical Association Conference, Darwin, 1991, pp.1-24.

60 *The Economics of Road Transport of Beef Cattle*, Canberra, Bureau of Agricultural Economics, 1959, p.74.

61 *The Australian Beef Cattle Industry Survey: 1962-63 to 1964-65*, Canberra, Bureau of Agricultural Economics, 1970, p.1.

62 *An Outline of the Australian Export Meat Industry*, Sydney, Australian Meat Board, 1969, pp. 2-3.

63 E.M.Crouch, Chairman's Address, Annual General Meeting, 22 October 1969.

64 'A Short History of the N.A.P.', Brisbane, 19 July 1973, E.M.Crouch Papers.

65 Minutes of Directors' Meeting, NAP, 12 June 1968.

66 The Heidelberg Restaurant was one of the tenants.

67 Francis Foster to Manager, Union Bank of Australia Ltd., 8 April 1929, Foster Family Papers.

Agnes MacGregor, a thrifty and efficient Scot who worked at NAP's head office 1936-1963. Peg Corbett

reasons. A call had been made by the Queensland National Bank for NAP to reduce its overdraft and the company favoured the stronger capital base of the Union Bank.[68] In 1951 when the Union Bank merged with The Bank of Australasia (both of which had branches in New Zealand), to form the Australia and New Zealand Bank Limited (ANZ), NAP went with the merger and is still with the ANZ forty years later.[69]

Douglas Fraser was ably assisted by his Brisbane office staff. The contribution of the Beardmore family has already been mentioned. William Doull, formerly Secretary of the Queensland Club, was appointed secretary of NAP in 1950. He was likeable but inexperienced in the economics of pastoral matters; he resigned in early 1956.[70] Another staff member, Scottish born Agnes Christain[71] Mary MacGregor was one of the most loyal and competent employees NAP has ever had. She retired in December 1963 after nearly thirty years service to NAP.[72] DM also depended on Bill Frith, the pastoral inspector. An Englishman from Somerset, he joined Collins White in 1919 and managed Eulolo and Strathfield. In the 1920s he became pastoral inspector for both NAP and Collins White, and was a director of both companies although not a shareholder. When Fraser became chairman of NAP in 1936, it was Frith's calm judgment and reasoning which was so often a healthy restraining influence on DM's intense likes and dislikes.[73]

Harold Wentworth (Jerry) Chambers was another who made a very significant contribution to NAP. Born on 19 May 1896, from 1930 to 1940 he was manager of the John Collins and Sons property Westgrove which provided the nucleus of the Alexandria and Coorabulka studs. He managed Yacamunda from 1940 to 1945, but in July 1940 was also appointed to supervise the trucking of NAP fats at Winton.[74] So began thirty-five years of service to the company.[75] In 1945 he was appointed pastoral inspector for NAP, succeeding Bill Frith. He was made pastoral inspector for the Collins White station, Glenormiston, in 1951, and for the other Collins White properties, Eulolo, Strathfield, Rangeview, and Imbi Imbi from 1954. His capability in planning and supervising improvements, fencing, yards, bores and dams was excellent and equal to his knowledge of livestock, both cattle and horses. It was said that, 'He'd arrive at turn-off time and was always willing to lend a hand with the drafting.'[76] He nearly always hired the drovers and could have as many as five mobs on the road at once. In Boulia they called him 'up and down the river Willy' as he traversed the stock route, checking on drovers and encountering occasional problems. As Ken McGuire recalls,

> You had to steer them [the drovers] past the Boulia pub, if they were going from Marion with fats to Winton, then the Hamilton pub, then the Macunda pub, then look out for the mailman bringing the grog down from Winton down the old Cork mail road when they were going up the Diamantina.[77]

Despite the rough roads, Chambers always seemed to arrive at his destination — generally covered in dust. Travelling around the properties advising and supervising the company's operations, he provided an important link between the

68 Memorandum, Minutes of Directors' Meeting, NAP, 18 November 1937.

69 Minutes of Directors' Meeting, NAP, Alexandria, 15 March 1952.

70 W.F. Alexander, Ninety Years of the NAP Co.

71 This differs from the usual spelling 'Christian'.

72 Agnes MacGregor died 10 April 1974.

73 William Chard Frith, in W.F.Alexander, 'The N.A.P.Co., 1878-1967'; Frith died 18 March 1958.

74 D.M.Fraser to H.W.Chambers, Yacamunda, 18 April 1945, D.M.Fraser File, April to October 1945, NAPR.

75 Minutes of Directors' Meeting, Monkira Pastoral Company, 24 July 1940.

76 Ken McGuire Interview, Bell, 5 February 1990.

77 *ibid.*

Weebollabolla Shorthorn Sales: Moree 23 October 1974. Left to Right: W.E. Young: Manager, Alexandria; E.M. Crouch: Chairman of Directors; K.W. Moore: General Manager; H.W. Chambers: Pastoral Inspector.

managers and Fraser in Brisbane. He was usually at board meetings and became alternate director for Francis Foster in 1945, when meetings were held at Alexandria. The replacement of droving with road transport, and improvement of communications with telephone contact and aeroplane travel, meant less need for a pastoral inspector. Ken Moore's job as general manager began to absorb this inspecting job, so that when Chambers retired on 1 March 1975 he was not replaced. He was invited to join the board as a director, and served in this capacity from 28 July 1971 until 1 March 1975. After he retired, he was not replaced on the board and died on 7 January 1977.

One person who should be mentioned at this stage, although more will be said of his contribution in a later chapter, is Kenneth Woodhead Moore. He was appointed assistant manager and secretary in April 1956, and in the same year was appointed general manager. He and DM formed a strong managerial partnership in the late 1950s and 1960s, just as Moore and Michael Crouch were to do after DM's death.

While Douglas Fraser's service to business and the community was outstanding, in 1935 his personal life was saddened by the loss of his 20-year-old son, Douglas, in an accident on a property in Victoria. Four years later, the family home at Mundoolun was burnt to the ground and his valuable library destroyed. Two decades of hard work followed until 1963, when his wife was killed in a car accident. His health was deteriorating by this time, although he continued as chairman of the board. Fraser's shaky signature on the minutes testifies to his declining health and, although ably assisted by Ken Moore, he wa unable to continue. He presided over his last meeting on 9 August 1967. His son Bill was acting chairman briefly until Michael Crouch was appointed chairman on 15 November 1967.[78] Douglas Fraser died the following May and is buried at Mundoolun. Fraser's working life was very largely devoted to

78 Meeting of Directors, NAP, 15 November 1968.

NAP and his record thirty-one years as chairman, twenty of these as managing director, clearly establishes him as one of the most influential figures in the company's history. Perhaps more than any other man, he can be seen as the father of the modern NAP.

Through all these years of development, Francis Foster was the other prime mover in policy matters, and he continued to play his part for another eleven years after Fraser's death. Foster's strong and established rapport with Michael Crouch and Ken Moore provided a great source of stability for the company during Douglas Fraser's declining years, and in the period following Fraser's death this stability provided a sound basis for the next phase of the company's development.

CHAPTER 8

Alexandria Consolidation: 1938-1971

Henry Heathcote Barnes: Manager of Alexandria 1938-1946. H.R.L. Barnes

The three managers of Alexandria between 1938 and 1972 all made significant contributions. They were:

Henry Heathcote Barnes	1938-1946
Rowand Hall-Scott Murray	1946-1955
William Ernest Hamilton Young	1955-1971

Henry Heathcote (Harry) Barnes was born at Dyraaba, New South Wales, on 6 March 1893. After leaving The Southport School in 1911 he was engaged in horsebreaking on the family properties, Dyraaba and Tabulam. He was employed as a jackaroo on Mitchell Downs in Queensland and was understudy to his uncle J.H.S.Barnes on Lyndhurst prior to enlisting in 1916 with the 5th Light Horse Regiment in World War I. Returning from the war, he went to Western Australia where for fifteen years he managed properties in the West Kimberleys. During 1936 he broke in horses for his younger brothers — Eric at Chatsworth who had left Coorabulka two years before; and Tom at Brunette who had previously managed Coorabulka and Chatsworth. Harry went contract-fencing in 1937 on Brunette Downs and so was in the vicinity when the Alexandria job became available. He was appointed manager of Alexandria in 1938, having known Douglas Fraser in the 5th Light Horse Regiment.

He was joined at Alexandria by his son Harry who worked as a jackaroo for five years before taking up the duties of bookkeeper when John Owen left. Mrs Margaret Weston, who had worked at Glennie School, Toowoomba, was employed as housekeeper. Father and son were at Alexandria for eight years, but Harry snr was forced to resign on 19 September 1946 after losing the sight of his left eye following a severe bout of shingles. Following a long convalescence, he returned to various station jobs in between numerous droving contracts. Never again, however, did he take on the task of managing such a huge station. He died on 29 April 1985, aged 92. The epitaph on his headstone — 'A True Bushman'— succinctly sums up Harry Barnes. He was completely at home in the 'Blue Room' as he described the outdoors and had a great affinity with horses. His main priority was his stock and anything else was secondary.[1]

He was at Alexandria during the difficult years of World War II when materials and labour were scarce. 'Black boys are our main stay. The few whites offering are not worth employing', said Harry Barnes.[2] But somehow work on the station had to continue: mustering, branding and the delivery of cattle to drovers. Improvements were out of the question then, and even basic repairs to bores and fences were difficult because of the lack of materials. Before 1941, the problem was exacerbated by the reluctance of the NAP board to contemplate major improvements until the renewal of leases was settled.[3] Barnes did manage, however, to have a bore on Soudan drilled and equipped (1946) and to build substantial yards at Lorne Creek for spaying, inoculating and drafting out weaners, which eased the pressure on the Soudan camp.

A summary of Alexandria's brandings from 1901 to 1971 illustrates how creditably Barnes performed, despite the difficult circumstances.[4]

	Years Inclusive	Total Brandings	Average
23 years	1901-1923 (R.Holt)	186,294	8,099
15 "	1924-1938 (CAY Johnston)	164,387	10,959
7 "	1939-1945 (H.Barnes)	102,482	14,640
10 "	1946-1955 (R.H.Murray)	118,966	11,896
16 "	1956-1971 (W.E.Young)	223,223	13,951

His brandings were the highest on average, despite working with the same number of bores as Johnston, and with none being put down between 1922 and 1946. Also, Murray and Young benefited from vast post-war improvements, such as more bores, fences, staff and better transport, although it must be conceded that Murray had to contend with the bad drought of 1952 which pulled his figures down, while Barnes had reasonable seasons.

Rowand Hall-Scott Murray: Manager of Alexandria 1946-1955. Michael Murray

Rowand Hall-Scott (Paddy) Murray was born at Brisbane on 17 July 1902. He was manager of a number of stations: Emu Plains, Sonoma, Carpentaria Downs, Forest Home, Wandervale and Jessievale. But Paddy Murray, as he was known, always had his eye on Alexandria. His wife Betty recalls that when they married in 1935, he pointed to Alexandria on a map and told her: 'I will never be happy until I go there. It's the biggest cattle station in the world.' The dream came true ten years later when he was appointed manager. The Murrays arrived with their two sons, Peter and Michael, and a governess, Miss Wiltshire. Betty Murray recalls the trip from Mount Isa: 'The heat was so bad that the asphalt on the road melted and we had to keep stopping to throw water on the tyres to keep the pressure down'.[5]

They arrived in time to oversee a vast increase in improvements; Murray was the right man at the right time. When the opportunity came to spend money on bores, fences and buildings, he did so with enthusiasm. Douglas Fraser seldom praised his employees but in 1949 he said he was amazed at the amount of work Murray had got through.[6]

1 H.R.L.Barnes, Typescript, Brisbane, 1990.
2 Harry Barnes to Directors, 22 February 1944, Alexandria Correspondence, 1944-1947, NAPR.
3 D.M.Fraser to Harry Barnes, 31 August 1938, Correspondence, 'A-C', D.M.Fraser's Room File, 1938-49, NAPR.
4 'Alexandria Station Pty. Limited', in W.F.Alexander, 'Ninety Years of the NAPCo., 1878-1967', p.4; Alexandria Stock Returns, 1968-71.
5 *Sunday Sun*, 6 November 1988.
6 D.M.Fraser to R.H.Murray, 13 January 1949, R.H.Murray File 1948-51, NAPR.

Six years later he even considered that Murray was 'proceeding too fast with [the] drilling arrangements'.[7] Significant repairs were made to the Alexandria homestead and buildings, while the homesteads at Gallipoli and Soudan were replaced. Murray could handle Aborigines and was able to get the job done despite what he saw as the retention of too much red tape by the Federal Government following the war years.

Mrs Murray was a trained nurse and a hospital was built at Alexandria to work in conjunction with the Flying Doctor. This put her at loggerheads with the local tribal 'doctor', until one night, with her husband away, the head stockman asked her to help bury the 'medicine man' who had just died. Because the Aborigines were too superstitious to do it themselves, she and the stockman had to dig the grave that night, to avoid the dreadful daytime heat.[8]

Murray was a very keen gardener and, in order to have rain water for the garden at the homestead, he built an over-shot dam across the Playford in 1948-49 using old steam engines to reinforce the earth banks. For the wings and spillway, since stone was scarce, he used bitumen drums.[9] The dam was unsuccessful, so Murray persevered with bore water and kept relocating the garden as each area became impregnated with minerals from the underground water.

Communication both on the station and with head office was somewhat improved although not always reliable. The main contact with Brisbane office was by surface mail although urgent telegrams could be sent and received through the Flying Doctor Base at Cloncurry.[10] Alexandria and Soudan had Traeger radio transmitting sets for this purpose and in 1955 two-way Crammond transceivers were installed for communication between the head station and Gallipoli and Soudan.[11] Not until the 1960s were portable (and initially cumbersome) transmitters fitted in vehicles.

Although direct communication with Brisbane improved after the telephone was connected at Soudan following construction of the Barkly Highway in 1943, the telephone at Alroy Station was sometimes used because it was closer to the Alexandria homestead than Soudan. Murray tended, however, to make decisions of his own instead of involving Head Office. When finances became tight after the drought of 1952 Fraser interpreted some station expenditure as extravagance. He also displayed dissatisfaction that in 1954 Murray corresponded with the Department of Native Affairs about establishing a school at Alexandria for the Aborigines. Fraser very soon made it quite clear that such discussions must be conducted through him.

The 1952 drought, and to a lesser extent the drought of 1954, were very difficult times and huge losses occurred. The pressures of management eventually affected Murray especially during and after a cattle-duffing case held in a four-day sitting at Alexandria in April 1955. Apart from a police patrol to the Nicholson country in 1943-44 which found some 70 head of stolen Alexandria cattle,[12] the Barkly Tableland had been one of the few places in northern Australia where cattle duffing was not usually prevalent. This changed with the drought of 1952, when Murray strongly suspected that Alexandria cattle, especially breeders, were being stolen from the outskirts of the property in order to establish breeding herds on smaller properties. Staff on Alexandria meanwhile, were preoccupied with moving cattle and keeping them alive.

Following one particular bout of cattle-stealing from the Barkly stations in June 1954 during the Brunette races, the Ranken policeman, Hughie Deviney, set out after the culprits with his party but returned with insufficient evidence. Not to be beaten, Deviney arranged for Soudan manager Ken McGuire to invite him to Soudan for Christmas, 1954 — over the two way radio 'galah session'[13] for all to hear. Deviney accepted the invitation and immediately set off north to find the offenders who had assumed he was safely at Soudan enjoying himself with the McGuires. After a three-months' journey of 2,100 kilometres during the wet, Deviney returned with 88 head of cattle. Sixty-five carried the Alexandria brand, although Murray estimated that another 500 head had been stolen. There was sufficient evidence for two men to be convicted of cattle duffing. Although the resulting sentences of three months hard labour for the offenders and fines of £100 each were said to be 'ludicrous and inadequate',[14] it was some deterrent

7 D.M.Fraser to R.H.Murray, 17 March 1955, R.H.Murray File, 1951-1955, NAPR.

8 *Sunday Sun*, 6 November 1988, p.15.

9 R.H.Murray to D.M.Fraser, 28 January 1949, R.H.Murray File, 1948-51, NAPR.

10 R.H.Murray to D.M.Fraser, 11 November 1950, R.H.Murray File, 1948-51, NAPR.

11 Correspondence D.M.Fraser and R.H.Murray, September — October 1954, R.H.Murray File, 1951-1955, NAPR.

12 Patrols 1943-44, File F1, 56/661, Australian Archives, Darwin.

13 Two-way radio communication between properties which can be heard by everyone.

14 R.H.Murray to D.M.Fraser, 29 March 1955, 22 June 1955, R.H.Murray File, 1951-1955, NAPR.

to others and attracted much attention in the media because of the precedent it set.

Fraser remained in Brisbane while the case was heard, which did not help communication between him and Murray, who once complained in 1949 that he had not seen Fraser at Alexandria for over fourteen months. 'That is too long considering the big programme we are engaged upon', Murray had said.[15] The stress became too much and Murray's health suffered with the recurrence of an ulcer. It is interesting to note that Harry Barnes had also developed a stress-related disease, shingles. Murray tendered his resignation, which the board did not accept at first, hoping that he would change his mind. But the demanding job of managing Alexandria had taken its toll and he left in December 1955. In one of his last letters he wrote:

> I am satisfied that 10 years on Alexandria is long enough for any one man. Alexandria requires a considerable effort to make it run . . . station affairs are constantly on my mind and extreme detail is essential each day.[16]

Jack Palmer was bookkeeper during Murray's time. He was a conscientious but slow worker; he would stay in the office at night until the lighting plant was turned off and then wend his way to bed in the dark. He is perhaps remembered most for the manner in which he left the company's employ. In all innocence, he wrote to head office stating his intention of getting married and living in one of the newly erected cottages. Apparently the cottages had been built for married engineers or mechanics, and Fraser would not allow it. Besides, he had Palmer in mind as the next bookkeeper at the Collins White and Company property Eulolo. After his wedding at Alexandria on 25 May 1953, however, Palmer accepted a job as bookkeeper at Alroy and his wife a position as housekeeper.[17]

In 1947 during Murray's time, Ted D'Arcy, a pumper at No. 21 bore on Gallipoli, died tragically. Doug Harris,[18] who had managed Gallipoli since serving with distinction as an officer in the RAAF in World War II, went on a Christmas binge at Camooweal and ended up in jail. While Harris was away, D'Arcy, apparently out of tobacco, began to walk into Gallipoli. Harris returned from Camooweal some days later and found D'Arcy

William Ernest Hamilton Young: Manager of Alexandria 1956-1971.

W.E. Young

dead, nineteen kilometres from Gallipoli. His grave is marked with a pile of stones where he was found by the side of the road. Murray partly blamed Harris for the death and as a result sacked him.[19] Harris later became head stockman at Marion Downs and finished his time with NAP as head stockman on Coorabulka in 1956.[20]

Another pumper perished in 1956 after setting out to recover a trunk of his personal effects which had been lost during his transfer from No.38 to No.7 bore. He spent many agonising hours before his death. Bill Young, the new manager, who found him, explains:

> Having no water with him it looks as though he must have panicked and by the tracks did quite some running about mostly in circles and as time went on, he shed his clothes, and finally died out on the plain . . . We buried him where we found him. From indications death by perishing is not a very nice way to go.[21]

William Ernest Hamilton (Bill) Young was born at Gloucester, New South Wales, on 11 January

15 R.H.Murray to D.M.Fraser, 29 August 1949, R.H.Murray File, 1948-51, NAPR.

16 R.H.Murray to D.M.Fraser, 2 August 1955, R.H.Murray File, 1951-1955, NAPR.

17 R.M.Murray and D.M.Fraser correspondence, R.H.Murray File, 1951-55, NAPR.

18 Brother of Herbert Harris who married Douglas Fraser's daughter, Margaret, Doug commenced as a jackaroo on Chatsworth and worked on various NAP properties prior to the war.

19 R.Murray to D.M.Fraser, 9 January 1947, Alexandria File, 1944-1947, NAPR.

20 Ken McGuire Interview, Bell, 5 February 1990.

21 Bill Young to D.M.Fraser, 17 January 1956, Alexandria Correspondence, 1956, NAPR.

1908. He finished school at 12 and undertook a business course for twelve months at a commercial college in Newcastle. But city life was not for him. He headed west and worked on cattle and sheep properties, learning various aspects of the pastoral industry. He tried to enlist in the army in World War II but was rejected because of diabetes. A series of managerial positions followed at St. Aubin's Stud, Scone; Ord River Station, Western Australia (1946); Strathmore for the Kidman/Angliss interests (1949) and Iffley (1950). When he started work on Alexandria on 1 December 1955, his wage was £2,000 per year with an increase of £1,000 after one year if he proved himself. This was one of the highest-paid manager's jobs in Australia but of course Alexandria was also reputed to be the biggest property.

Bill Young was once described by Ken Moore as memorable for

> the little red boots, the white moleskins, the blue shirt, the grey hat, the cigar, a lot of white hair, and above all, the man himself.[22]

He was a big man — 183 centimetres tall (6 ft) and over 105 kilograms (16.5 stone). He was a most generous person but his temper earned him the name, 'Willie the Rager'. It was said that he summoned his employees 'in a voice which would make a sensitive heifer on the furthest corner of his vast domain start nervously'.[23] To begin each day he would gather his staff around the front steps of the office and issue instructions. Of the more than 150 people who lived on Alexandria at that stage, more than half were Aborigines, which made his handling of them a significant part of the job. Indeed, Rowand Murray was correct in his predictions that 'The new man will have to understand Blacks and will stand or fall on his ability to work them'.[24]

As with other station managers of the time, Young could be described as treating the Aborigines 'with an earthy kind of aloofness or gruffness' which was in keeping with the character of the 'hard, tough' Territorian.[25] His time at Alexandria began before the era of equal wages but if he thought any Aboriginal workers deserved higher wages than the award, then he paid them more. Finding them more reliable than whites as pumpers, he increased the number of Aborigines checking the bore pumps. But he was determined always to be one step ahead of his Aboriginal workers:

> Whenever I went into the stock camp, I did the branding and cutting calves and got on the broncho horse and challenged the black fellow, see. And if you could get about two calves on him — you had him. But don't let that black b . . . get two calves ahead on you.[26]

It is significant that it was disillusionment with the changing Aboriginal situation that contributed to the termination of his employment on 31 December 1971. In 1962, Aborigines received full voting rights for Commonwealth and Northern Territory elections.[27] Full white wages to Aborigines and equal drinking rights had come in the late 1960s and the pattern of Aboriginal life was changing. Young explains:

> The trouble was — with your black labour — when they got the rights to go on drinking where and when they wanted and [when] the old bush fellows who controlled the camp died out . . . the young fellows had no one over them to steady them up . . . Things were getting no good. You might wake up on Monday morning and find half of them were away at a pub somewhere. And white labour was getting worse and worse.[28]

Bill Young was a legend in his own time and was highly regarded in the company. Many humorous stories can also be told of his days on Alexandria. In 1961 the directors had told him to be on his best behaviour when Sir Herbert and Lady Ingram visited the station from England. Not only were they substantial shareholders, they were not accustomed to his rough ways. But it made little difference to Young. At the time of the Ingrams' visit there was a rat plague at Alexandria and while chatting after a meal in the kitchen, Young casually tossed a piece of meat into the corner of the room where rats emerged from cracks in the wall. Picking up his shotgun, he fired. 'Good, I got three of the bastards,' he exclaimed, and carried on talking, to the Ingrams' astonishment.

His letters to head office were very colourful. Of the monsoonal heat in January 1971 he wrote:

> In the heat wave believe this or not, but I have the proof, a pair of my boots just burned to pieces, and I can hardly walk for sore feet so the bottom of a Land Rover must get pretty warm . . . Coming from

22 Ken Moore to Bill Young, 23 December 1971, File 1, Correspondence, Alexandria, 1971, NAPR.

23 *Courier-Mail*, 6 April 1964.

24 R.H.Murray to D.M.Fraser, 2 August 1955, R.H.Murray File, 1951-1955, NAPR.

25 Ann McGrath, *Born in the Cattle*, Sydney, Allen and Unwin, 1987, p.99.

26 Bill Young Interview, Casino, 26 January 1986.

27 A. Powell, *Far Country: a short history of the N.T.*, Melbourne, Melbourne University Press, 1982, p.234.

28 Bill Young Interview.

> the Darriel (sic) Gate home I ran through 5 storms and you could smell the paint blistering . . . No wonder calves die.[29]

The English shareholders were sent some of Young's reports and he was highly amused to learn that when he had written that 'some galah' had been lighting fires, the shareholders were trying to imagine how a bird could be responsible.

Other members of the Young family also worked on the station. Daughter Barbara was the bookkeeper, son Nelson was manager of Soudan between 1960 and 1967 and George MacQueen, son of Bill Young's wife, Pat, was manager on Gallipoli from 1958 before being appointed manager of Monkira in 1961. MacQueen returned to manage Alexandria in 1976. Mrs Young did the cooking for the staff for a long time. 'We got fed up with a few cooks — they couldn't cook, and rows and fights and brawls and I don't know what', Young said.[30] Being a family member meant no special privileges; in fact he probably expected more of them.

The management program on Alexandria through this period is illustrated from Douglas Fraser's instructions in 1941 to Harry Barnes, the manager. Barnes was told to:

1. Brand at three camps — at Gallipoli, Alexandria and Soudan — to be completed by the end of October if possible, to avoid handling calves in the heat,
2. Spay 2,500/3,500 cull females and make arrangements with Kemp Bates[31] to carry this out,
3. Muster for sale — 1,250 males and 1,250 females (spayed),
4. Receive from drover J.McCawly 100 stud bulls and distribute on the run,
5. Make repairs and alterations to the kitchen and the men's dining room,
6. Survey the western boundary between Alexandria and Alroy.[32]

Before the three stock camps went out each season, horses had to be mustered, drafted and shod. Foals were branded and yearlings taken off. With 240 horses needed at any one time, 1,000 were needed for the season. No.1 stock camp was the largest with approximately twenty-five men, most of them Aborigines. Apart from a break for the races, the mustering, branding, ear-marking, castrating, spaying and inoculating continued as the groups moved around, camping at bores and drafting out cattle to send to the Channel properties. Because the Adder block was close to the Rocklands and Avon boundaries, it was mustered by camping beyond the lead of the cattle and mustering back. Until a bore was established there in 1957, it was a dry camp where water and firewood had to be carted.

By November, it was too hot to work the cattle and storms heralded the beginning of the wet. Bill Young describes the pattern:

> If you got early storms in November — you got a good start with your grass. If you got storms late, well, you were behind a bit. Well if you got rain then through December and January and February and if you got rain in March, you'd have a really good season. Six weeks after it rained you'd have a crop of button grass . . . and my God! breed flies . . . Then your Flinders and Mitchell came on top of that and in a good season if you went through with a hay making machine you could have made millions of bales of hay.[33]

The management policy during these years was consistent. Breed on Alexandria and fatten on the Channel properties, Marion Downs, Monkira and Coorabulka. Aged and culled stock were walked to Mount Isa to be railed to coastal meatworks for slaughter. With fencing consisting of only small holding paddocks and some boundary fences, it was considered essential to run steers with the breeders who knew the property and the various watering points. Steers and spayed cows were then separated from the rest of the herd and taken to Soudan pending delivery to drovers. Cattle were walked down the Georgina Stock Route to NAP's Channel properties for fattening. In the reasonable seasons of the early 1940s the turnoffs from Alexandria were:

1939/40	2,500 to Marion
	6,250 to Monkira
1940/41	5,075 to Marion
	3,779 to Monkira

Numbers varied, however, with 12,000 steers leaving Alexandria in 1943, including 4,000 from Marion Downs, which had been on agistment at Alexandria.[34]

After road trains were introduced in the 1950s, droving declined and cattle were trucked from Soudan. In the 1968 season, for instance, of 10,000 steers that went from Alexandria to the Channel

29 W.R.Young to General Manager, 28 January 1971, File 1, Correspondence, Alexandria, 1971, NAPR.

30 Bill Young Interview, Casino, 26 January 1986.

31 He also spayed on NAP's Channel properties.

32 Minutes of Directors' Meeting, NAP, Alexandria, 5 July 1941, NAPR.

33 Bill Young Interview, Casino, 26 January 1986.

34 Minutes of Directors' Meeting, NAP, Alexandria, 27 April 1943.

Mustering on Alexandria 1960s. Left to Right: Jack Draper, Bill Young and Patrick Draper. W.F. Alexander

properties, 6,000 went on the hoof and 4,000 were trucked.[35] Having been dipped at the Ranken, they were unloaded at Camooweal and dipped again before being trucked direct to the Channels. In one huge lift of 3,011 head during Bill Young's time, thirty-eight transports were required.[36]

Drovers who took NAP cattle to NAP's Channel properties, markets and railheads, included Tom, Jack and Jim Laffin, Joe and Peter Craigie, Jack Morris, George Girdler, Fred Barlow, Don Booth, John Jessop, Jim Dwyer,[37] Laurie Hansen, Harold (Paddy) Fogarty, H.W.(Mick) Horn, Ian McBean, Clarrie Pankhurst, Laurie Pit, Owen and Brian ('Rawhide') MacNamara, Malcolm Thomson, Peter and Billy Pedwell, Ernie Jones, Bill Sharpe,[38] Bill Cousens, Pic Willetts, the Richardson brothers (Reg,[39] Ken and Les), Jack Stewart and later, Clive and Glen ('Smiley') Teece. There were also three Howard brothers (Dan, Jack and Jimmy) who, in earlier times, brought store steers from Soudan to Marion, Coorabulka and Monkira. They would then pick up a mob of fats and take them through to Marree or Fareena to be railed to Adelaide for slaughter.

Usually seven men were required in a droving plant; the head (or boss) drover, a cook, horsetailer and four assistants. The number in a mob was about 1,500 with up to an extra five for 'killers' to be eaten along the way. In a typical droving job in 1951, Jack Laffin left Soudan on 25 April and seven weeks later had delivered 1,498 bullocks in good order to the Monkira boundary. At fifteen kilometres a day, Marion Downs could be reached from Soudan in five weeks. There were often delays and losses, however, through redwater disease and poor conditions on the stock route. Fred Barlow was delayed for two weeks at Carandotta in June 1951 because of redwater, delivering only 1,364 out of his original 1,500,[40] and in 1959 Les Richardson lost over twenty head because the cattle ate fuchsia bush because of the scarcity of feed on the route.[41] Rates were fairly standard throughout the industry. In 1956 NAP paid:

35 Minutes of Directors' Meeting, Marion Downs Pty. Limited, 12 June 1968.

36 Bill Young Interview, Casino, 20 January 1986.

37 Later manager of Glenormiston 1971-1986.

38 After his retirement he lived at Mundoolun as caretaker until his death in 1988. He is buried at St. John's Church, Mundoolun.

39 Reg Richardson died in 1954.

40 Droving Book, 1951 season, NAPR.

41 Ken Moore to R.Gunther, Manager, Monkira, 17 April 1959, File 1, Correspondence, Monkira, 1959, NAPR.

Branding at Gallipoli, 1958. George MacQueen

Rate per 100 miles for stores travelling 63 miles per week [42]

4s 6d	per head per 100 miles for mobs	1,500
5s 3d	" " " " " "	1,250
6s	" " " " " "	1,000
7s 8d	" " " " " "	750

Depending on the availability of cattle, the size of mobs has increased in recent years. The combination of quieter cattle and increased droving costs has made it more economical to do so. Thus in April-May 1991, Smiley Teece walked a mob of 1,803 (three for killers) from Soudan to Monkira in seven weeks.

From 1959, Miner's Meat Supply in Mount Isa was an excellent outlet for approximately 4,000 head per year. A good relationship developed with manager Jimmy Moy and the proceeds of these sales more than paid the running expenses of Alexandria. After the purchase of Miner's Meat by Smythe's Butchery in 1966, sales of Alexandria cattle continued until the end of 1973.

A stud of good quality Shorthorn cattle was introduced to Alexandria in 1943. It had originated on the Collins property, Westgrove, where a stud had been in existence since the 1880s. When Westgrove was sold in 1940, the stud of 391 head was moved to Coorabulka which had just been purchased by NAP. In 1943, the stud was moved on from Coorabulka to Alexandria. By this time, it had increased to 496 but fifteen young bulls and eleven females were lost on the trip. By 1945 there were 480, of which 400 were females. A bad outbreak of pleuro-pneumonia over the next few years, however, lowered herd numbers to 294 by 1956. The stud was first run in the Buchanan Paddock but, with No.27 bore completed in 1949,[43] the cattle were shifted into the three stud paddocks erected in 1950. Over 400 selected herd females were added in 1950 but the devastating drought of 1952 caused the paddocks to be eaten out and the stock were turned bush to fend for themselves. Many died, but after subsequent additions to the herd there were 1,200 breeding-age females in the stud and weaner paddocks by 1960.

When the stud was moved to Alexandria in 1943, Len Carrington, the manager of Coorabulka, kept eight selected cows and one young bull. With purchases of stud bulls from Royal National Shows, the Coorabulka Stud increased. During the drought on Coorabulka in 1958, the stud was shifted for a year, first to Sandy Creek paddock on Herbert Downs and then on to Eulolo, a Collins

42 H.W.Chambers to Bill Young, 8 March 1956, Manager, Alexandria File, 1956-1965, NAPR.

43 No.27 bore was later replaced by No.53 bore.

White property in the McKinlay district. The stud then returned to Coorabulka and gradually, during the 1950s, its progeny were sent to Alexandria. The remaining portion of the stud, 112 females and 48 bulls, was transferred to Alexandria in 1960 and became known on Alexandria as Top Stud, with the stud already on Alexandria called the Commercial Stud. It took some months for the Coorabulka stud to get to know the locations of the Alexandria bores. They were trying to hunt storms[44] as they had done on Coorabulka. Bill Young explains:

> About 150 left 49 bore and I found them in 3 corners of the paddock on Tuesday morning pretty dry, so we just had to cart them water as the heat is bad, very steamy and burning. I was not game to try and drive them back to the bore. They are giving them another sip this morning and we will shift them into the bore tonight.[45]

All brands in the Northern Territory had to include a 'T', and as all the stud cattle from Coorabulka were branded QIF on the off ribs, the QTF brand was granted for use in the Top Stud in 1960. The remaining Alexandria herd had used the brand 676 since 1912 when a Brands Directory was compiled containing all the brands registered in the Territory by the South Australian Brands Office.[46] When the now famous [A] symbol brand ('box A') was granted in 1940 to replace ⊡ (dot within a square) which had been used since 1922,[47] the Commercial Stud retained the 676 brand and the main herd was gradually branded [A] on the off rump. Another connection with Coorabulka is in the earmarks. The [A] and QTF branded cattle now have the Coorabulka stud earmark as a result of an administrative move by the company in 1947 to have these cattle owned by Alexandria Station Pty. Limited.[48] The remaining herd branded 676 continued to use the old earmark (see Appendix for diagrams).

Breeding stud bulls on Alexandria was a saving to the company but both before and after the studs were established it was necessary to purchase bulls to improve the herd and to keep up the numbers to service such a huge herd. In 1933, 280 bulls came from Kianga and Gracemere studs in Queensland. In 1937, 522 were purchased and 250 in 1938. Other bulls came from Chatsworth and Westgrove Stations and the Royal National Shows. By far the largest number came from the A.G.F. Munro partnership, which owned Goodar, Banool and Weebollabolla in New South Wales. In the twenty-six years from 1940, NAP purchased 2,754 Shorthorn bulls from the Munro properties. These purchases continued until the change of breeds in the 1980s. Many Shorthorn breeders endeavoured to produce stud stock which followed the pattern of the low-set Scottish Shorthorn, but Rowley Munro refused to follow this pattern and maintained that a beast with the scale, bone and conformation of the Old English Shorthorn needed to be an upstanding animal to carry its massive frame, cover the distances of the larger properties and still be able to vigorously carry out the duties of a sire.

After a time, the stud Shorthorn bulls purchased at the various shows were considered by NAP management to be too short-legged with not enough scale and it was decided in 1962 to acquire only big, top-class Munro bulls for the Top Stud. From that time, the best stud bulls from Top Stud, which had 200 cows, went into the Commercial Stud which had 1,500 cows. The resulting bulls, usually 500 per year, were put into service in the main herd at about two years of age and at eight years they were sent to meatworks.

Alexandria cattle also had to be able to withstand the stress of severe heat and limited feed. For three months after the wet there was good surface water and the cattle had the benefit of good shade along the creeks. For the rest of the year they were on bores on open downs.[49] Calf shelters were erected in the 1960s, being just low enough for the calves. They were not successful, however, because the cattle ate the cane grass used in the roofing. Although there were problems with rats, grasshoppers and heat, Bill Young also began planting trees at the bores within the fenced-off areas of the turkey's nests.[50]

The average percentage of bulls to breeders was 4 per cent in the fenced stud areas and eight percent on the outside run. Through the spaying of culled cows, there was a gradual improvement in the quality of the herd. Young describes what he did:

---

44 Cattle instinctively hunt storms by walking to areas where rain is falling.

45 Bill Young, to Ken Moore, General Manager, 15 November 1961, File 1, Correspondence, Alexandria, 1961, NAPR.

46 Report of the Administrator of the NT, 1912, *CPP*, 1913, p.118.

47 Deputy Registrar of Brands to Ken Moore, 23 October 1972, Brands File, NAPR.

48 H.W.Chambers to Register of Brands, Northern Territory, 7 May 1947, H.Chambers and D.Fraser File, April 1945-October 1949, NAPR.

49 *Queensland Country Life*, 14 April 1966.

50 Bill Young to Ken Moore, 7 August 1968, File 1, NAP stations, 1969; 1 January 1969, 28 January 1969, Correspondence File 1, NAP stations, 1969, NAPR.

> I'd brand up to 70% in the stud herd and I used to cull the bulls 33% and the heifers 27% — spay them and cut the cull bulls . . . I'd have 400 really good bulls to put out on the run.[51]

Having averaged 43,358 between 1929 and 1939,[52] stock numbers on Alexandria gradually increased to over 60,000 by the 1960s. Brandings varied depending on the season but averaged between 10,000 and 20,000 by that time. Turnoffs also varied. For instance in 1961, 24,000 were turned off to the Channel properties, the result of two years' brandings and the holding back of stock while the Channel properties were in drought. Stock were again held back in the mid-1960s, with the result that the Alexandria herd increased to 75,000, somewhat above the property's optimum of approximately 60,000 to 70,000.[53]

Lease 2121 and part of lease 2148 were used as a bullock paddock to hold steers after weaning until they could be put on the road for the Queensland properties. At times there were up to 7,000 in the paddock, as well as spayed cows on transfer.[54] Breeding was essentially on Leases 2123 (Homestead Block), 189 (Gallipoli), and 2114 (Ranken).

Ticks and redwater recurred at various times, although not to the same extent as in earlier years. Gallipoli was always a troublesome area and heavy rains would wash the ticks down and affect the rest of the property. This was certainly the case after the 1974 flood, and there were also outbreaks in 1954-55. Dipping was still employed to contain the spread and by the late-1970s, dips were constructed at No.16 bore for the studs, at the Alexandria homestead, at Soudan, No.36 bore, and at Gallipoli.

Just as ticks had been a major problem on Alexandria in the early days, so too was tuberculosis (TB) in more recent times. Detection of this disease in the herd in the mid-1950s resulted in significant changes to management policies. In the belief that TB was spread by bulls introduced into the herd, the company flew Brisbane vet Jack C.Maunder to Alexandria in 1956 to test the stud herds. He reported the high proportion of 8.5 per cent of animals reacting positively. Maunder had been to Alexandria on occasions before 1956 to inoculate during pleuro-pneumonia outbreaks, but from 1956 he tested the stud herds and all introduced cattle each year for TB. Any cattle showing reactions were slaughtered. Testing in early April or June before the young bulls went out into the herd meant that they were tested twice before joining providing an added margin of safety. And because all Alexandria's cattle were fattened and slaughtered off the property, it was possible to keep a close check on the incidence of the disease in the commercial cattle.[55] By the late 1960s, however, TB was on the increase and a major testing program was commenced in the early 1970s during the wider National Brucellosis and Tuberculosis Eradication Campaign. Of major significance, the national testing program will be dealt with in a later chapter.

Bush fires were an occasional problem although the run was so large that the fires generally burnt themselves out. In 1951 fires kept breaking out along the stock route and as there was a rat plague at the time, Murray concluded that fires were ignited when rats chewed on loose waxed matches dropped by the drovers. As a result, he requested all stores and stations on the stock route to sell only safety matches.[56] Electrical storms were often the cause, although fires could also start at the bores from sparks from the engine belt. Fires were sometimes inadvertently started by tourists passing through; sometimes they were believed to be deliberately lit if someone had a grudge against the station. An engine-driven jetter spray, together with hand-held jetter sprays were used to fight fires, and fire-breaks were made with the station's grader.

An important pre-condition for successfully working a station such as Alexandria is that there should always be spare country available for relief during the dry times and especially in a drought. Good wet seasons obviously assist with such a strategy. In January 1939, for example, the news was telegraphed from Alexandria:

> Playford highest since 1922 Buchanan ran banker . . . all mails held up . . . rain general . . . Rankine River running strong at store and through Soudan have closed all pumping plants abundance feed and water and season looks promising.[57]

At the other extreme, the 1952 drought was the worst in all the Fraser years and one of the worst droughts Alexandria had ever experienced.

51 Bill Young Interview, Casino, 26 January 1986.

52 D.M.Fraser to Minister for the Interior, 30 August 1939, Foster Family Papers.

53 W.E.Young, Evidence given to the Commonwealth Conciliation and Arbitration Commission, Cattle Station Industry (Northern Territory) Award, 1951, Alexandria, 6 October 1965, p.1005.

54 D.Fraser, 18 April 1940, Foster Family Papers.

55 *Queensland Country Life*, 15 July 1965.

56 R.H.Murray File, Alexandria Station, 1951-1955, NAPR.

57 Telegram, H. Barnes to D.M.Fraser, 26 January 1939, Foster Family Papers.

No. 16 Bore, Alexandria, 1964. W.F. Alexander

Eighteen months of continual pumping followed and during the first six months of 1952, 14,660 head perished, with estimates as high as 35,000 (more than half the herd) for the number lost during 1951-52.[58] At one point, virtually the whole of the station herd was held on Gallipoli. Subsequently, many cattle were agisted on the upper Playford, Carrara, Muswellbrook, Highland Plains and Nicholson areas. Murray liaised with Rocklands manager Robey Miller[59] as Alexandria cattle passed through Rocklands on their way to agistment. The loss of breeders and calves was particularly heavy, which was devastating for the future herd. As an economy measure, and to help its recovery, Gallipoli was virtually closed down for two years after the manager Dick McCullagh left in 1953.

There could be great regional variation in the effects of drought; Brunette did not lose cattle in the early 1950s but the situation was reversed in 1959 when thousands were lost on Brunette, while Alexandria lost none. The precaution was taken in 1959, however, to sell 6,500 steers which were suitable for the American market, direct from Alexandria.[60] Later, the drought on Alexandria in 1970 accounted for losses of some 7,000 even though bores had been put down in the dry country towards Alroy.[61]

No bores were put down on Alexandria between 1922 and 1946 because of uncertainty over the extension of the leases and shortages during the war years. From the 1930s, diesel engines were phased in to replace the earlier steam-driven units. In 1937 thirteen Lister diesel engines were purchased,[62] though Ronaldson Tippet engines were also used. No.14 bore was the last bore to be replaced with a diesel engine; the steam engine was still being used there in 1944.[63] In May 1951,

58 R.H.Murray to D.M. Fraser, 15 June 1952, R.H.Murray File 1951-55, NAPR.

59 Robey Miller's son Philip has been the manager of Soudan since 1989.

60 Directors' Annual Report, NAP, 30 June 1959.

61 Directors' Annual Report, NAP, 30 June 1971.

62 NAP to Minister for the Interior, 26 September 1938, Foster Family Papers.

63 Inspection of Alexandria and Marion, 1939, Minutes of Director's Meeting, NAP, 6 June 1939; H.H.Barnes to Directors, 22 February 1944, Alexandria Station Correspondence, 1944-1947, NAPR.

the government bore, No.7A, on Soudan was purchased for £694.

The 1960s were a busy time for the sinking of bores with rotary drilling used for the first time in 1963 making the process speedier and more efficient. The company used its own bore drilling plant on Alexandria in the late 1950s and early 1960s but then found it was more economical to employ contractors with their own plants. Twenty-seven bores were put down between 1956 and 1971, bringing the total to 70 working bores, all equipped with Lister engines and either Pomona, 'A28' Oilbath, Bulwheel or Mono pumps. The Mono pumps were revolutionary because of the lightness of their water columns and the increased amount of water that could be brought up. In order to improve efficiency by having spare parts readily available, it was decided to standardise and bores were to use only Lister engines and Mono pumps; Land Rovers and Bedford trucks became standard vehicles.

Since it was no longer necessary to have a pumper at each bore to keep the steam engines supplied with firewood, six pumpers equipped with Mini Moke vehicles could travel between bores, doing the work previously done by 30 men. Roads were graded and cottages were built at some bores. Each man was allotted his own area and was responsible also for watering the trees at the bores. Sometimes they took on the task of trapping and poisoning dingoes with strychnine which was used until 1969 when the N.T. Administration allowed the use of 1080 poison.

World War II brought social change to Alexandria. A military camp was established two kilometres from No.19 bore, with water being pumped to the camp.[64] Coupons were required for tea, sugar, butter and Aboriginal clothing, and the associated paperwork was a nightmare. 'Miss MacGregor was the best backstop we ever had,' Harry Barnes Jnr recalled. The rationing of petrol and diesel was felt, as was the shortage of tobacco and cigarettes. Maintenance supplies were always delayed with a three-year wait for fencing materials.

Army personnel were sent to Alexandria to train everyone in guerilla warfare. Staff were taught how to set live demolition charges and how to throw grenades. Machinegun training was held on the aerodrome at the edge of the Playford desert, three kilometres north of the homestead; any sightings of aircraft had to be reported.[65] From December 1942, cattle were sold to the military authorities to feed the troops with whom there were often social contacts. Harold Lloyd, an Alexandria well-borer recalls the Americans who travelled on the new Barkly Highway:

> Their first night camp was at Number 19 Bore on Soudan, 8 miles west of Avon Downs . . . They had a telephone line to Avon Downs from Number 19 and when a convoy arrived the officers would ring up and say 'What's on for dinner, there are three of us and we will be over soon' . . . They were really a fine bunch of men.[66]

The war brought home to pastoralists the potential of machinery and rapid changes were to follow. Alexandria had relied on an International utility truck and a few lorries. The war brought other possibilities — four-wheel drive vehicles, larger trucks, semitrailers with stock crates, D4 bulldozers, graders and horse floats, as well as lighting plants, cold rooms and refrigerators. Gone, or on their way out, were the packhorses, German wagons, carbide lights, Coolgardie safes, steam engines and the like. By 1970 there were 4,800 kilometres of graded roads on Alexandria, many more vehicles had been purchased, and air mail, radio telephone services and the Royal Flying Doctor Service were well established. The company did not, however, own an aircraft. As Bill Young said, 'If we wanted to go anywhere or look at anything we used to grab the mail plane for a couple of hours'.[67]

One of the major influences on cattle management programs in the Fraser era was the construction of the Barkly Highway from Mount Isa to Tennant Creek in the early 1940s. 'The Bitumen', as it was called, was completed in 1943 and passed within 100 metres of the Soudan homestead. Although the stock route still passed through the Ranken, the new road decreased the significance of the Victoria River/Alexandria/ Ranken route for vehicles. This was compounded in the late 1960s when the Beef Road from the Barkly Highway to Borroloola via Alroy was sealed. With the arrival of the road trains in the 1950s, trucking from Soudan was facilitated and trucking yards were built on the highway near the Soudan homestead. An all-steel yard, reported to be one of the largest cattle yards in Australia, was constructed in the late 1970s to combat the problem of white ants.

After hostilities ended in 1945, NAP undertook a major building program on Alexandria. At the

64 H.Barnes to D.M.Fraser, 10 November 1944, Alexandria File, 1944-1947, NAPR.

65 H.R.L.Barnes. Typescript, Brisbane, 1990.

66 H. Lloyd in C. Wagstaff, *Avon Downs NT 1882-1982*, Armidale, Australian Agricultural Company, 1982, p.19.

67 Bill Young Interview, Casino, 26 January 1986.

homestead complex, the house kitchen and bulk store were replaced with easily constructed Sydney Williams' buildings; two cottages, an engineer's and machine shop building were also constructed; the men's quarters were renovated; the butcher's shop was replaced; a lighting plant was constructed to house a new 240-volt generator; a saddle room was provided adjacent to the stock yard; and single quarters were built for Aboriginal workers.[68] A cold room was constructed in 1955 to avoid a wastage of meat and, in an effort to attract more permanent staff, three cottages for married couples were built in 1965.[69]

In 1948 Soudan saw another extensive building program. A new homestead was built in front of the store and existing men's quarters, renovations to the kitchen included the laying of a concrete floor, a new store was built, and a central septic tank installed, as well as a steel building constructed to accommodate the men from the stock camp. The latter was built at right angles to the rear of the other buildings so that the rooms would benefit from the south wind.[70]

The buildings at Gallipoli also needed replacing. Barnes commented in 1944 that the homestead, with its bore casing for uprights and its earth floor, was uninhabitable because a windstorm had blown off the whole roof. Furthermore, the blacksmith's shop, which was built out of old troughing and tanks cut in half for the roof, had caved in on one side.[71] A new homestead was completed at Gallipoli in 1945, but not without some problems. It was built at ground level with a concrete floor but, being in white ant country, was attacked by these pests even before the carpenters had finished. Subsequently, a new homestead was built on stumps but this time there were difficulties with the carpenter who was described as 'a slow, cranky old devil . . . each alteration he has to make is something fresh to growl about'.[72]

Finding reliable skilled labour was a real difficulty for the managers. Men from Brisbane could not imagine what it would be like on a station. As Murray commented of an Englishman who arrived in 1953: 'He just couldn't camp out or drink bore water, so decided to make back to the city'.[73] Of another British migrant employed in 1950 as a pumper, Murray wrote to head office:

> He could not make an edible damper or cook anything at all, in fact he had the greatest difficulty in lighting a fire. I put him off before he started walking in and dying on the track.[74]

During the war those men who were not serving with the Armed Forces could obtain well-paid jobs with Mount Isa Mines or with the Allied Work Council which was responsible for the building of the Barkly Highway. Drovers and tradesmen were scarce. Even after the war, the cattle stations of the Barkly had to compete with the higher wages of the mines. Murray complained in 1948 that unskilled labourers could get £14 10s a week at the Mines plus £5 a week lead bonus. First-year jackaroos were paid £2 9s 10d a week plus keep.[75] Furthermore, to be competitive with Queensland, Alexandria generally had to pay the Queensland State Award wage for stockmen and pumpers which was higher than the Northern Territory award, although during the 1952 drought pumper wages reverted to the lower Territory rate. As a result, most pumpers left and Murray employed lower-paid Aborigines.[76] An unusual labour relations problem confronted Murray in 1947 with the absence of tobacco supplies. He complained to head office: 'If no tobacco arrives before end of month, I will lose additional men. Other stations have plenty so causing discontent here'.[77]

Jackaroos were usually engaged from head office in Brisbane. As Jerry Chambers said to Bill Young about the 10 jackaroos he sent at the beginning of the 1964 season:

> Some of them may not turn up, and the most likely looking ones are often the duds. You can only try them out and cull them.[78]

There was a high turnover of jackaroos but if they came back for a second year they were usually given full wages. Murray was to comment in 1955 at the end of his 10 years on Alexandria that Rod Bellette was 'the first of any of our Jackeroos who

68 Minutes of Directors' Meeting, NAP, Alexandria, 12 July 1948.

69 Minutes of Directors' Meeting, NAP, 19 February 1969.

70 Minutes of Directors' Meeting, NAP, Alexandria, 12 July 1948.

71 H.Barnes to D.M.Fraser, 2 April 1944, Alexandria File, 1944-1947, NAPR.

72 H.W.Chambers, Pastoral Inspector, to D.M.Fraser, 6 May 1947, D.M.Fraser File, April-October 1949, NAPR.

73 R.H.Murray to D.M.Fraser, 11 April 1953, R.H.Murray File, 1951-55, NAPR.

74 R.H.Murray to D.M.Fraser, 14 October 1950, R.H.Murray File, 1948-51, NAPR.

75 R.H.Murray to D.M.Fraser, 13 April 1948, Alexandria Correspondence File, 1948, NAPR.

76 R.H.Murray to D.M.Fraser, 24 October 1952, R.H.Murray File, 1951-55, NAPR.

77 R.Murray to D.M.Fraser, 14 March 1947, Alexandria File, 1944-1947, NAPR.

78 H.W.Chambers to Bill Young, 5 February 1964, Green Copies, Marion Downs Pty. Ltd., November 1963-February 1964, NAPR.

Jack Draper, head stockman Alexandria Station.

has made the grade and is worthy of promotion'.[79] Young always referred to them as 'roos' and, when the fashions changed in the 1960s, sometimes as 'long haired louts'.

With relatively little fencing in the 1940s, segregation within the herd was limited. To maintain a quality herd and to avoid overstocking, many cows were culled and spayed, which also allowed weight gain which in turn made for a more marketable animal, as did the castrating of male calves. Open bronchoing and broncho panels in yards were used for branding and castrating; spaying was generally carried out in a crush in holding yards, while hessian panels were used to contain the cattle for the treatment of pleuro-pneumonia. More crushes and yards were required and gradually installed, although external fences were virtually still unknown. Tendering, therefore, continued to be a necessary part of mustering. A program of fencing 360 kilometres of grazing land was begun after the war.

One of Alexandria's best-known characters during this period was Jack Draper, a big man who arrived in 1945 at the same time as Murray, and remained on the property until his death in 1983. As head stockman, he was at the No.1 stock camp until he retired on 30 April 1980. He related well to Aborigines and had a wonderful capacity to get the work done despite the huge numbers of stock to be handled. As Bill Young wrote in a letter to head office in July 1969, when Draper was doing an excellent job of branding and inoculating in record time, 'If Jack Draper was a racehorse I feel quite confident we could collect the Melbourne Cup'.[80] He seemed able to adjust to most situations, whether it was lack of manpower or the absence of a cook. Like most stockmen he was partial to drink but he still did his work. Bill Young tells of the time the Chairman of Directors, Michael Crouch, was visiting:

> Old Jack was there — and he was half full. So, I said to him 'Listen here, you be bloody sober in the morning because Michael wants to come down and see some branding.'
> 'Righto, what time are you going to be here?'
> 'I'll be there at daylight.'
> We were there at daylight — and he had about fifty calves branded when we got there.[81]

Jack loved his cattle and his broncho horses and was especially proud of two beautiful, grey half-drafts. On one occasion he had them in Gidyea Yard between the Ranken and Soudan just as the directors drove up. 'Now I'll show these bastards [the directors] a bit of class', he remarked.[82] He was not happy with the disease control program and TB testing during the 1970s and felt it was a dreadful waste to see so many cattle slaughtered. Cattle were his prime concern and he would get very agitated if he had weaners in the yard and the trucks were late in arriving to take them away.

His Aboriginal wife Myra was camp cook while their children were small. Their marriage ceremony is spoken of with good humour. It was illegal for white men to consort with Aboriginal women, so when Myra became pregnant the Native Affairs Officer suggested that Draper marry her. It also meant her other children would not be taken away by the authorities. At the ceremony in the stock camp kitchen, which was witnessed by John Ohlsen and George MacQueen, both later managers of Alexandria, Jack, as usual, wore a singlet and sandshoes; a cigarette hung from his mouth. When Myra was asked 'Wilt thou take . . .' she giggled shyly and scratched the back of her leg with her big toe. Jack gave her a shove and said, 'Say bloody yes can't ya?'.[83]

---

79 R.H.Murray to D.M.Fraser, 9 March 1955, R.H.Murray File, 1951-1955, NAPR.

80 W.E.Young to Ken Moore, General Manager, 18 July 1969, Correspondence File 1, Alexandria, 1969, NAPR.

81 Bill Young Interview, Casino, 26 January 1986.

82 E.M.Crouch, Typescript, Brisbane, 1990.

83 Ken McGuire Interview, Bell, 5 February 1990; Bill Young Interview, NTRS 226, TS 434, Northern Territory Archives Service, Darwin; George MacQueen Interview, Alphadale, N.S.W., 26 March 1991.

Throwing a weaner foal for branding at Old Horse Yards, Alexandria, 1974; head stockman Jack Draper in charge.

On his retirement, the company gave Jack a shotgun and allowed him the use of the Land Rover he had been driving at the stock camp. He was also allowed to stay on at his home on Alexandria. He eventually died of cancer on 29 April 1983.[84]

Another interesting character, who spent practically all his life on Alexandria from the 1920s, was Fred Deihms. Teamster, saddler and later pumper at No.9 bore, he purchased his first car after the war and after obtaining his licence from the Ranken policeman, Hughie Deviney, headed for No.9 bore. As a later manager of Alexandria recounted, he 'went flat as a strap and Hughie chased him, thinking he might have an accident'. When he finally stopped, Deviney asked him why he was doing 60 miles an hour on such a rough road. Fred put his finger on the oil gauge and said in all honesty, 'No, it's only doing thirty-five'. He put the car away and it was never driven again until he left the bore and retired to Soudan 10 years later.[85]

The use of private vehicles gradually became more widespread but in one hilarious incident in 1945, Florence Mofflin, wife of the Ranken policeman at the time, tells of her experiences in driving home to the Ranken with author Ernestine Hill and Ernestine's son, Bob.

> There was this place on the side of the road. They had a house and a lot of chooks in the back yard and the lady was hanging out the washing. So Bob drove in there, and then he found he couldn't pull the car up ... she dropped the washing and the basket and the chooks jumped out of their pens ... The husband raced out of the house to see what was going on — oh, you've no idea the commotion. Ernestine and I couldn't stop laughing ... we were going round and round and round. He [Bob] hurt his head hanging out the window, saying 'How do we get on the road to the Ranken River Store?' She'd shout something out, and the husband shouted something, so we'd go round again. The chooks were flying everywhere ... I wondered if we were ever going to arrive at the Ranken River, but eventually we did.[86]

Mick Hill was another long-serving employee on Alexandria. An American, he was a qualified steam engineer and was particularly valuable during the steam-engine days. He finished up as a pumper at No.7 bore and ran a very clean camp. In the dry

84 Ken McGuire Interview, Bell, 5 February 1990.
85 *ibid.*

86 Florence Mofflin Interview, NTRS 226, TS 435, p.70, NTAS.

year of 1952 he pumped for a mob of 3,000 for ten months straight without any problems. His wife, Ethel, was a good cook, and Jerry Chambers the pastoral inspector always made sure he was in the vicinity at meal times. Meals were eaten in the pump shed because, although it was noisy and the fumes tainted the meat hanging there, at least the flies were kept away.[87]

Entertainment at Alexandria still revolved around the annual race meeting which was transferred from the Ranken to Brunette Downs in 1949.[88] Bill Young relates a humorous incident when the vehicle carrying the Mount Isa Pipe Band, to play at the races, broke down somewhere near Alexandria.

> We went down and got them and took them up to the station and give them a big feed. There was about twenty-five of them. All done up in their kilts and bagpipes. They drank a case of whiskey and three cases of beer and the more they drank the more they played. The blokes from Brunette were looking for them and they were on to me [over the radio] and I'm telling them I haven't heard of them and I haven't seen them but they could hear those bagpipes. Eventually we got them over there.[89]

Visitors were always passing through, some of whom were quite prominent. Former Australian Prime Minister W.(Billy) Hughes called at Alexandria when Barnes was manager, and the Governor-General, Sir William Slim, stayed four days in Murray's time. The American ambassador of the day visited Alexandria during Bill Young's time as manager, when Young commented that he had a job to see the ambassador's staff for 'brass buttons, ribbons and braid'.[90] In October 1965 John Kerr, later Australia's Governor-General, accompanied the Arbitration Commission delegation concerned with the employment of Aborigines and their living conditions. Bill Young gave evidence at the hearing held at Alexandria.

Until the 1960s the Ranken Store was a hive of activity. But as Chambers pointed out,

> there are hundreds of man-working hours that have been wasted there and good workers that left or got sacked for getting drunk there.[91]

It was claimed that heavy stock losses at Lorne Creek in 1945 were partly because of an employee getting drunk and not taking oil to the pumper at No.11 bore which was near Lorne Creek. Jack Charlton, a drover at the time, recalls how cattle became bogged in the creek.

> Others climbed up over them trying to get to the water. Over 3000 cattle died of thirst in that tragedy. Some perished in the bush. Hundreds of horses perished.[92]

Murray described the situation in 1953 when a case of overproof rum came into the head station from the Ranken, causing the inevitable fights, brawls and general trouble.

> To hold some semblance of order I had to fire five of them, including my Head Stockman, Brian Egan, who has the Irish tendency to fight when drunk. The Rankine is responsible for a lot of capable men losing their jobs, and is one of the main difficulties in management. Large doses of 130 proof rum together with 109 degrees temperature produces quite remarkable changes in the character of men.[93]

During 1950, 57 white men were either sacked or left Alexandria head station, as well as 17 from

Johnny Rankine, road train driver, born and raised on Alexandria.

87 Ken McGuire Interview, Bell, 5 February, 1990.

88 J.M. Holmes, *Australia's Open North*, p.228.

89 Bill Young Interview, Casino, 26 January 1986.

90 Bill Young to General Manager, 22 June 1960, File 1, Correspondence, Alexandria, 1960, NAPR.

91 H.W.Chambers to D.M.Fraser, 15 November 1947, Chambers and Fraser File, April 1945-October 1949, NAPR.

92 Jack Charlton, 'Moleskin-moulded in the Murranji', *Stockman's Hall of Fame Magazine*, Volume 42, March 1992, p.7; H.W. Chambers to D.M. Fraser, 15 November 1947, Chambers and Fraser File, April 1945-October 1949, NAPR.

93 R.H.Murray to D.M.Fraser, 5 December 1953, R.H.Murray File, 1951-55, NAPR.

Soudan and 11 from Gallipoli.[94] It was inevitable that the men would congregate at the Ranken, with the dip still in use there, a new one being erected in the late 1960s. The bore water, however, was of poor quality and drinking water had to be carted from No.9 bore five kilometres away to supplement the tank supply.

It was the people associated with the Ranken who made life so interesting. Policeman Hughie Deviney and his sister Jeanie have already been mentioned. George Watson, the first store-owner at the Rankine, had by an Aboriginal woman several children who seem to have taken the name Rankine as their surname.[95] One son, George Rankine, became a trusted employee of NAP and was a boreman on Alexandria between 1955 and 1972.[96] Previously he lived and worked at Gallipoli, being out-station manager for a time in the mid-1940s. He raised a large family, several of whom worked on Alexandria. Noel Rankine was a boreman and his brother Johnny still drives the road trains for the station. Johnny is married to Carol, daughter of Jack and Myra Draper, and they have four children. Like his father, Johnny is a trusted employee as is indicated by the general manager in 1973: 'this man is an ideal worker who will carry out his duties at any hour without a complaint but with enthusiasm and reliability'.[97]

RANKEN STORE-OWNERS

George Watson/Rankine	1904
Herbert and Gwen Lloyd and son Harold	1929
Jimmy and Ella Fowler	1941
Joe Barton	1948
H.Creighton	1950
Dick Carter	1952
NAP	1968
Store closed	1970

The Lloyds sold out in 1941 to Jimmy Fowler after Herbert Lloyd died in January 1941.[98] Harold Lloyd, Herbert's son, continued to work on Alexandria as a well-borer using a steam engine. He was bore engineer of Soudan in 1952 before George Schultz went there as manager; he then continued as bookkeeper[99] but left after Ken McGuire arrived to manage Soudan in 1954. Jimmy Fowler had been a head stockman in CAY Johnston's time. When he married Ella, they thought she and her mother could carry on at the store so that Fowler could work as head stockman. But the demands were too great and Fowler returned to work there. Ella was also secretary of the ABC Race Club. Fowler died tragically in 1947 from peritonitis, while Clarrie Hudson, the Camooweal to Borroloola mail contractor, was desperately trying to get him to Alice Springs during a huge wet. Ella subsequently married Hudson and after selling the store, they moved to Alice Springs. The subsequent owner Joe Barton ('Commo Bob') was jailed in 1950 for supplying liquor to Alexandria's Aborigines.[100] NAP had the opportunity to purchase the store in 1947 and in 1951 when Fowler and Creighton respectively decided to sell. Douglas Fraser, however, was not interested[101] and this was perhaps one of the worst decisions he ever made. All managers complained about it; for example, Murray said in 1947:

> This hotel has caused me more worry and trouble than anything else on the Station and unless owned by us will remain a danger point for all time.[102]

Dick Carter,[103] a Scandinavian, had been a fencing contractor for Murray and ran the store for sixteen years until 1968 when it was finally sold to NAP. By then the company was eager to purchase the store and close it down, especially after the supply of alcohol to Aborigines was legalised. Carter was another of the Barkly's characters and Michael Crouch recalls the first time he (Crouch) took Sir Herbert and Lady Ingram to the Ranken. Carter insisted on their having 'morning tea' with him and, although the temperature was well over 100 degrees, they all drank sparkling burgundy. The Ingrams, not wishing to be impolite, drank their full share.[104]

In 1969 the company was granted a special purpose lease for an area at the Ranken, which was converted in the following year to two special purpose leases in perpetuity.[105] Jim Dwyer ran the

94 R.H.Murray to D.M.Fraser, 4 January 1951, R.H.Murray File, 1948-51, NAPR.

95 At the time, the spelling 'Rankine' was still being used and although the family has kept the spelling, the Ranken area has reverted to the original spelling.

96 George Rankine died at Alexandria 27 October 1972.

97 Ken Moore to Ken McGuire, 4 June 1973, File 1, Correspondence, Alexandria and Glenormiston, 1973, NAPR.

98 Rankine River Police Journal, 1941, F300, (4), NTAS.

99 R.H.Murray to D.M.Fraser, 5 December 1952, R.H.Murray File, 1951-55, NAPR.

100 R.H.Murray to D.M.Fraser, 1 May 1950, R.H.Murray File, 1948-51, NAPR.

101 D.M.Fraser to R.H.Murray, Manager of Alexandria, 11 May 1951, R.H.Murray File, 1951-55, NAPR.

102 R.Murray to D.M.Fraser, 16 November 1947, Alexandria File, 1944-1947, NAPR.

103 His full name was Richard Carter-Clausen. He died at Mullumbimby in 1969.

104 E.M.Crouch, Typescript, Brisbane, 1990.

105 Minutes of Directors' Meeting, NAP, 15 July 1969, 18 March 1970, NAPR.

Clarrie Hudson's Mail Coach, afterwards the 'Soudan Camp Truck' until 1955. June Blaschek

store until he became manager at Soudan and, after several changes of staff, NAP finally closed the store in 1970. With the decline of droving and the absence of other traffic, the Ranken was no longer important. The police station had moved to Avon Downs in 1963 and there were dips at Alexandria and later at Soudan, if required.

It was the end of a colourful history for the Ranken. A pumper now lives in the cottage and, except for a few fowls nesting on the shelving and gear scattered around, the store is deserted. There were discussions in 1971-72 with the Shell Company of Australia about the possibility of NAP's establishing a roadhouse on the Barkly Highway between Camooweal and Soudan, near bore No.22A, but this did not eventuate.

The last mail contractor for Alexandria was Clarrie Hudson, who made thirteen mail deliveries per year between 1940 and 1946. He used a truck and a rubber-tyred 'chariot' in the dry, while in the wet the 'chariot' was pulled by three horses which he kept at No.1 bore. Connellan (Connair) won the mail contract when it came up for renewal in 1946 and a fortnightly airmail service was introduced from Alice Springs to Tennant Creek via the various stations. A new all-weather airstrip, capable of landing larger aircraft, was put down in 1946, ten kilometres south of the station. Previously there had been a strip on an ant-bed surface at the edge of the Playford Desert, three kilometres north of the homestead, but access over the black soil was very difficult in the wet.

In April 1950, Alexandria was included on a Trans-Australian Airlines (TAA) domestic route between Brisbane and Darwin. The Douglas aircraft arrived on Tuesdays from Brisbane and called again the next day on its return from Darwin. People from the Barkly district used the service, and Alexandria's bookkeeper Jack Palmer, as the booking agent, received a commission from TAA. From 1955, the station was changed from a weekly to a fortnightly stopover, with Anthony Lagoon being approved as the other landing place on the Barkly. From 1959 Brunette Downs was preferred to Alexandria because of its more central position to the other Barkly properties. The Brunette strip was also located at the homestead which did not pose a problem for transportation in wet weather.[106]

Serious health-related problems were treated by the Flying Doctor base at Cloncurry, although it was difficult to land at Alexandria in the wet. In 1950, the Gallipoli manager Dick McCullagh following his wife's illness, built a landing strip at the out-station for the Flying Doctor.[107] An epidemic of diphtheria and whooping cough on the Barkly Tableland in the mid-1950s resulted in the death of a number of Aboriginal children and of the manager's daughter at Lake Nash Station. Mrs Murray inoculated the Aboriginal children on Alexandria, with the result that there was only one death there from whooping cough.[108] A few Aborigines suffered from leprosy in Bill Young's time and they were removed from the station. Death was regarded philosophically, indeed almost casually, as indicated in Young's telegram to head office in January 1971: 'Jack Neilson passed away early hours Sunday morning at Rankin . . . What about sending some spuds etc.'.[109]

Aborigines made up a significant part of the Alexandria community but, while they were an important section of the workforce, not everyone could relate to them. George Schultz's inability in this respect contributed to his dismissal in 1953 after only one year at Soudan as manager.[110] After

106 Manager, TAA, Queensland to Manager, Marion Downs Pty.Ltd., 6 March 1959, File 1, Correspondence, 1959, NAPR.

107 R.H.Murray to D.M.Fraser, 19 April 1950, R.H.Murray File, 1948-51, NAPR.

108 R.H.Murray to D.M.Fraser, 19 September 1955, R.H.Murray File, 1951-1955, NAPR.

109 W.Young, Soudan, to head office, 18 January 1971, Correspondence, File 1, Alexandria, 1971, NAPR.

110 R.H.Murray to D.M.Fraser, 3 July 1953, R.H.Murray File 1951-55, NAPR.

Ranken Store in 1976 used as a stock camp following its purchase by NAP in 1968. George MacQueen

receiving Christmas presents, all the blacks would go on walkabout each wet and return about March in time for mustering. As well as the local tribes, stockmen would come for the season from missions at Borroloola, Yuendumu and Phillip Creek[111] although by the mid-1950s they came only from Borroloola. One Borroloola Aborigine was Mick Dajahai who arrived at Alexandria with Cec Teece, a horse-breaker who had owned Rosie Creek Station. Mick was a good roughrider and in his later years did odd jobs around the homestead. Another Borroloola Aborigine, Willie Reilly, worked for O'Reilly, a well-borer in the 1940s, after whom O'Reilly's bore on Soudan was named. Willie became an off-sider for George Rankine and his knowledge about bores was of great assistance to the Alexandria managers after Rankine died.

In order to comply with instructions from the Department of Native Affairs, a number of huts were built in 1951 for the Aborigines at both Alexandria and the out-stations.[112] It was not until 1963, however, that a more permanent camp was established at Alexandria. Situated about 400 metres from the homestead, the camp comprised fifteen galvanised huts built in a circle. Nine were erected in July 1963, three more arrived that November, and three were added later. The huts had one large room with a dirt floor, which was later concreted, and an extended roof at the front provided a veranda. Water was obtained from a bore put down in 1962 and there were showers, toilets and a laundry. The Aborigines who worked on Alexandria were fed at the station. The rest received food which they cooked for themselves.

It is only since the 1970s that whites have outnumbered Aborigines on Alexandria. In 1945 there were 42 whites and 32 male Aborigines, with at least another 50 Aboriginal wives and children. In 1965 the number of Aborigines remained much the same. Of a total of 90, 27 were working men, 17 were working women, 11 were pensioners and there were 35 children, although numbers fluctuated as they moved between Alexandria, Brunette and Anthony Lagoon. There were 10 Aborigines at Gallipoli but none at Soudan. Bill

111 George Schultz, Manager of Soudan, to D.M.Fraser, 1 March 1953, Soudan File, 1953, NAPR.

112 R.H.Murray to D.M.Fraser, 4 January 1951, R.H.Murray File, 1948-51, NAPR.

Young pointed out, 'You cannot keep the natives there on the bitumen because they are here today and gone tomorrow'.[113]

Aborigines were slow to acquire a sense of the value of money, which made them vulnerable to hawkers. The stations therefore preferred that they spend their money at the station store. James White, part-owner of Brunette Downs, describes the situation there in the 1950s:

> Then hair oil became the fashion. So we bought Yardley's brilliantine in gallon jars . . . Instead of putting a little dab here, they used to get the whole bottle and pour. Then gramophones became popular . . . , then jewellery . . . , and they were dripping in diamond combs and things that we used to get from Woolworths, rings and bangles and beads, they were really characters, laugh their heads off . . . They used to laugh at us and we used to laugh at them, it was really magnificent repartee between black and white.[114]

From 1962, while the law called for a minimum wage of £2 8s 3d[115] a week for Aboriginal males with the company supplying food and clothing, Bill Young preferred to pay them £3 10s a week with any additional clothing purchased out of that wage.

113 W.E.Young, Evidence given to the Commonwealth Conciliation and Arbitration Commission, Cattle Station Industry (Northern Territory) Award, 1951, Alexandria, 6 October 1965, pp. 1002-3.

114 James White, Oral Interview, NTRS 226, TS 140/1, NTAS.

115 F. Stevens, *Equal Wages for Aborigines*, p.23.

On some properties, Aborigines, indeed, received as much as £7 10s a week, but Young followed his policy of paying less cash, in part because he did not believe Aborigines were capable at that stage of properly handling their payments.[116]

Following subsequent pressures to increase payments to Aborigines, an award was negotiated between the Federal Government, the NT Administration, the NT Cattle Producers Council and the North Australian Workers' Union, which took effect from 1 November 1966. The Queensland rate for stockmen of \$37.70 per week plus keep was paid to whites on the Barkly in order to attract staff. The Aboriginal rate from November 1966 was tied to the NT rate for white stockmen of \$23.25 per week plus keep although there was variation within the award; from those regarded as 'fully efficient', to those regarded as 'not fully efficient', which included gardeners, female domestics, and aged and infirm workers. Alexandria implemented the rates so that three men were on the full award or better, 15 were on wages between the minimum and full award and 22 were in the 'not fully efficient' category which attracted a lower rate. There were seven non-working Aborigines and 30 children bringing the total of Aborigines to 77.

The company did not object to the award and found that running costs on the station were not

116 K.Moore to Bill Young, 26 September 1960, File 1, Correspondence, Alexandria, 1960, NAPR.

No. 1 Bore and Turkey's Nest, Alexandria. The first to be drilled (1894-95), the bore was in use by 1900 and is still in good working order.

Alexandria Station ca. 1964, showing the Playford River and the Aboriginal huts in the foreground. George MacQueen

adversely affected. Some improvements had to be made to the Aboriginal accommodation including concreting the floors of the huts and improving the cooking and messing facilities.[117] As a result, the company received a certificate from the Welfare Department stating that the accommodation was up to standard. 'Personally I think it is the only one in the Territory', said Ken Moore.[118]

From 1 December 1967, Aborigines were granted a wage increase according to the same categories. It was noted in correspondence between the Alexandria manager and head office that Paddy O'Keefe was receiving higher than the award[119] of $25.78 because of his ability. Income tax was deducted which increased the paperwork — and forced social change on the Aborigines. For the first time, full names had to be submitted which led to a request from head office to clarify the names of Left Hand Dave, Johnny Picaninny, Sandy, Keithy, Joe Mick, Barny, Tobacco Jack, Ling Norman, Major, Gilbert, Reggie and Maudie. They were to join others already listed, names such as Chungaloo Janama, Puddin Yagamari, Sybil Nangala, Bessie Ninama and Lizzy Nurulama.

Aborigines were granted full white wages from 1 December 1968[120] but problems often remained. While the stations would sell clothing and boots to their employees and deduct the cost from their wages, workers would often gamble the remaining money away at cards or dissipate it in other pursuits, especially after the supply of alcohol to Aborigines was legalised. The number of Aborigines employed on cattle stations gradually declined after 1968 and by the mid-1980s, the huts which had accommodated the Aborigines on Alexandria were deserted. The few Aborigines who still worked on the station shared the same accommodation as the white employees.

117 Ken Moore to Directors, 18 November 1966, File 4, Directors, 1966, NAPR.

118 Ken Moore to Bill Young, 8 December 1967, File 1, Correspondence, Alexandria, 1967, NAPR.

119 *NT Government Gazette*, 20 December 1967.

120 Under the Cattle Station Industry (NT) Award and the Federal Pastoral Industry Award; Ken Moore to Bill Young, 3 December 1968, Green Copies, Letter File, October-December 1968, NAPR.

CHAPTER 9

The Channel Properties: Marion Downs, Monkira, Coorabulka

The inland river system of western Queensland is characterised by streams which in their lower courses spread into a network of shallow channels. Fed by tributaries having their sources in zones of relatively high monsoonal rainfall, these rivers periodically flood on average once every three years,[1] causing a widespread distribution of slow-moving water in the channels and inter-channel areas. The Georgina and Diamantina are the major rivers for NAP's channel properties: Glenormiston, Marion Downs, Coorabulka and Monkira which benefit also from the Hamilton, Mulligan and Burke rivers and numerous smaller creeks. The focus in this chapter is on the three properties which NAP acquired before World War II. Glenormiston which was purchased much later, in 1968, is discussed in a subsequent chapter.

The soils of the river channels of south-west Queensland and the surrounding swampy areas are of heavy, grey clay which contracts and cracks during the dry periods between floods. These soils are moderately fertile, there being comparatively little leaching of nutrients to depths beyond the reach of plant roots. Bluebush and lignum dominate within the limits of the floodwater distribution, but while bluebush has some feed value, lignum is inferior as a stock feed. The most effective fattening feed in the channels themselves consists of native sorghum, neverfail, pepper grass, channel blue grass and some Flinders grass together with a variety of good herbage. In the Georgina channels, clover and native sorghum are the dominant channel feeds, while clover does not grow in the Diamantina channels. Native sorghum grows vigorously to a height of nearly two metres and is of greatest value when green; the cattle show less interest in it as it dries off. Vast sorghum fields can emerge in good flood seasons, and, in the earlier days, sorghum hay was made on Monkira (at Neuragully). Sorghum and the other grasses respond best to summer floods, while with later flooding in March or April, the herbage response is much greater. The dominant timbers are coolibah, black wattle and eremophila, which occur mainly as thin, fringing belts along the channels. Gidyea trees also grow along the gullies which run off the pebbly country away from the channels.

Beyond the heavier swampy channel country extend level plains over which the flood waters disperse. Normally these flood plains are only submerged for a few days at a time, which is sufficient to soak the ground thoroughly and produce a good pasture response. But, as Bob Gunther, a former manager of Monkira explains:

> It is very disheartening to witness a flood, that for want of a two foot rise which would inundate the surrounding country for miles, remains stationary for days just below flooding level and then falls away without doing any good at all.[2]

Button grass and Flinders grass predominate on the flood plains, and provide excellent fattening feeds, while neverfail grass and herbage from winter rains are also valuable. There are some bare claypans, especially on Monkira, though these tend

1 P.J. Schmidt and N.T.M. Yeates, *Beef Cattle Production*, (second edition), Sydney, Butterworths, 1985, p.182.

2 R. Gunther to Douglas Fraser, 12 March 1947, Monkira File 1947-50, NAPR.

The Diamantina River at Monkira during the 1974 floods.

not to be extensive. The soils in these areas are usually silty, almost impervious to water, with a smooth, almost polished, surface.

Beyond the flooded country extend gravelly and pebbly downs, where the most important vegetation is undoubtedly Mitchell grass, with Flinders and button grass after good summer rain. Even if it is dry enough to be only powdery lick-up feed, Flinders grass will still fatten cattle. The area between the Georgina and Diamantina immediately south-east of Boulia near Coorabulka comprises approximately 15,600 square kilometres of such country, with practically no channel or flooded country.

The other outstanding feature of the lower reaches of the Georgina and Diamantina River is the sandhill country forming the eastern fringe of the Simpson Desert. This area is roughly bounded by Glenormiston, Marion Downs, Cluny (near Bedourie), Monkira and Mooraberree stations. The sandhills vary in height from low ridges with very loose sand at their summit to imposing hills of loose sand up to 30 metres high, the summits of which are entirely bare of vegetation. Such sandhills generally run from south to north and may continue for several kilometres without a break, with the distance between consecutive formations varying from one kilometre to several.[3]

MARION DOWNS, with its out-station Herbert Downs, is the largest of NAP's Channel properties — 12,700 square kilometres (4,885 square miles) including Breadalbane[4] — and also the earliest acquired. The property is watered by the Hamilton, Georgina, Burke and Mulligan rivers, and by the Cottonbush, Sylvester and Bellevue[5] creeks. Between the river channels are poorly grassed open plains on the eastern boundary, open plains and sandy forest country on the north-east, and high, stony downs on the western side of the

3 *The Economics of Road Transport of Beef Cattle,* Canberra, Bureau of Agricultural Economics, 1959, pp.80-3.

4 Breadalbane at 2,600 square kilometres is part of the Coorabulka lease but for convenience is worked as part of Marion Downs. The Marion Downs lease is actually 10,100 square kilometres but with Breadalbane the working area totals 12,700 square kilometres.

5 Spelling also Bellview.

Sandhills at Glenormiston.

Georgina. There are five good waterholes, lasting fifteen to eighteen months after flooding, and by 1928 just prior to NAP's acquisition, there were twenty-four bores, some equipped with windmills.[6] The Herbert Downs section is watered by holes lasting up to eighteen months — Fourteen Mile hole and Charlie's hole to the west, Old Station hole and Paravituari.

The longest flood on Marion Downs was from December 1949 until August 1950, while the biggest floods during NAP's ownership have reached heights of 7.42 metres (1974), 7.11 metres (1977), 6.45 metres (1972) and 6.35 metres (1953).[7] But the lack of water could be just as devastating, with droughts in the late 1920s, the early 1960s and from 1979-80. In the drought of the early 1960s, for instance, every waterhole on Herbert was dry except for Yellow Waterhole.[8]

The first run, Marion Downs No.1, on what became Marion Downs Station, was taken up by Edward Rowland Edkins of Mount Cornish in July 1876. It was named after his wife Edwina Marion Edkins, who was known as Marion. Frank Scarr, William Bucknell, John Brown and Harry Grantham Duffield also took up runs which were subsequently included in Marion Downs. Although this property was declared stocked in 1877, Edkins forfeited his leases in the same year to the Queensland National Bank and in 1879 most of the Marion Downs leases were taken up by Andrew Tobin of St Kilda, Victoria. In 1888 Tobin transferred all the leases to his new partner, Daniel Mackinnon[9] of Yallock near Camperdown, Victoria, who transferred the leases again in the following year to his three sons, William Kinross, James Curdie and Kenneth John Mackinnon. Thus began more than four decades with the Mackinnons as lessees.[10]

The out-station Herbert Downs had been taken up in July 1876 by Colclough Kirwan for William Wentworth Bucknell of Cooks River, New South Wales. The taking up of Herbert Downs had tragic consequences for Kirwan who had also taken up Glenormiston in 1876. Historian Hudson Fysh relates the story:

> He [Kirwan] started his team away to Burketown, five hundred miles distant, for rations, and expected

6 Marion Downs, LAN AF 289, QSA.

7 Marion Downs Flood Levels, 1934-1990, NAPR.

8 Rhondda Alexander, 'Sam Hill: 50 years of service', Typescript, Brisbane, 31 March 1981.

9 One of Victoria's celebrated horse racing events, the Mackinnon Stakes is named after another family member, L.K.S. Mackinnon.

10 Gregory North Pastoral Holdings, LAN N43, QSA.

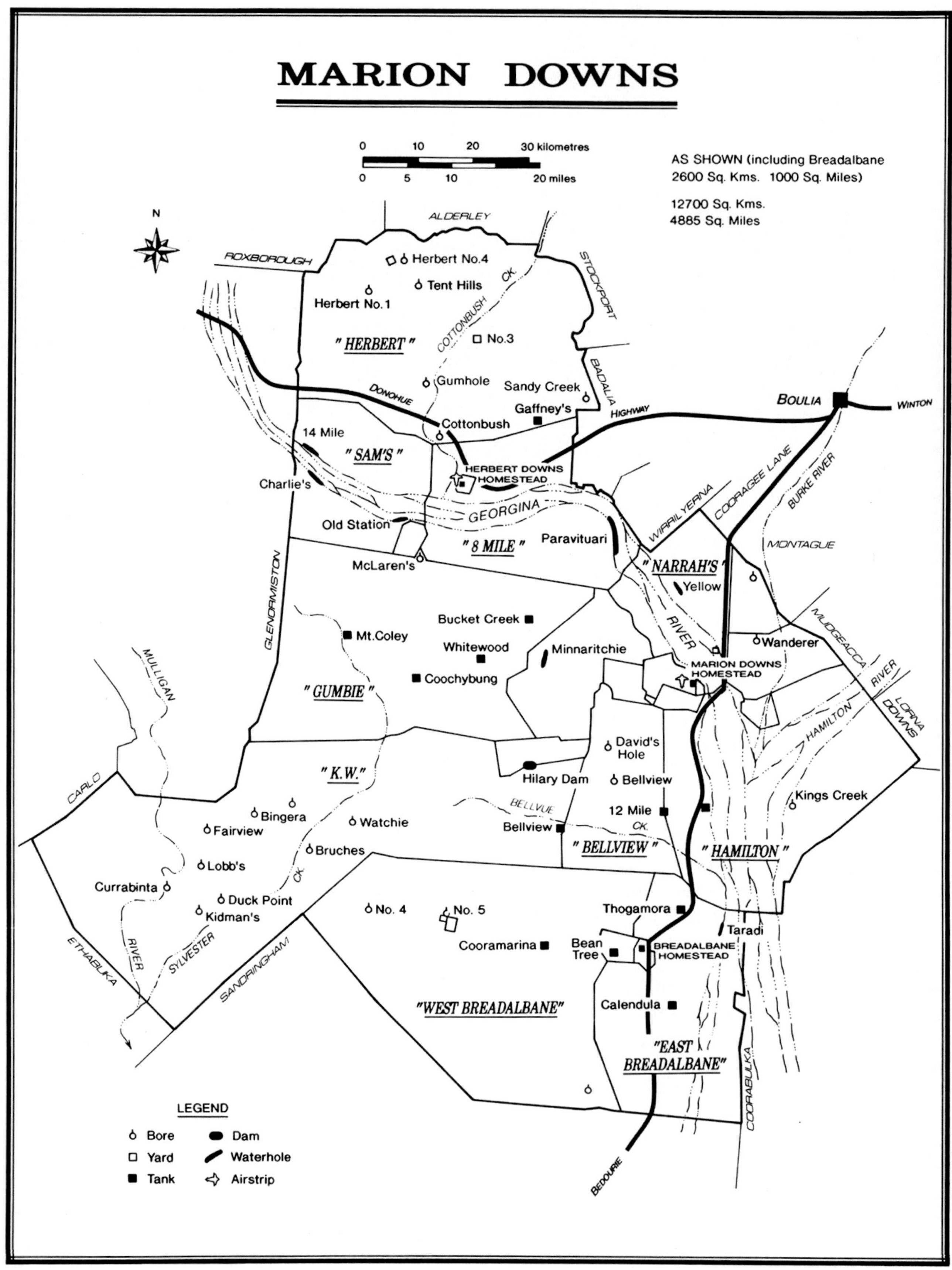

Map of Marion Downs.

its return in six months' time. No road existed, and held up continually by rain and floods in the coastal regions, it was actually two years before the team again pulled into Herbert Downs. Kirwin (sic) was dead, having died from starvation, his constitution not being equal to living on native food for a long period.[11]

Herbert Downs was transferred to John Brown in 1877, to George Augustus Mein and Pultney Mein in 1885, and in 1898 to the Mackinnons who needed more feed for their stock.

James Mackinnon managed Marion and Herbert on behalf of his brothers and in 1901 declared proudly that the herd had been 'worked up at great expense and was probably the best Shorthorn herd in Queensland'.[12] In 1909, for instance, Victoria River Downs Station in the Northern Territory purchased 69 cows, 70 bulls and 48 bull calves from Marion Downs.[13] But the droughts of 1890-91, 1898-1902 and the late 1920s were devastating and only a few hundred head survived each time. The Mackinnons restocked and carried on with the financial support of their other business interests in Victoria. But the drought years were heartbreaking for them, as William Mackinnon commented in 1930:

> It is now eight years since we had sufficient rain out there to make feed off the frontage; in fact the bulk of the country is a dust heap . . . We estimate having lost fully £100,000 on these Georgina holdings during the last 8 years, and with the present outlook we are seriously considering abandoning these properties.[14]

By the early 1930s the Mackinnons had had enough, and manager Joe Coghlan, together with drovers Dan Howard and the four Oldfield brothers, took 2,500 male cattle to South Australia. Coghlan was paid off by the Mackinnons and Marion was put on the market. Its purchase by NAP afforded the company greater flexibility in stock management, so that in dry times cattle could be either marketed or sent off to Alexandria or Monkira (after 1939) for agistment. Alternatively, cattle could be sent to Marion at times when other NAP properties were dry.

The individual runs held by the Mackinnons on Marion Downs and Herbert Downs included Belle Vue, Cannonbar, Belmont East and West, Mylrea Downs, Blackall, Kasouka, Coradilla, Eton Vale and Tent Hill. But when NAP took over in 1934, the leases had been consolidated as follows:

Marion Downs	1,176 sq. mls.	3,058 sq. km.
Headfort	161 " "	419 " "
Amaru	635 " "	1,651 " "
Mount Whelan	515 " "	1,339 " "
Marion Downs South	42 " "	109 " "
Herbert Downs	1,685 " "	4,381 " "
TOTAL	4,214 " "	10,957 " "

In August 1958, the company decided that Breadalbane, an area of 2,600 square kilometres which was one of the Coorabulka leases, should be run in conjunction with Marion Downs rather than with Coorabulka.[15] This made a total working area on Marion of 13,557 square kilometres, although resumptions have since decreased this area to 12,700 square kilometres. The main reason for this arrangement was management flexibility; to service the Breadalbane homestead, particularly during flood time, the Coorabulka manager had to go through Marion Downs.[16] Douglas Fraser, however, argued against the transfer of responsibility; there was always a certain amount of rivalry and jealousy among the managers and he believed it would upset the Coorabulka manager if part of his country were handed over. Such jealousy occurred particularly when stock were transferred. Any fat stock wandering from one NAP property to another would be pounced upon and sent in with a mob of fats, while in dry times the opposite occurred and the lean strays would be smartly turned back.

Because of problems with flooding, the original Herbert Downs homestead near the Georgina River was moved to its present site near Cottonbush Creek in 1899, one year after the Mackinnons had purchased the run. A bore was put down in 1899 but the water was salty and drinking water was obtained from the creek instead. When the creek dried up, water had to be carted from Cottonbush Well, thirteen kilometres away. The problem was finally solved, however, when a bore was put down in Cottonbush Creek in 1985, and the plentiful water supply that resulted

11 Hudson Fysh, *Taming the North*, Sydney, Angus and Robertson, 1950, p.110.

12 Daniel Mackinnon, South Yarra, to Secretary of Lands, 25 February 1901, LAN AF289, QSA.

13 Stock Inspector, Palmerston, 20 January 1910, Government Resident's Report, 1909, *SAPP*, 45 of 1910, Volume 3, p.17.

14 W.K.Mackinnon, quoted in letter from Gordon Graham and Co. to Secretary, Land Administration Board, 1930, Herbert Downs, LAN AF 281, QSA.

15 Minutes of Director's Meeting, Monkira Pastoral Company, 21 August 1958.

16 Coorabulka, Ken Moore Typescript, Brisbane, 1990.

Marion Downs homestead, 1960s. W.F. Alexander

has helped to transform the homestead surroundings.

The stone for the new Herbert homestead in 1899 was transported from the site of the old. Records show that another homestead was built on Herbert for £500 in 1917,[17] which possibly refers to the present stone and pise building located in front of the 1899 stone building which was demolished in 1960.[18] The present homestead was remodelled in 1958; the ceiling was lined and the veranda concreted and enclosed. Further upgrading was carried out in the early 1980s.

The original homestead on Marion Downs was at Old Station Creek where a dam was built. A huge wet washed the dam away, and at the turn of the century the homestead was relocated to its present site on the highest point near the Four Mile waterhole on the Georgina River.[19] The first construction at the new site was a slab building of bush timber with an ant-bed floor, an iron roof and three rooms. It was eventually demolished by manager Ken McGuire after a big wind storm in 1958. After completing the 1899 homestead on Herbert Downs, the same German stonemason moved on to Marion where he built the present homestead which is 80 kilometres from the Herbert house. Of stone and pise, it is similar in construction to the Monkira homestead, although Monkira has four large rooms off a central passageway whereas Marion has five. Both houses have wide verandas which have been progressively enclosed to provide additional living and working areas. Dust storms were always dreaded because of the clean-up involved; if a strong wind blew from any direction after a storm, there was another internal dust storm, because dust which had lodged on the half metre thick wall-tops rained down upon the inside of the house. Although less cleaning work was required at Marion Downs after the house was ceiled in the late 1930s, the dust still poured through the open gauzed verandas and the three McGuire children would pile the dirt up and play with match box cars.[20]

One of the requirements of the lease, which ran to 1959 was that the Marion boundary be fenced, an almost impossible task because at least 160 kilometres traversed flooded and sandy country. A dingo barrier fence on Herbert Downs was erected

17 Herbert Downs, LAN AF 281, QSA.

18 Ken Moore to Manager, Marion Downs, 6 May 1960, File 1, Correspondence, Marion Downs, 1960, NAPR.

19 Ken McGuire Interview, Bell, 5 February 1990.

20 Lenore McLaren Interview, Monkira, 18 November 1990.

Shorthorn bullocks at Gibbers Waterhole, Marion Downs, 1978. Rhondda Alexander

George Girdler (1983): Drover, horse breaker, brumby shooter and dingo trapper. Ken McGuire

in 1956 on part of the northern boundary, but there was concern that wild brumbies in the south-west would gallop straight through any fences.[21] Brumbies had become an enormous problem on Marion, coming in from the Simpson Desert and breeding into the thousands. A common method of combating this nuisance was to build a 'hide' at a watering place, wait until the horses came to water and then shoot them. If a mob with mares and a stallion was met in the open, the strategy was to shoot a mare, whereupon the stallion would halt his harem. Other mares would then be shot, leaving the stallion until last. In 1950, 4,000 brumbies were shot and in the 1950s Jim Dwyer killed between 1,200 and 2,000 a year, but this hardly kept up with the natural increase, so great were the numbers. George Girdler[22] effectively resolved the problem when he shot 6,200 brumbies in an eighteen-month period in the late 1950s.

Other than a few horse paddocks and two stake yards of Coolibah timber built by the Mackinnons — one at the 18-mile yard in the middle of the Georgina Channel and another on the western side

21 Douglas Fraser to Secretary, Land Administration Board, 25 July 1956, LAN AF 289, QSA.

22 Also a horse breaker, drover and dingo trapper.

of the Hamilton Channels[23] — there were no significant improvements apart from bores on Marion until after the war when men and materials (especially wire) became available. As Bob Kirk said, even in the early 1950s:

> You were flat out seeing one post sticking up that was the boundary . . . You could ride from Marion to Birdsville and you wouldn't see a decent fence.[24]

When developments did start, gidyea trees suitable for fence posts were not readily available, so large quantities had to be purchased from Blackall and Barcaldine.[25] Cattle and horse yards were originally of timber, but white ants limited their life span. In 1969, forty-six tonnes of sugarcane tramlines were purchased for distribution to Marion Downs and other properties for use in fencing. Despite initial opposition from the older men, this was a most successful move and, with rust no problem in the far western areas, maintenance costs were minimal.

The Tent Hills bore and the Gumhole bore were put down in the northern region of Herbert Downs in 1945. Bill Alexander, a later manager of Marion Downs, relates how the site for one of the best flows in the area, Herbert No.1, originated.

> Mr Chambers picked the site . . . He drove up there with [the overseer] Sam Hill and asked, 'How far are we from the boundary?'
> 'About five miles,' said Sam.
> 'Pull up!' said Chambers and with that he drove a steel post into the ground. 'That's where the bore goes. I'm sick of bouncing over this rough ground.'[26]

In this area there was some country carrying caustic bush, and stock losses were reported at times. As discovered later, the problem could have been gidyea poisoning. Tank-sinking was undertaken by A.E.Griffiths at Bucket Creek and Bellevue and a tank was built by the Gaffney Brothers on Herbert. Since the property now had a number of ring tanks and dams, a Fowler diesel track tractor with scoop was purchased to keep them in good working order. Manager Ken McGuire began making roads on Marion and Breadalbane with a one-cylinder Field Marshall tractor and a tow grader; Douglas Fraser was so impressed with the results that a new D4 crawler was purchased in 1959. McGuire describes the first time the D4 was used:

> They even sent a man up to deliver it and show how to drive it. That's how we came to put the road straight down the hill across the river. We dozed it down the very first job to get the hours up so the man could come back and do his second service.[27]

New roads were developed and a policy established of keeping them in good repair. Travelling time between the Herbert and the Marion homesteads fell from a full day to a couple of hours. Another new road gave much easier access to the Mulligan country than the alternative route past Hilary Dam. It was also proved that the gibbers on the ridges on the original track to the Sylvester were only surface stones; a grader made the road reasonable.

Stock horses were still of great importance. Horse drafting yards were built at Hilary Dam in 1957 in a paddock keeping the spell horses, a much needed improvement which saved a tremendous amount of work. Brumbies were still in the vicinity at this time and those caught were either destroyed or branded with the station brand.

Comfort at the homestead was gradually enhanced with the provision in 1958 of a single room air-conditioner. Improvements for the Marion Downs homestead approved by the board in late 1962 included a 240 volt lighting plant, a cold room capable of storing three bodies of beef, and an outboard motor to use during floods.[28] During the wet, a vehicle would be left on the other side of the river and the boat used to cross the channels in order to reach Boulia.[29] In 1967 cold rooms were also approved for all company out-stations including Herbert Downs and Breadalbane.[30]

In the 1960s extensive improvements were undertaken on Marion Downs following recommendations by manager Ken McGuire. These included:

1. Drilling of new bores : Currabinta[31] (1960); in Tent Hills area (1963); and one at Old Station Creek (1964) which was subsequently abandoned (no water);
2. Replacing Cottonbush Well (1961);
3. Building of broncho yards at Tarridy[32] waterhole and Cottonbush well (1963); and Gaffney's tank (1967);

---

23 Refer to page 184 for photo of manager Bill Alexander at these yards.
24 Bob Kirk Interview, Herbert Downs, 14 November 1990.
25 Ken Moore Typescript, Brisbane, 1990.
26 Bill Alexander Interview, Marion Downs, 15 November 1990.
27 Ken McGuire Interview, Bell, 5 February 1990.
28 Minutes of Directors' Meeting, Marion Downs Pty.Ltd., Brisbane, 13 December 1962.
29 Lenore McLaren Interview, Monkira, 18 November 1990.
30 Minutes of Directors' Meeting, NAP, 13 December 1967.
31 Also spelt Carabinta.
32 Also spelt Taridi.

4. Replacing the Breadalbane homestead bore which had collapsed;

5. Dividing of Wanderer paddock (1963);

6. Erecting a boundary fence with Sandringham on a share basis;

7. Erecting a turkey's nest at Sandy Creek bore;

8. Fencing of Top Horse Paddock (1964);

9. Building the Bucket Creek earth tank between Bucket Creek Bore and Whitewood earth tank (1964);

10. Equipping of No.4 Bore Herbert Downs (1964) and re-equipping the Tent Hills bore with a Mono pump (1970);

11. Replacing the station Land Rovers with new models;

12. Extending the Breadalbane cattle yards (1967);

13. Erecting guest quarters — called the 'motel' (1967).[33]

Communications gradually improved, with the first airstrip put down on Marion in the 1930s. Another was built in 1958. There was then a fortnightly service of DC3 aircraft and occasional light charter aircraft.

In December 1957, Basil Essam installed a Traeger transceiving set at Marion Downs which enabled Ken McGuire to communicate daily at 7am and 1pm with Herbert Downs. Glenormiston also joined the morning session while Coorabulka and Monkira participated in the Springvale sessions at 7.15am and 1.15pm. Breadalbane joined the Marion Downs session in 1961 but not until 1976 was a fixed radio band available for all NAP's channel properties to communicate daily (7am and 1pm) with each other, the stock camps and the grader drivers.[34]

When communication with the Flying Doctor improved, Breadalbane's Crammond two-way transceiving set was replaced, in 1960, by Marion's Traeger set so that the Charleville Flying Doctor Base could be reached from Breadalbane. Marion Downs then received a new Traeger Transceiver 60T25 which was connected to the Cloncurry base, thus allowing a wide communication span.[35] Later, telegrams could be sent through the Flying Doctor Base at Mount Isa. By the late 1960s there were portable sets in the managers' vehicles. In December 1975, a trunk line telephone was installed at Marion Downs,[36] the only NAP property in the Channels to have a telephone until 1991, when an STD telephone service was connected to each property except Monkira.[37] Until the STD connection, with the exception of Marion Downs, the NAP Channel properties had to book calls through the community 'rat phone' connected to the Mount Isa Flying Doctor Base for Glenormiston and to Charleville for Monkira and Coorabulka.

On occasions various dignitaries visited the station. The McGuires hosted a visit from the Queensland Governor in 1964 as did the Alexanders in 1976. Before the 1976 visit, Ken Moore commented:

> Sir Colin [Hannah] is a whiskey and rainwater man, although a bit of muddy water from the Georgina might make him know what you fellows have got to put up with.[38]

A successful field day was held at the station in July 1972, a year after Bill Alexander was appointed manager. Opened by the Minister for Lands and attended by beef producers and government officials as far south as Brisbane and as far north as Mount Isa, it was the first of its kind in the area.[39]

The size of the herd during the Mackinnons' ownership averaged 9,000 and when NAP took over in 1934 there were 6,457 head[40] on the property. Although stocking with Alexandria cattle soon increased this to a high of 21,665 at 30 June 1936,[41] numbers always fluctuated with the seasons, 1965 and 1966 being especially dry when only 74 and 73 millimetres fell respectively. In a reasonable year such as 1963, however, 2,000 bullocks and 2,000 cows were turned off, being transported by road train to Winton.[42] On a big open run like Marion, the cattle would spread out in the showery weather so that it was impossible

33 Minutes of Directors' Meeting, Marion Downs Pty. Ltd., Brisbane, 22 September 1960, 14 March 1963, 20 February 1964, 7 December 1966.

34 Ken McGuire Interview, 26 February 1992; Ken Moore to Bill Alexander, Marion Downs, 21 June 1976, File 2, Queensland properties file, 1976.

35 Minutes of Director's Meeting, Marion Downs Pty.Ltd., Brisbane, 19 May 1960; Ken Moore to Secretary, Royal Flying Doctor Service, Brisbane, 28 July 1960, File 3, Miscellaneous, 1960, NAPR.

36 Minutes of Directors' Meeting, NAP, 17 December 1975.

37 Connected to Monkira in 1992.

38 K.W.Moore to W.G.Alexander, 7 May 1976, Green Copies, 1.1.76-8.7.76, NAPR.

39 E.M.Crouch, Chairman's Address, Annual General Meeting, 18 October 1972.

40 Herd Size, Marion Downs, LAN AF 289, QSA.

41 Directors' Annual Report, Marion Downs Pty. Ltd., 30 June 1937.

42 Minutes of Directors' Meeting, Marion Downs Pty.Ltd., Brisbane, 9 May 1963.

to muster cleanly,[43] the Mulligan country being especially difficult to muster, and since it consisted of sandhills and saltpans, cattle became adept at escaping the muster there, and over the years bred up. Trapyards were established at the main waterholes to retrieve them, but until four-wheel drive vehicles, motorbikes and a helicopter were used, all the cattle could not be successfully mustered.

Marion Downs was in a good strategic position to supply markets as far apart as Townsville, Bowen, Rockhampton (Lakes Creek), Gladstone (Swifts), Brisbane (Cannon Hill), Adelaide and Melbourne, and, as early as 1918, Marion stock were being walked to Butru, north of Dajarra, and railed from there to Townsville. The railway from Townsville reached Butru in 1915 and Dajarra in 1917.[44] NAP was fortunate initially in being able to send their fats away from Marion with cattle from other Collins family interests — John Collins and Sons, Collins, White and Company and, after 1925, the Monkira Pastoral Company owned by Foster and Fraser. Adelaide was a major market, but once more than 1,000 were taken per week prices dropped. To avoid price fluctuations, the combined numbers of cattle offered from the companies each week never exceeded 500.[45] In 1935 Sir Herbert Ingram was anxious that NAP become acquainted with the Vestey family who owned the meatworks at Lakes Creek, Rockhampton, because he was friendly with Sam Vestey (Lord Vestey's eldest son), and felt the company could benefit from a closer association with them. While Phil Forrest acknowledged that it was better to be with than against Vesteys, NAP preferred to sell on an open market and no special arrangements were made.[46]

MANAGERS

MARION DOWNS	
1889 - 1890s	James Curdie Mackinnon
1890s - 1895	Gordon McLeod
1902 - 1934	Joseph R. Coghlan
NAP Ownership 1934	
1934 - 1936	Tom Bennett
1936 - 1939	Mr Macdonald
1940 - 1955	Thomas Harold Cook
1955 - 1971	Ken McGuire
1971 -	Bill Alexander
HERBERT DOWNS OUT-STATION	
1940 - 1945	Michael Naulty
1945 - 1947	Dick McCullagh
1947 - 1980	Sam Hill
1980 - 1982	Don Smith
1982 - 1989	Peter McLaren
1990 -	Bob Kirk

James Mackinnon was the first manager of Marion Downs, followed by Gordon McLeod, who committed suicide in 1895 by cutting his throat with a razor, the aftermath of a Christmas binge on rum.[47] Joe Coghlan who had been on Herbert, then served for many years as manager on Marion although Tommy Bennett had taken over by the time of the NAP purchase in 1934. He was replaced two years later by Macdonald, and Bennett found a job with the Winton stock and station agent Herbert Venness. On returning to Winton after war service, Bennett acted as agent for the NAP road trains operations.[48]

Macdonald and his wife had no children, but they did have some prized little dogs. On the first visit of Douglas Fraser and Bill Frith to the Macdonalds at Marion, they were greeted in the sitting room before dinner with a snarl and a snap from one of these dogs already occupying a chair. The dog proved not to be as important as the managing director and pastoral inspector and was promptly removed.[49]

Macdonald enlisted in the armed forces after the outbreak of war in September 1939 and was killed in action.

Thomas Harold Cook replaced Macdonald and stayed for fifteen years from 1940 until 1955. He came from Bradshaws on the Victoria River in the north-west of the Northern Territory, and was described by Jerry Chambers as 'a first class cattleman, conscientious, hard working and an economical manager'.[50] The first few years of the war were difficult for a manager because of the rationing of petrol and food, and the shortages of materials and good staff. But when materials became available after the war, improvements were undertaken. Cook's brother-in-law did contract fencing on the place and once took charge while Cook was away. Various disagreements between Cook and Fraser came to a head in April 1955 when Cook left for holidays without the permission of head office. His employment was terminated and Ken McGuire, who had been

43 D.M.Fraser to Moreheads, 31 August 1936, Monkira File, NAPR.

44 John Kerr, *Triumph of Narrow Gauge: A History of Queensland Railways*, Brisbane, Boolarong, 1990, p.224.

45 P.Forrest to Herbert Ingram, 15 December 1934, Herbert Ingram File, NAPR.

46 P.Forrest to Sir Herbert Ingram, 14 March 1935, Herbert Ingram File, NAPR.

47 Ken McGuire Interview, Bell, 5 February 1990.

48 Bennett died December 1967.

49 Marion Downs, W.F.Alexander, 'Ninety Years of the N.A.P.Co., 1878-1967', p.1.

50 H.W.Chambers, Pastoral Inspector, to Chairman of Directors, Marion Downs Pty. Ltd., 20 December 1948, H.W.Chambers File, NAPR.

Kenneth Milton McGuire: Manager, Soudan 1954-1955; Marion Downs 1955-1971; Alexandria 1972-1976.

Ron Spedding

Sam Hill who retired from Herbert Downs on 31 October 1980 after a record 50 years of loyal service.

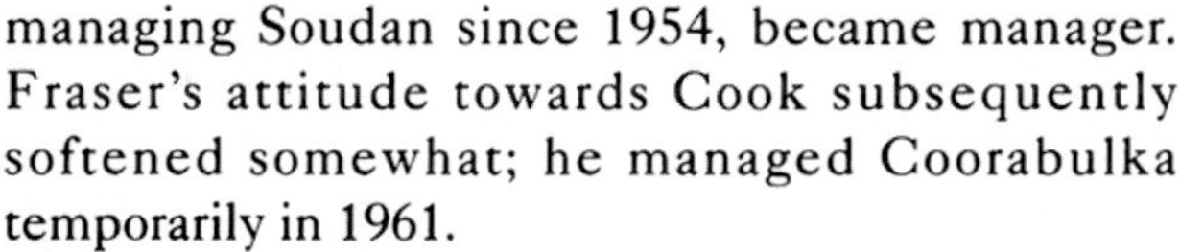

managing Soudan since 1954, became manager. Fraser's attitude towards Cook subsequently softened somewhat; he managed Coorabulka temporarily in 1961.

Kenneth Milton McGuire was born at Kyogle in 1925, one of fourteen children. He left home at 16 and headed for the bush, spending the next twelve years on cattle and sheep stations in Queensland and the Northern Territory, including Kamileroi, Mount Doreen, Lake Surprise, Plainby and Pine Creek. For a short time in 1950 he was a warder at the Alice Springs jail and in January 1954 joined NAP as manager at Soudan. He managed Marion Downs for sixteen years from 1955, and for part of that time was councillor on the Boulia Shire Council and president of the Boulia Race Club. A keen racing enthusiast, he rode in many country race meetings. He had an authoritarian approach to his work and while he was a good cattleman and very conscientious, his staff were sometimes at the receiving end of his fiery temper.

Fraser, Moore and later Michael Crouch were pleased with his results at both Marion and Alexandria (1972-1976) and only ill health forced him to retire from Alexandria, after his being there during the important period when the homestead complex was rebuilt. His wife Lorna was a trained nursing sister and she helped him considerably, especially in medical emergencies. The family's interest in NAP continues with one of their children, Lenore, who is married to Peter McLaren, manager of Monkira. Ken McGuire's letter of resignation to the chairman of directors illustrates his depth of feeling for the company. In concluding he wrote:

> it gratifies me to know I have been associated for such a lengthy period with a bunch of good men. I state here and now the altogetherness of this Company — yourself — the Directors — the General Manager — our late Pastoral Inspector — Head Office Staff and fellow Managers is something we can be proud of.[51]

James Michael Dwyer was a long-term employee of NAP who started on Marion as a stockman and brumby shooter in 1947.[52] He left NAP in 1949 and returned three years later. As a brumby shooter, he received two shillings and sixpence per head for the first 100 brumbies, rising by sixpence per head per hundred. In the trucking season he mustered fats for the road trains, and spent part of 1955 at Mundoolun. He returned to the west to run the stock camp at Glenormiston, and in the late 1950s went droving with Bob Kirk, another long-serving employee. In 1958 these two, together with Bob Graham, purchased Carlo Station but, unfortunately, there was no rain for the eighteen months they had it. 'All I could see was dust', recalls Bob Kirk.[53] From 1964 until 1968, Dwyer managed Eulolo for Collins White at a time when

51 K.McGuire to Chairman of Directors, 31 October 1975, Ken McGuire Personal Papers, Bell.

52 Born 30 September 1921 at St. George.

53 Bob Kirk Interview, Herbert Downs, 14 November 1990.

Glenormiston 1974. Left to Right: Pilot B. Newsome, Manager Jim Dwyer, Head stockman Bob Kirk, General Manager Ken Moore, Chairman of Directors Michael Crouch.

Glenormiston cattle were being agisted there during a drought. He was always willing to go where he was needed; he managed the Ranken store from 1968 until he became manager of Soudan out-station in 1970-71. He returned to manage Glenormiston, which had been acquired by NAP in 1968, from 1971 until his retirement in 1986.

Dwyer was ably assisted by Bob Kirk[54] in many of his endeavours. Bob started at Marion Downs as a jackaroo in 1950 and has worked in the area ever since. He was at Glenormiston from 1953 to 1956 and returned to Marion as head stockman in 1956-57. This was in the days when stock camps were a constant round of mustering, branding and turnoffs. The cattle then needed to be tailed and watched, a different story from today. Bob explains:

> You'd be in the camps twelve months and you wouldn't see the station . . . You'd only go in for tucker. These days it's a picnic mustering. You got a bike and a plane and there's plenty of water . . . I don't think anyone would know how to watch[55] a bullock these days.[56]

Bob spent time at Eulolo with the Glenormiston cattle in the 1960s and has since been overseer at Marion Downs, Glenormiston and Herbert Downs.

There have also been some notable overseers of the out-stations. Samuel Thomas Hill, who managed Herbert Downs for thirty-three years from 1947 to 1980 was born at Charters Towers on 12 August 1915. His father had worked on Chatsworth from 1918 and later on Breadalbane. Sam started work at Coorabulka in 1930 and in 1931 was one of the youngsters who helped Les Stretton move all the cattle from drought stricken Coorabulka to Chatsworth. After a year at Chatsworth, he returned to Coorabulka until 1939, the year it was taken over by NAP. After the war, on 1 January 1947 he started work at Herbert Downs and stayed until his retirement on 31

54 Born 1934; Townsville.

55 To watch cattle is to mind them at night during musters or droving trips.

56 Bob Kirk Interview, Herbert Downs, 14 November 1990.

December 1980.[57] A quiet, efficient bushman, he made the Herbert Downs section of Marion extremely productive. Sam was heavily involved in the management and movement of stock, but gidyea poisoning eventually restricted the options in his area, especially when fences were built during the BTEC program in order to segregate stock.

Conditions were difficult for the Hill family as they endeavoured to raise five children at Herbert. Monday was washing day at Cottonbush Creek where Mrs Hill spent all day with the children, washing clothes in the creek and then spreading them over bushes to dry. All wood had to be pulled over by a broncho horse and dragged back to the house to be cut up by axe and stockpiled. Water was carted from Cottonbush Creek and in summer Mrs Hill would wet the large grey flagstones forming the floor to keep the house cool; the floor in the kitchen was dirt. Carbide lights provided lighting in the early days, and, with no refrigeration, salted meat would last about a month. Supplies came from Marion every six to twelve months and the mail was delivered from Boulia by packhorse each fortnight. During the wet, Sam would ride to Boulia for supplies, and if there was an emergency such as a child's illness, he would ride to Badalia Station, thirty-two kilometres away. Three of his daughters married into the company — Rhondda to Bill Alexander, Valerie to Bob Kirk and Rosemary to Don Smith, manager on Herbert from 1980 to 1982.

Horses were the only means of transport on Herbert until 1952 when a Bedford three-ton truck was purchased. Sam was a fine horseman and his expertise with stock horses, especially those considered unmanageable by other riders, was well known. On one occasion he rode 120 miles (193 kilometres) in the one day using three horses. He left Herbert at 5 a.m. to ride to the Glenormiston boundary to count Alexandria steers through. He returned to Herbert, changed horses and rode to Tent Hills with meat for the pumper, returned again and saddled another horse to ride to Sandy Creek. He was back at Herbert just on dark. In those days, it took Sam ten days to ride the Herbert Downs boundary, using four riding horses and two pack horses.[58] He was one of the most respected employees NAP has had; and in 1968, knowing he had a deep interest in cricket, the company paid all the expenses for a trip to Brisbane so that he could see the Test match between Australia and the West Indies.[59]

Henry Marion and Billy Brady, Marion Downs, 1947.

Jim Dwyer

Michael Naulty was another long-serving stockman on Herbert and Marion. Born at Clare, South Australia on 10 August 1879 he was head stockman on Marion Downs in the Mackinnons' time and took charge at Herbert Downs from 1940 after Cook took over at Marion. Following his death on 7 January 1945 alone on the run, NAP erected a headstone over his grave near the Herbert homestead which reads 'Erected to the memory of a trusted and faithful headstockman'. Other graves are to be found on Marion Downs — such as that of Dan Richardson who died after the policeman from Boulia paid a friendly visit to the Marion Downs homestead. As Dan had stolen some food, he thought the police wanted him and he cleared out in panic, became stranded, and perished. His grave is between the Gibbers and Narrahs on the eastern side.[60]

An Aborigine, Yummpi Billy, known as Billy Brady, spent nearly all his life — seventy-five years — on Marion Downs. As a child aged about eleven, he was captured from his tribe by 'Cracker' Jimmy

57 Rhondda Alexander, 'Sam Hill: 50 years of Service', Typescript, Brisbane, 31 March 1981.

58 Bill Alexander Interview, Marion Downs, 15 November 1990.

59 Ken Moore to Sam Hill, 21 October 1968, File 4, Directors, 1968, NAPR.

60 Rhondda Alexander, 'Sam Hill: 50 years of service', Typescript, Brisbane, 31 March 1981.

These cattle died from gidyea poisoning at Herbert Downs, November 1980. E.M. Crouch

Clark, a head stockman on Marion during the Mackinnons' time. Although he retained some traditional skills and behaviour, he also learned the white man's ways — 'a white man through and through', said Sam Hill.[61] Although he was illiterate, he was expert in accurately determining cattle numbers. He worked in many capacities with the cattle: mustering, tailing, droving, tendering and as cook. For the last fifteen years of his life he was caretaker of the Wanderer Bullock Paddock, fifteen kilometres from the station. It was always his wish that he be buried where he lived after his retirement, at the Wanderer Waterhole; but there was a big wet on in 1962 when he died and he was buried at the Boulia cemetery instead. The company erected a headstone over his grave.[62]

Although it was essential to have an out-station at Breadalbane in the early days to efficiently service the property, it eventually became sufficient to have a man there to check that drovers and stock did not linger or stray from the stock route. Between 1946 and 1959 Breadalbane was occupied by Billy Gordon, a well-educated Englishman who from the mid-1920s was head stockman on Monkira. On Breadalbane, it became his habit to go to Bedourie each Friday afternoon with the mailman, Eddie Miller, on his way from Boulia. Having satisfied his thirst for liquor and companionship, he would return the next day. After Gordon developed an unsightly cancer of the nose and face, it was decided that he should retire to Birdsville. The following week, however, he cleaned the house, put out the kerosene refrigerator, placed his personal papers on the kitchen table, laid his swag out in the bow shed so as not to make a mess in the house, and shot himself. The deed was timed for the mailman to find him dead and in order to inconvenience no one. He is buried at the Bedourie cemetery.[63]

Marion's head stockman during McGuire's time was Colin Albert (Bill) Cockerill. He had been on Soudan from 1948 and in the late 1950s transferred to Marion Downs. Ken McGuire described him as

> one of the best horsemen I've ever seen. So kind breaking young horses; a good man on a broncho horse; always liked the pack horse best; said it was the most efficient way to get a camp transported about. I've heard him complaining at times when the truck had been bogged, 'Christ boss,' he'd say,'if we had the pack horse we'd be miles and miles in front.'[64]

61 Sam Hill, in Rhondda Alexander Typescript, 1981.
62 Ken McGuire Interview, Bell, 5 February 1990.
63 Ken McGuire Interview, Bell, 5 February 1990; Ken Moore Typescript, Brisbane, 1990.
64 Ken McGuire Interview, Bell, 5 February 1990.

Waterhole on Monkira, 1964. W.F. Alexander

With the Diamantina River running north to south through the centre of the property, MONKIRA is fortunate to have a plentiful water supply, including four permanent waterholes with several others lasting eighteen months. The 3,730 square kilometre (1,435 square miles) property consists chiefly of country flooded by the Diamantina River, with some sandhill country and various creeks. In addition there are some lightly grassed low hills on the eastern side running out to the adjoining property, Palparara, and approximately 260 square kilometres of downs country to the north-west. The station has the use of a portion of Palparara, which adjoins the eastern boundary and which has been fenced into Monkira by mutual agreement with the owners.[65] The largest coolibah tree in the world is at Neuragully, one of the permanent waterholes. Toonka waterhole had never been known to be dry until the mid-1960s although silting has occurred in the watercourses — not from overstocking, but rather from droughts, duststorms and the moving of sandhills.

The first runs on what became Monkira station were taken up by John Costello in May 1875. Named Mackhara and Willpally, they were situated in the middle of the main Diamantina Channel. 'Mackhara' was first used as the name for the station although an alternative spelling, 'Mackarra', was also used.[66] It seems likely that the name 'Monkira' was a derivative of 'Mackhara'. Declared stocked in October 1875, the leases were transferred in 1877 to Frederick Peppin and John Webber of Kyabra; in 1887 to the partnership of Edwin Thomas Smith, George Leonard Debney[67] and Edward Laughton; and in 1895 to Sir Edwin Thomas Smith of Adelaide.[68]

Sidney Kidman purchased Monkira from Smith in 1902, with John Forrest as managing director of Moreheads acting as Kidman's agent. By 1914, Kidman also owned other stations in the district and while he ran Monkira separately, the other stations were controlled by a manager at Glengyle and head stockmen at Sandringham and

65 During earlier times, Francis Foster endeavoured to purchase 970 square kilometres from the owners of Palparara but the price asked was too high.

66 N.Macdonald to John Collins, 14 August 1876, Collins Family Papers, privately held by Joan McKean, Brisbane.

67 The present manager of Glenormiston, Malcolm Debney, is a great grandson of George Debney's brother, Frederick.

68 Gregory North Pastoral Holdings, LAN N/42, QSA.

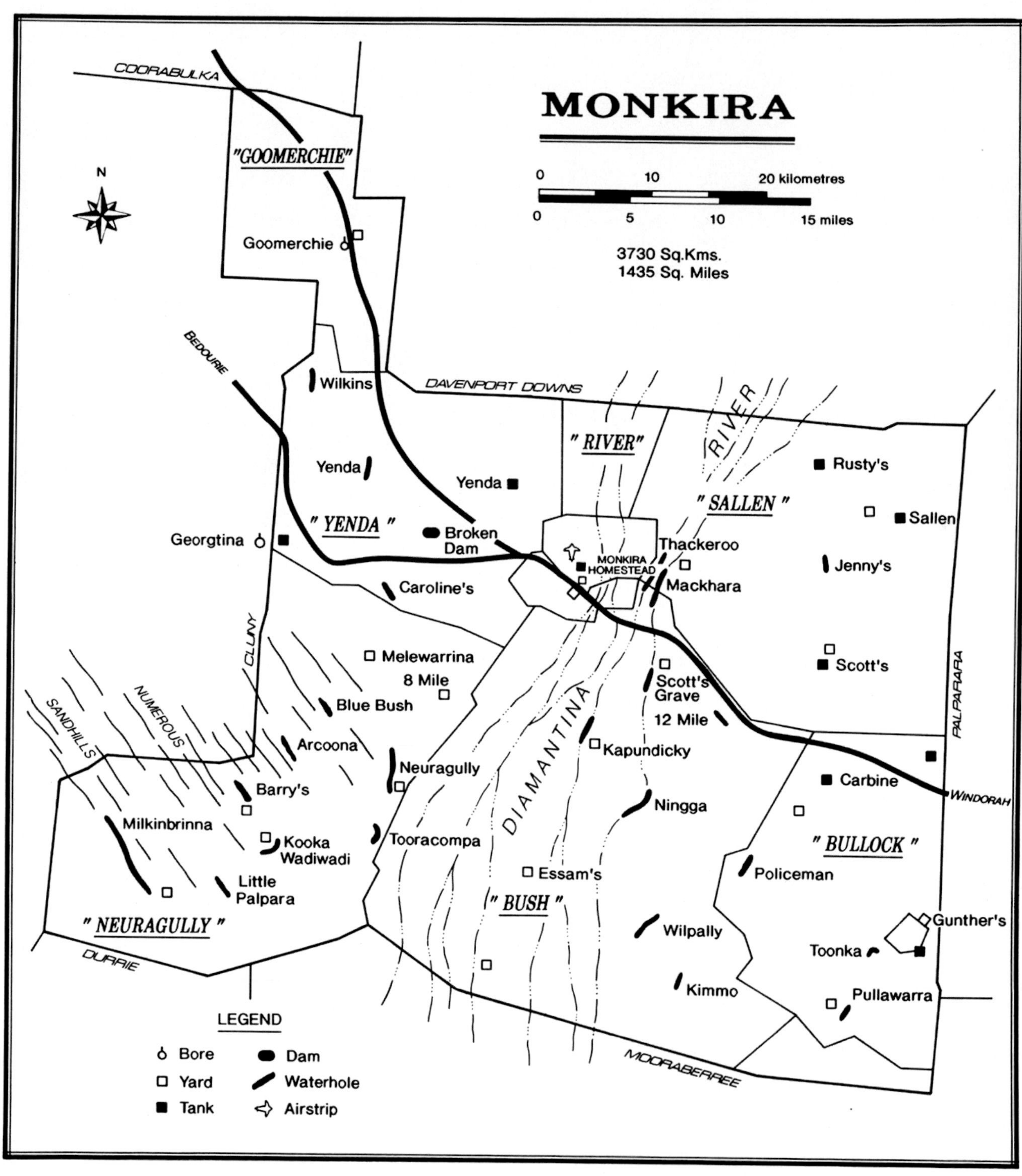

Map of Monkira.

Annandale.[69] Monkira's rent, at twelve shillings per square mile, was the highest at this time and Annandale's the lowest at seven shillings.[70]

Although Monkira presently has the use of the Georgtina bore[71] situated on Cluny just metres from the boundary, the only bore on Monkira, Goomerchie Bore, was put down in Kidman's time by the Inter Colonial Boring Company. Started in 1919, the bore had to be recased and was completed in 1921 at the enormous cost of £7,704.[72] Kidman was disillusioned with the considerable effort involved, complaining of the trouble in transporting the bore casing over the sandhills and the necessity of initially having a flowing bore.[73] The depth was 730 metres giving a little over 3 million litres per day.

Although Kidman was concerned about water, he does not seem to have paid the same attention to other aspects of Monkira's management. Scathing reports by the Land Commissioner in 1921 criticised Kidman — his choice of staff, the fact that he did not build fences but waited for neighbours to do so, and his neglect in carrying out improvements. The Commissioner concluded:

> I left Monkira and as I rode over the treeless downs of Coorabulka with its flowing bore drains I reflected over the 10,500 square miles under the blighting control of a man knighted for his services to the empire with mingled feelings of anger and disgust.[74]

'Costello's Camp', marked on an early map north-east of the present Monkira homestead, was the first place of white habitation.[75] An early homestead with walls of clay mixed with chopped cane grass and two feet thick, was erected on the present site, and while George Debney was manager for the Smith, Debney, Laughton partnership, the present house of sandstone and pise was constructed in the late 1880s.[76]

Situated conveniently in the middle of the block on the Diamantina from which it draws water, this homestead was built at a time when, it is said, conflict with the Aborigines made the provision of small rifle holes in the walls necessary, in case of attack. Until the 1930s the floor was of compacted fat and ashes and the ceilings were of calico. Improvements were carried out, although there were complaints in 1949:

> The verandah under which the house kitchen and bathroom is, needs renewing. When it rains the water runs down between the house wall and the end of the verandah, and floods these two rooms out; a new laundry is needed; on the West side of the house a verandah is needed to protect that side of the house, as the mortar is beginning to scour out from between the stones.[77]

The house was burnt down in 1977, although the stone walls remained. In the reconstruction by Bunny Clissold, a long-standing building contractor for the company, one wall showing the fire damage has been left as an interesting feature wall for the new homestead, which is now one of NAP's most spacious and comfortable residences.

In 1925 when Foster and Fraser purchased Monkira, improvements consisted of various stone buildings, drafting yards, Goomerchie bore, a few hundred kilometres of fencing and galvanised iron barracks for the stockmen, which Kidman had built under the Workers' Accommodation Act. Goods arrived once a year by camel from Marree in South Australia and although there was a vegetable garden, there were seven deaths from beri-beri in 1919, and other deaths from typhoid including that of a Chinese cook.[78] Bedourie, 116 kilometres away, with its hotel, store, police station and one residence, was the nearest town while the closest doctor was at Winton, 480 kilometres to the north.

Fencing was added during the administration of the Monkira Pastoral Company. When NAP took over the property in 1939 the northern boundary fence was in good order; the southern boundary was marked by a government fence which the station had to acquire; the eastern boundary fence was not in good order and adjoined a considerable area not used by neighbouring Palparara, and the western boundary between Monkira and Cluny was incomplete. The average capacity was 6,000 head with an annual turnoff of 3,000 fats. Being near to Marion Downs, and on a different river and

69 Other Kidman properties in the vicinity included Davenport Downs, Diamantina Lakes, Durrie, Roseberth, Morney Plains, Retreat, Dubbo Downs, Kaliduwarry, Carcoory and Thargomindah.

70 Assessing Commissioner, 27 December 1913, Monkira, LAN AF 290, QSA.

71 Known originally as the Georgina or Gerara Bore, it was put down in 1954 to a depth of 2,700 feet.

72 Sidney Kidman to Under Secretary of Lands, 23 December 1921, Monkira, Pastoral Holding No.384, LAN AF 290, QSA.

73 Sidney Kidman to Minister for Lands, 13 April 1922, Monkira, LAN AF 290, QSA.

74 W.W.Williams, Land Commissioner, to Under Secretary, Lands Department, 20 July 1921, Monkira, LAN AF 290, QSA.

75 Monkira, Pastoral Holding No 384, LAN AF290, QSA.

76 *North Queensland Register*, 17 January 1970.

77 H.W.Chambers, Pastoral Inspector, to Chairman of Directors, Monkira Pastoral Company, 22 November 1949, H.Chambers File, NAPR.

78 The graves can be seen near the end of the airstrip.

Monkira homestead after 1977 fire. E.M. Crouch

Monkira homestead after the rebuilding. E.M. Crouch

watershed, added greatly to management flexibility, as cattle could be moved from one property to the other as circumstances dictated.

The disastrous drought of 1927-29 had overwhelmed the Monkira herd and although some 3,100 head were removed before the stock routes were entirely closed, most of the cattle (which were cows and calves) perished. At the end of 1929 a bangtail muster showed there were only 752 head left alive on the run, of which 466 were aged bullocks and the balance dry cows and steers. The calamity showed clearly the dangers of carrying too many breeders on Monkira, and as a result a successful policy of spaying and purchasing store steers to fatten was adopted. Good seasons followed, Fraser commenting to Foster in 1933: 'This has been the best rain for ten years and I really think it looks like the turning of the tide'.[79] The company improved financially, and in 1938 had a herd of 5,600. In 1931 the old lease was surrendered and a new one obtained on the more favourable terms of eight shillings per square mile.[80]

Although, as elsewhere, there were shortages of materials for improvements during the war years, wind-driven electric generators were installed in 1941 both at Monkira and Coorabulka. Seasons were good then, and the quality of cattle turned off was excellent. In August 1940 a mob of Monkira and Coorabulka bullocks delivered to Ross River meatworks averaged 347 kilograms (766 pounds) and 368 kilograms (813 pounds) respectively.[81] Improvements followed in the post-war years, with broncho yards for branding erected at watering places, new men's quarters built in 1951 and a cottage for Basil Essam, the mechanic, constructed in 1953.[82]

Improvements at Monkira in 1957 included two 23,100 cubic metre tanks (Sallen and Yenda),[83] a new saddle room and a heavy machinery shed.

79 D.M.Fraser, Mundoolun, to Francis Foster, 18 February 1933, Foster Family Papers.

80 Francis Foster to Managing Director, NAP, 27 January 1938, Amalgamation File, NAPR.

81 Meeting of Directors, Monkira Pastoral Company, 21 August 1940, NAPR.

82 This building is presently the head stockman's residence.

83 Also spelt Yendah.

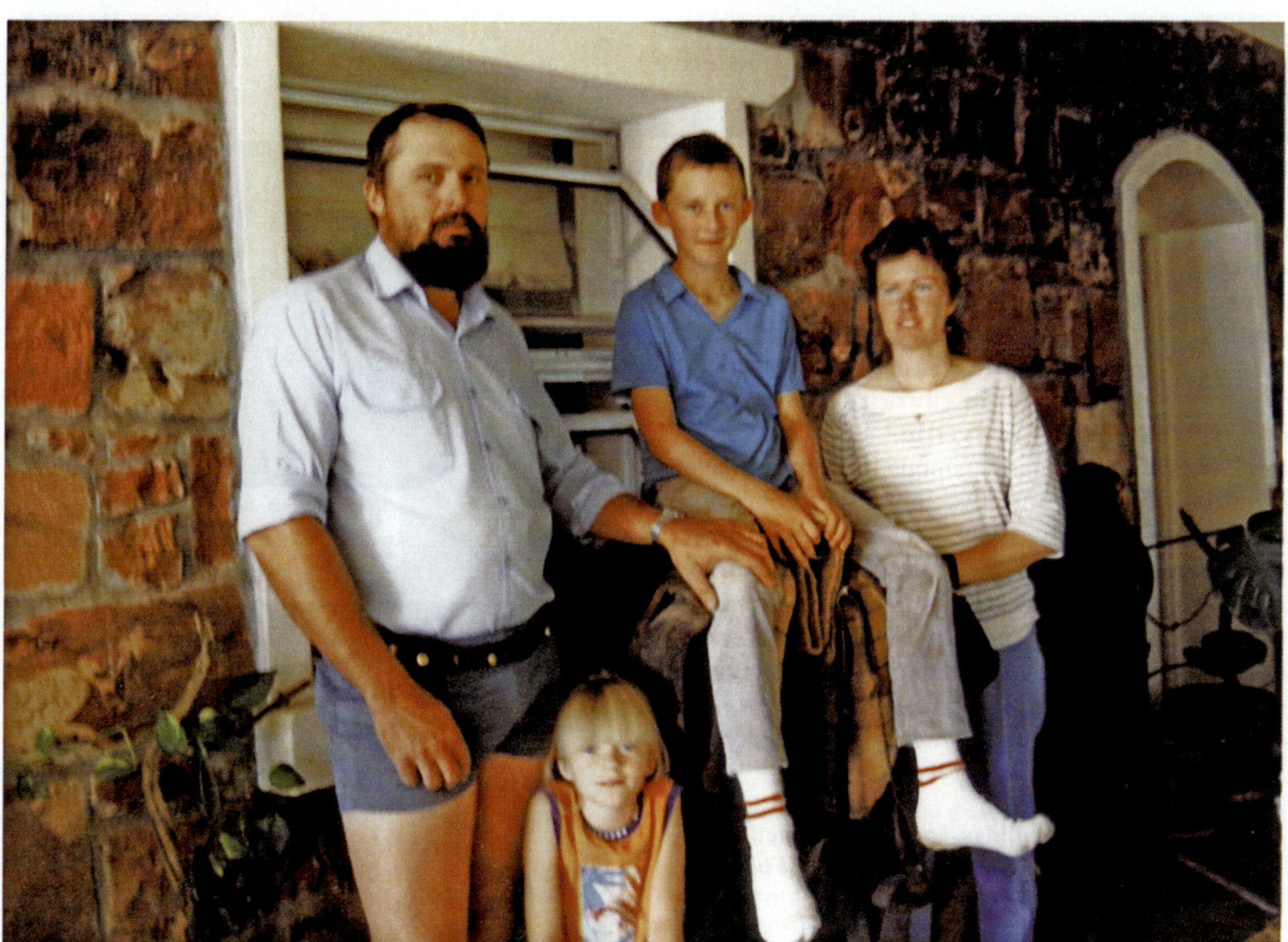

Manager of Monkira, Peter McLaren, his wife Lenore (née McGuire) and their children Dwayne and Helen, 1990. In the background is the feature wall left after the fire in 1977.

Bigger lighting plants were installed on Monkira and Coorabulka in 1957; not only the homestead but all the buildings including station kitchen and quarters, were then lit. The smaller plants were moved to Breadalbane and Herbert Downs. New vehicles were purchased, including Land Rovers for Monkira and Coorabulka in 1959. In the next decade more earth tanks were built to facilitate watering, and major watering holes were cleaned out to remove the silt. In 1961, a caterpillar grader was purchased for making and maintaining the roads on Coorabulka and Monkira. As a result the approximate total length of graded track within the company at that stage amounted to:

Alexandria	5,000 miles	8,050 kilometres
Marion Downs	3,000 "	4,830 "
Coorabulka	1,000 "	1,610 "
Monkira	1,000 "	1,610 "
	10,000 miles	16,100 kilometres

NAP's road trains were used from the 1950s to truck Monkira cattle to the railheads, although droving continued whenever there was feed on the stock routes. Carbine tank was a government-owned stock route watering facility which NAP later purchased as the stock route became redundant, and at the same time the company also purchased the Thogamorra[84] and Calendula tanks on Breadalbane.

Monkira was central for a number of main markets with mobs travelling to either Sydney, Brisbane, Adelaide or Melbourne. In earlier times, Adelaide via the Birdsville track was the usual destination but once railways were established fats would also be walked to the railhead at Winton, travelling from there to the coastal meatworks at Townsville, Bowen and Rockhampton. The most famous bullock to leave Monkira, the 'Monkira Ox', held the record for more than forty years as the heaviest bullock ever slaughtered in Australia. Bred by Smith, Debney and Laughton, after walking to Adelaide its live weight was 1,378 kilograms (3,042 pounds) and dressed weight was 902 kilograms (1,992 pounds) when slaughtered in 1894.[85] Douglas Debney, son of the manager and part-owner, George Debney, recalls:

> I was given a half holiday and accompanied my father to see the mob start on their journey. The big bullock, a beautiful roan had been a milker's calf and was very quiet and allowed us to rub his back.[86]

Of the first ten years of NAP ownership, only in 1947 were there any brandings on Monkira and Coorabulka. Subsequently, it was decided to hasten the stocking program and both properties were used for breeding purposes. Seventy Boonal bulls were purchased from the Munro family at Weebollabolla, continuing a long association in the purchase of Shorthorn bulls. Monkira brandings increased to a peak of 3,354 in 1955.[87] The breeding program was discontinued in 1964, although a certain number of station-bred cattle were always necessary to lead newcomers from Alexandria to the property's watering places.

Being on the edge of one of the main channels of the Diamantina, Monkira serves as a weather station and records flood heights. In 1934 it became the second cattle property in Australia to receive a pedal wireless from the Flying Doctor Service.[88] Monkira was connected to the Cloncurry base until 1956, when NAP asked for a transfer to the Charleville base because of their doctor's

84 Also spelt Thogomora and Thogomarra.

85 Jill Bowen, *Kidman: the Forgotten King*, North Ryde, Angus and Robertson, 1987, p.135.

86 *North Queensland Register*, 17 January 1970.

87 Monkira, W.F.Alexander, 'Ninety Years of the N.A.P.Co., 1878-1967', p.7.

88 This set was replaced in 1947 by a later model.

Traegar Pedal Wireless at Monkira, 1934. Henry Foster

Monkira, early 1950s. Left to Right: Bob and Norma Gunther, Annie and Henry Bruxner. Richard Purvis-Smith

monthly routine service.[89] People would attend from surrounding stations and these visits were made easier because Monkira's two airstrips were virtually all-weather strips. In the 1950s and 1960s the station was on a regular service flown by the TAA DC3 aircraft. Mail arrived fortnightly on the Sunday and the pilot would stay the night at the station. The manager would promptly answer any correspondence that night so the mail would catch the plane leaving the next day.[90] Air services declined in the early 1970s when government freight subsidies were discontinued.

With the lease expiring on 30 December 1959, negotiations on lease conditions were entered into with the Lands Department. The new leases that resulted were thought to be most satisfactory. The main terms included the enclosing of a bullock paddock at Gunther's Yards and the erection of seven sets of broncho yards at various places, including the Sallen and Yenda dams.

Monkira's management has been significant for its stability. There were only two managers in the Fraser years — Bob Gunther and George MacQueen.

MONKIRA MANAGERS

pre-1927	P.D. Edwards
NAP Ownership since 1939	
1927 - 1961	Bob Gunther
1961 - 1976	George MacQueen
1976 - 1985	John Crane
1985 - 1989	Bob Wales
1990	Jamie MacFarlane
1990 -	Peter McLaren

Gunther worked at Monkira for a record forty-one years, thirty-four of these as manager. Born in Adelaide, he arrived at Monkira in 1920 as bookkeeper and became head stockman in 1925. He married Norma, daughter of the then manager,

89 Manager, Monkira Pastoral Company to Manager, Monkira, 20 September 1956; Manager, Monkira Pastoral Company to The Postmaster General, Brisbane, 17 September 1956, Green Copies, Letter File, July-September 1956, NAPR.

90 Ken Moore, General Manager to George MacQueen, Monkira, 4 August 1961, Green Copies, Marion Downs Pty. Ltd., May 1961-August 1961, NAPR.

Monkira Homestead on the Diamantina River, 1991.

P.D. Edwards, in 1926; and when Edwards left to manage Norley the following year, Gunther took over as manager at Monkira. He worked for three owners — Kidman, the Monkira Pastoral Company (Foster/Fraser) and NAP — and was well-liked by all three; 'a wonderfully good chap', said Kidman when he sold Monkira.[91] On 30 July 1961, he died at the age of 61, while still manager. Leaving Monkira after Gunther died was extremely difficult for Norma Gunther; as she said, 'I don't know what I'll do. I have never had another home'.[92]

The Gunthers were very proud of Monkira and were gracious hosts. Bob was one of far-west Queensland's best known identities and leading figures. He was chairman of the 93,600-square-kilometre Diamantina Shire Council for fifteen years and shire councillor for many more. As a keen advocate of beef roads, one of his final acts as chairman was to raise finance to extend the road from Charleville to Currawilla.[93] A lasting tribute to Gunther on Monkira, the Gunther yards used for trucking, were built in 1961[94] to replace the old stake yards; they were improved in 1968. Originally it was thought that Bob Gunther's son Robin might take over from his father, but Robin married a city girl and they decided that the outback life was not for them. George MacQueen, the son of Bill Young's wife, followed as manager of Monkira in 1961. MacQueen continued with NAP until 1981; he will be discussed in a later chapter.

Basil Essam was an exceptional man. Left by a hawker on Monkira as a young boy in the late 1930s, he was illiterate, but manager Bob Gunther encouraged him in mechanical skills. He became an expert mechanic, watchmender, radio repairer and movie operator. After marrying and finally leaving Monkira, he worked on Marion Downs for a few years until he took up a position with the Boulia Shire Council.[95]

Donald McKenzie was another loyal station hand during Gunther's time. He stayed on after his retirement in 1957, a weekly payment being made by the company to meet his needs. McKenzie was remembered for stories concerning his teeth. Once when he had toothache, he tied one end of a string to his saddle and the other end around the offending tooth. He gave the horse a hit on the rump, the horse galloped off leaving the tooth dangling.[96] Some time later he lost his false teeth;

91 Sidney Kidman to Moreheads Ltd., 14 August 1925, Foster Family Papers.

92 Ken McGuire Interview, Bell, 5 February 1990. Norma Gunther died August 1982.

93 *Queensland Country Life*, 10 August 1961.

94 Monkira Development Book, 1960, NAPR.

95 Ken Moore Typescript, Brisbane, 1990.

96 Monkira Pastoral Company Pty. Ltd., W.F.Alexander, 'Ninety Years of the N.A.P.Co., 1878-1967', p.5.

Yellow Top, Coorabulka, 1978. Steve Millard

Unloading steers from Alexandria at Coorabulka, 1984. Steve Millard

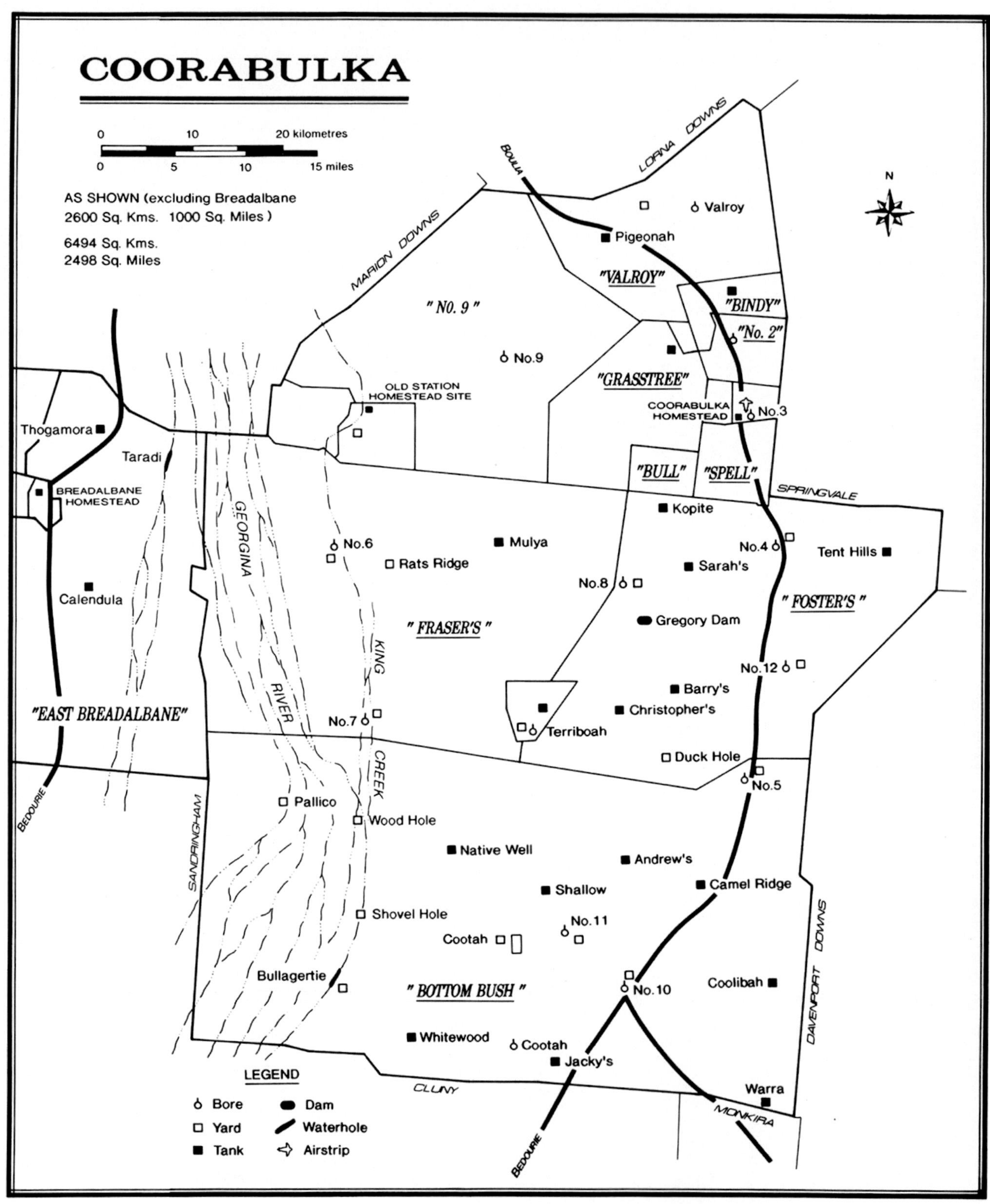

Map of Coorabulka.

he asked the mailman to get him a new set in Winton, saying they could judge the size from the fact he took a seven and one-eighth inch hat and a size 10 boot.[97]

A young Aboriginal woman, Annie Daley, known as 'Topsy', arrived at Monkira in the 1920s and worked in the Gunther household. She was the only source of female companionship for Norma Gunther; she cared for the children and worked in the house. One of her tasks was to pedal furiously on the pedal wireless so that the Gunthers could contact Cloncurry. Born at Glenormiston in 1907, she later married Jack (Snap) Hansen who was head stockman on Glenormiston for twenty-eight years.[98] She died in August 1988.

Situated 113 kilometres south of Boulia, COORABULKA adjoins Davenport Downs, Springvale, Monkira, Cluny, Sandringham, Marion Downs and Lorna Downs. Its total area of 9,094 square kilometres (3,498 square miles) includes Breadalbane (2,600 square kilometres) which is worked as part of Marion Downs. Nearly one-quarter of Coorabulka is flooded country from the Georgina River and King Creek with Flinders, sorghum, yellowtop, bluebush and herbages growing there. The balance of the property consists mainly of open rolling downs with Mitchell and Flinders grasses, and (on Breadalbane and Mount Tarley) ironstone, pebbly country, lightly covered with saltbush. Ken Moore describes how such country

> produces shimmery hazes, so much so that some miles from the homestead on very hot days the white buildings appear to be upside down some 10 to 15 feet off the ground. In those hot dry times it looks like the end of the world.[99]

The four leases comprising Coorabulka station — Coorabulka, Mount Tarley, Breadalbane and Bittoorong — were taken up by various people in the 1870s. Prout and McGuigan took up Coorabulka, while Duncan Campbell of Mount Margaret Station took up Breadalbane in 1876. The next year, the Breadalbane lease was transferred to Duncan Campbell and William Wright Richardson and in 1881 to Richardson alone. He forfeited the lease in 1890 to the Australian Joint Stock Bank. By 1900 the leases were with the Collins family in the name of John Collins and Sons, and remained so until the purchase by the Monkira Pastoral Company and the later resale to NAP in 1939.[100] The property had gone to auction in 1933 and 1935 during attempts to wind up John Collins and Sons, but no sale eventuated.[101]

In the early years, Coorabulka relied almost entirely on local rain, and to alleviate this dependence, bores were put down. By 1906 seven were in use,[102] and by 1913 there were ten flowing bores, all artesian, Nos 4 and 10 being equipped with windmills, tanks and troughing.[103] Only four more bores have been drilled since — two in 1956 and two in 1981-82 — but nearly thirty earth tanks and desilted waterholes have been added to provide a permanent water supply. In 1970-71 manager Bill Alexander had a busy time supervising the construction of ten of these.

When NAP took over Coorabulka the bores were in need of repair, but by the end of the war the situation was deplorable. In many, the casing had rotted away below ground and water was coming up outside the casing, creating a waterhole at the borehead. In the mid-1940s manager Len Carrington had some success in containing and cementing the boreheads, but some of the waterholes formed were too big, making it difficult to rig a boreplant over the hole. A large centrifugal pump was then used to keep the water away, a task made more difficult because of the scalding temperature of the water. No.4 bore gave particular trouble. Because it was on a knoll, the flow ceased with the lowering of the water table in the artesian basin; in 1967 a replacement hole was drilled and equipped with a windmill. While No.8 bore had been redrilled during John Crane's management in the mid-1970s, it was not until the 1980s in Steve Millard's time as manager that bores were successfully renovated (Nos 2,3,5,9,10) and in some cases redrilled — No.6.

There was an interesting quirk of fate in the drilling of Coorabulka's Terriboah bore in late 1981. The drillers were down to 610 metres to no avail, and head office staff agonised over whether the drilling should be abandoned. The decision to stop was finally made and, as the driller was using up his final length of drilling rod, water was struck at 628 metres.[104] It is the deepest bore on the

97 Ken Moore Typescript, Brisbane, 1990.

98 Marian K.Dent, 'The Life of 'Topsy' ', *Mimag*, September 1976, pp.11-6.

99 Coorabulka, Ken Moore Typescript, Brisbane, 1990.

100 Coorabulka, Pastoral Holding No.349, LAN AF266, QSA.

101 Auction Information, Coorabulka, 31 March 1933, 27 November 1935, Foster Family Papers.

102 'William Collins', *Pastoralist's Review*, 15 February 1906, p.942.

103 Advertisement for Auction of Coorabulka, Chatsworth and Noranside, 31 March 1933, Brisbane; Francis Foster, Diary of Western Queensland Trip, 1913, Foster Family Papers.

104 General Manager's report, No 12, 14 December 1981, NAPR.

No. 6 Bore, Coorabulka, 1913. Left to Right: Jim Nicholls (Pastoral Inspector for Collins White Pty. Ltd.); Francis Foster (sitting); Ivy Nicholls.

Henry Foster

property — situated in a dip in the artesian basin — and the luckiest to be in existence.

In good seasons and cooler weather Coorabulka was considered to be one of the best Mitchell-grassed properties in the west. The weights of the Coorabulka bullocks supported this, with one draft averaging 902 pounds dressed weight at the Townsville meatworks in the early 1950s.[105] A comparison of cattle sent from the Channel properties in 1945 also makes the point:[106]

	BULLOCKS		COWS	
Station	No.	Av. wt	No.	Av. wt
Marion D	2260	766.8lb	65	514.7lb
Monkira	1115	707.9lb	10	680.6lb
C'bulka	1932	827.1lb	273	579.6lb

The country and the seasons on Coorabulka are nevertheless always prone to extremes. In 1938 the herd totalled 6,700 with an annual turnoff of 3,000 and in 1941 stock numbers peaked at 10,498. There were 8,776 head on Coorabulka in 1957, but after 1958, drought conditions necessitated destocking to the extent that in 1960, there were only 981 left. With dependence on local rain for grass cover, water alone from the bores was insufficient to keep the cattle alive — they needed feed. In 1958, the stud herd was moved to Herbert Downs, up to 3,000 head of Coorabulka males were moved on to Monkira, while the excess remained on Coorabulka.[107] By 1959, however, Monkira was also feeling the strain and as many as possible were turned off. They were loaded at Butru instead of Quilpie, since the stock route between Monkira and Windorah was eaten out. Meanwhile only 204

105 Coorabulka, Ken Moore Typescript, Brisbane, 1990.

106 Particulars of Cattle sent off Station, 1945, H.Chambers File, 1945, NAPR.

107 Minutes of Directors' Meeting, Monkira Pastoral Company, Brisbane, 19 June 1958.

Coorabulka Shorthorns, 1974.

Dairy Shorthorn bull purchased for Coorabulka by H.W. Chambers 1969; Senior and Grand Champion at the Sydney and Melbourne Royal Shows.

were turned off Coorabulka. With the Channel properties unable to provide feed for store cattle, 4,000 were sold direct from Alexandria to meatworks.[108] Although a few reasonable seasons followed 1960, drought in the mid-1960s again almost destocked Coorabulka.[109] Manager Bill Alexander later commented:

> For the first three years I was there [from 1964] all that the cattle lived on were the bullrushes growing in the flowing bores.[110]

The company successfully applied to the Commissioner of Taxation to have sales from Coorabulka declared 'forced sales' thus allowing profits to be spread over five years for income tax purposes.[111]

The old station homestead and yards had been built at King Creek on Coorabulka East run. Although there was a good supply of natural water lasting up to twelve months, the threat of water difficulties resulted in the moving of the homestead. In the 1902 drought No.3 bore was put down for the station's new homestead on Station Creek at the north-eastern end of the property. This provided an excellent water supply, both hot water direct from the bore and cold water pumped to the house from a cooling tank.

Unlike the other homesteads on NAP's Channel properties, the one at Coorabulka is made of wood. Although spacious, it does not have the charm of the stone structures at Monkira, Marion and Glenormiston. Furthermore, it has had subsidence problems and seems to be continually plagued by rats. In 1962, Harold Cook, the manager commented that 'one needs a gas mask to live in the house these days, dead rats underneath . . . They are going through the store like a bush fire'.[112] As well as the homestead, other improvements when NAP took over included men's quarters, necessary out-buildings, drafting yards and some external fencing. In 1960, the men's quarters were burnt down and new ones erected.

As mentioned, there was a stud on Coorabulka from 1940 until 1960, when it was transferred to

108 Directors' Annual Report, NAP, 30 June 1960.
109 Cattle Statistics, Coorabulka, 1939-1967, NAPR.
110 Bill Alexander Interview, Marion Downs, 15 November 1990.
111 Directors' Annual Report, Monkira Pastoral Company Pty. Ltd., 31 December 1964.
112 Harold Cook to H.W.Chambers, 25 December 1962, Manager Coorabulka File, 1953-1962, NAPR.

Coorabulka Station, 1920s. John Oxley Library

Alexandria to form the Top Stud. In the dry years of the late 1950s the stud was moved first from Coorabulka to Sandy Creek and then agisted on the Collins White property, Eulolo, until late 1959.

Disease has not been a major problem on Coorabulka. Three-day sickness broke out in the district in 1955 and horses were affected by Birdsville disease at various times. But there was never the trouble with ticks that occurred at Alexandria. In 1960, a bullock infected with pleuro-pneumonia found its way from the Springvale herd into the Coorabulka herd of cattle for market, resulting in both stations being quarantined.[113] The quarantine lasted for seven months until Coorabulka cattle were allowed into all parts of Queensland except the south-east. As a result, the board decided that future policy would be to inoculate all calves on Coorabulka and Monkira.[114]

While Monkira had only two managers between its acquisition by NAP and the death of Douglas Fraser, Coorabulka had nine.

COORABULKA MANAGERS

1909 - 1910	C.S.Delpratt
1911 -	H. Afford
1920s- 1930	Tom Barnes
1930 - 1934	Eric Barnes
1934 - 1939	Jack Clanchy
NAP Ownership since 1939	
1940 - 1942	Les Stretton
1942 - 1948	Leonard Carrington
1948 - 1949	Ron Carrington
1949 - 1951	Tommy Groom
1951 - 1952	William Kriedemann
1952 - 1961	Tommy Groom
1961 - 1963	Harold Cook
1963 - 1964	Peter Donkin
1964 - 1971	Bill Alexander
1972 - 1976	John Crane
1976 - 1985	Steve Millard
1985 - 1989	Jamie MacFarlane
1990	Peter McLaren
1990 - 1991	Dieter Jordan
1992 -	John Rickertt

Before Coorabulka joined NAP, it was managed by Tom Barnes and then by Eric Barnes, both brothers of Harry Barnes, later manager of Alexandria. Tom went to Chatsworth and then to Brunette Downs, with Eric following him to each place. When NAP took over Coorabulka, Les Stretton was appointed manager from January 1940, replacing Jack Clanchy, whom Douglas Fraser had sacked because he had been allegedly branding NAP cattle as his own. On one occasion, Les Stretton nearly 'did a perish' following the breakdown of a station vehicle in poor condition because of war shortages. Stretton is also remembered by Camel Ridge between the homestead and the No.10 bore where he shot twenty-seven camels. Two years later Len Carrington was appointed manager, but he left in 1948 to manage Inverleigh in the Gulf, preferring breeders to fatteners. For a short time, his nephew Ron succeeded him on Coorabulka.

An old servant of NAP, Tommy Groom, relinquished his position as manager of Soudan to go to Coorabulka in 1949. Both he and Len Carrington were keen horse-breeders. Groom returned to Soudan in 1951 but was back at Coorabulka the following year after the shortlived management of William Kriedemann. Groom died in 1962. Harold Cook was there temporarily until Peter Donkin took over, also for a short time. Bill Alexander, who had been head stockman on Eulolo which was being resumed, followed as manager of Coorabulka in 1964. He will be discussed in a later chapter.

113 General Manager, NAP to Department of Primary Industries, 23 December 1963, Green Copies, Marion Downs Pty. Ltd., November 1963-February 1964, NAPR.

114 Monkira, W.F.Alexander, 'Ninety Years of the N.A.P.Co., 1878-1967', NAPR.

CHAPTER 10

The Road Trains

Alexandria's Road Train, 1991. Christine Taylor

In Australia, droving had always been the accepted practice of moving cattle to markets. This changed with the advent of the railways, when cattle walked only as far as the nearest railheads for transport by train to meatworks. Despite grand plans for a railway through Queensland to the Northern Territory, the economics of development in these remote parts could not justify the capital expenditure and one did not eventuate. A further advance — in motor vehicle transport — gave impetus to the cattle industry to combine the two forms of transport: rail and road haulage. In the Northern Territory, NAP played a major role in the development of the transport of cattle by road, and in Queensland it was the first company to establish a fleet of road trains. In the early days of NAP's road trains, the real benefit came from trucking cattle from the Channel properties, to railheads. Very few store cattle were moved this way from Alexandria to the Channel properties, but with improved roads and faster vehicles this also gradually changed; although even today in good seasons cattle are still walked.

There had been some road transport of beef cattle in the Meekatharra area of Western Australia in the late 1920s but such ventures were usually economically unsound because of the relatively small carrying capacity (about fifteen head), and the excessive wear and tear on the vehicles because of the poor roads and unsuitability of vehicles.[1] In 1945 Philip Ward of Banka Banka station in the Northern Territory began moving 1,000 fats annually by road to the railhead at Alice Springs in weekly lots of twenty; his alternative was to walk them 560 kilometres to the railhead at

1 *The Economics of Road Transport of Beef Cattle*, Canberra, Bureau of Agricultural Economics, 1959, p. 5.

Dajarra.[2] On a larger scale, in 1946 Kurt Johannsen began transporting cattle by road in the Alice Springs area using a Diamond 'T' 201 b.h.p. prime mover pulling three trailers.[3] In the following year he increased the number of trains to three. In September 1948 he trucked an experimental draft of 74 bullocks from Anthony Lagoon to Alice Springs. The cattle continued by rail to Adelaide, the whole operation taking eight days.[4] NAP's operations began soon after in November 1948, and at the same time Tancreds trialled the transport of cattle from south-west Queensland to the firm's works at Bourke. In 1952 Vesteys converted two huge army tank carriers to move cattle from their Wave Hill Station.[5]

The benefits to the industry and specifically to NAP were fully endorsed by General Manager Ken Moore, when he spoke at the Animal Production Society's dinner on 31 October 1962.[6] Stressing the saving in time, he pointed out that it took stock six weeks to walk the 525 kilometres from Marion Downs to Winton or the 500 kilometres from Monkira to Quilpie. By road transport it was a 16-hour trip to Winton or a 14-hour trip to Quilpie. Cattle retained their softness and did not become road-weary or muscle-bound on the journey. Any bruising occurred mainly during loading and unloading, due to handling by inexperienced men, bad yards and faulty unloading facilities. Moore also noted that road trains made it possible to get stock away, even if a stock route was bare of grass and water; during drought, stock might otherwise perish. And stock could be turned-off at their peak; in the Channel country, this was normally at the end of July and August, while Alexandria's peak came a little later — in mid-August to mid-September.

Moore quoted figures showing that although road transport at seven shillings per mile per K railway wagon was more costly than droving, the saving in weight largely compensated for this, and with a company operating its own equipment, there was a definite advantage in using road transport. He gave the comparison of 1,400 head sent from Marion Downs to the meatworks. One mob sent to the Winton railhead by road transport dressed 349 kilograms (770 pounds); a similar mob sent by drover dressed 319 kilograms (703 pounds).

There were other points in favour of road trains — such as the fact that cattle sent by road did not have to be inoculated for pleuro. Stock could also be marketed younger. Walked out, they had to be four to five years old; by transport they could be turned off at two to three years. Furthermore, in contrast to walked cattle that took time to settle and tended to 'keep walking', trucked cattle settled down quickly. On account of the heat, the latest cattle could be worked was the end of October. A six-weeks' walk to railheads meant a mid-September start, so mustering had to begin at the end of August. Using road trains, mustering did not need to begin until mid-October. The flexibility of road transport allowed consignments to be spread so as not to overload the market in any one day and so push prices down. Drafts could be dispatched two or three times a week, which also suited the meatworks better than having a flush of cattle during the peak killing season.

By the 1950s drovers were also becoming more scarce and droving costs were rising — £200 a week for a mob of 700 to 1,000 head. Other advantages of road trains were that mixed mobs of steers and cows could be marketed, which was not otherwise practicable, and road transport ensured that all possible stock were marketed. With droving, those cattle unable to withstand the walk could never be sold and would die on the property.

But road transport also had its disadvantages: a lack of feed around loading points; extra work for the station manager in dispatching cattle over a longer period; difficulty in arranging for buyers to inspect small mobs compared with one large mob; and transport delays in wet weather. Spelling of cattle both before and after trucking was necessary to decrease the chance of bruising, which could also be minimised by good roads, careful driving, loading animals of uniform size, and loading them sufficiently tightly to prevent undue movement. Drivers endeavoured to prevent any cattle falling down, because once one fell more could follow and be trampled to death.

The number of NAP cattle being trucked to railheads increased enormously in the early 1950s. In 1952, 1,728 head were transported from Marion Downs, while in the following year 6,204 bullocks and 94 cows went from Marion to Dajarra. In return, from Dajarra, 100 bulls for breeding were trucked to Marion, further illustrating the

2 *Queensland Country Life*, 20 July 1961.

3 J.H. Kelly, *Report on the Beef Cattle Industry in Northern Australia*, p.197.

4 *Adelaide Chronicle*, 9 September 1948.

5 'Road Transport of Cattle in Central Australia', *The Meat Producer and Exporter*, December 1948, p.8 D.M. Fraser File A to F, 1946-49, NAPR; Alan Powell, *Far Country: a short history of the NT*, p.220.

6 K. Moore, Speech to the Animal Production Society, Brisbane, 26 October 1962, in *Queensland Country Life*, 1 November 1962.

NAP road train drivers and co-drivers leaving Brisbane, Leyland Hippos ca. 1957. Rear Left to Right: Cam Doglione and Richard Purvis-Smith.

Richard Purvis-Smith

usefulness of the road trains.[7] Similarly, in the Northern Territory in 1956-57 only 13 per cent of cattle movements were handled by motor transport; by 1960-61 the proportion had grown to 45 per cent, and by 1974-75 to 99 per cent.[8]

The earliest official mention of road trains in NAP was at a directors' meeting on 13 April 1938, when Francis Foster suggested the purchase of diesel semi-trailer trucks for the transport of fat cattle to the railhead by road.[9] Douglas Fraser undertook to investigate the position in relation to road tax and competition with the railways. No further progress was made then, because the war effort strained resources. In November 1946 the board authorised Fraser and Foster to continue their investigations and, if they saw fit, to purchase on behalf of the company a lorry and trailer of whatever make and design they considered most suitable.[10] Douglas Fraser was encouraged by his son Bill, who during his years in the Army had seen the advantages of heavy transportation by road. In the south, Foster interviewed the International Harvester Company and the Fowler construction group, while in Queensland Fraser investigated the Leyland company.[11]

Following negotiations, two Leyland Diesel Hippo livestock trucks were purchased, and these arrived in Australia from England in late 1947. They were driven from Sydney to Brisbane where crates for the chassis and two trailers were built. In late 1948, the two units were successfully tested by Frank Russell, the first person in charge of the

7 W.F. Alexander, 'The Road Trains', in 'Ninety Years of the N.A.P.Co., 1878-1967', p.3.

8 Tom L. McKnight, *The Long Paddock*, Armidale, University of New England, 1977, pp.117, 120, 123.

9 Minutes of Directors' Meeting, NAP, Brisbane, 13 April 1938.

10 Minutes of Directors' Meeting, NAP, Brisbane, 27 November 1946.

11 F.Foster to D.M.Fraser, 17 February 1947, Alexandria Station Pty.Ltd. File, NAPR.

NAP's road train fleet, April 1961 following the purchase of four new Mack prime movers, two new trailers and modification of the six existing trailers.

NAP road trains, and Bill Fraser, on trips from Marion Downs to Dajarra, a distance of 225 kilometres. With each lorry and trailer holding approximately 27 head, three trips were required to fill the special trainload of 165.[12] At the time, the road was just a track and creek crossings were very bad; 'jack-knifing' was common, and Bill Fraser commented:

> The roads were really bad from Boulia right through to Dajarra . . . We'd load up late afternoon at Marion and we'd go through the night and we'd get to Dajarra next morning about half past eight. You had to do that because it was so hot. It was 110 degrees at ten o'clock at night at Boulia.[13]

Despite some bruising and the deaths of five head during transportation, the directors were enthusiastic about the long-term advantages of road trains.[14] Fraser claimed that the annual turnoff of 9,000 to 10,000 fats could be boosted by 20 per cent.[15] There were also advantages for Alexandria. In 1949, doing two trips per week of 640 kilometres each, the Hippos trucked 2,700 spayed cows from Soudan to the Mount Isa railhead.[16] Made easier by the bituminised military road, the trip took twelve hours; by contrast it would have taken four weeks to walk the cows into Mount Isa. Advantage was taken of the return trip to Alexandria to transport eight new stud bulls to the station from Mount Isa.[17]

The board decided in December 1948 to purchase four more Leyland Hippos, two each for Marion Downs Pty.Ltd. and the Monkira Pastoral Company. The company now had six lorries and six trailers valued at £40,000.[18] They were modelled on the railway cattle wagon and loaded from the side. Depending on the size of the cattle, the truck would hold about 12 head and the trailer 13. The whole fleet was equivalent in capacity to a special railway stock train of eight K wagons[19] and one KB wagon.[20]

12 Douglas Fraser to Thomas Cook, 26 August 1948, H.W.Chambers and Drovers File, NAPR.

13 Bill Fraser Interview, Mundoolun, 15 September 1990.

14 D.M.Fraser to T.Cook, Marion Downs, 15 December 1948, T.H.Cook File, November 1948-June 1952, NAPR.

15 *Courier Mail*, December 1948.

16 D.M. Fraser, Brisbane to D.G.Stokes, London, 22 February 1950, NAP Companies File, F to L, NAPR.

17 R.H.Murray to D.M.Fraser, 8 October 1949, R.H.Murray File, 1948-51, NAPR.

18 Minutes of Directors' Meeting, Monkira Pastoral Company Pty.Ltd., 3 December 1948, 185 Mary Street, Brisbane; *North Queensland Register*, 29 January 1949.

19 Particular classes of railway wagons which transport cattle. K wagons are approximately 10 metres long and each carry around 17 bullocks or 20 cows depending on the size of the animals. A KB wagon includes compartments to accommodate the drover and the guard.

20 'The Road Trains', W.F. Alexander, 'Ninety Years of the N.A.P.Co., 1878-1967'; *North Queensland Register*, 29 January 1949.

Bogged in the sand near Charleville; Cam Doglione and Keith Haughton. Richard Purvis-Smith

Although primarily for the transport of cattle, three of the lorries carried fencing materials from Newcastle to NAP properties in 1950-51.[21] A seventh truck, a Leyland Beaver was added by 1953. In late 1960 with the Leylands over 10 years old, it was decided to replace four of them with Mack diesel prime movers, and to purchase two new trailers built by the McGrath company and modify the six existing trailers.[22] With each of the four Mack prime movers pulling two dogs (towed trailers), 236 head could be carried by the fleet in one consignment (15 in each of the four prime movers and 22 in each of the eight trailers), compared with 140 to 150 in a Leyland consignment.[23] The outlay for each Mack truck after trade-in was £45,000 and delivery was taken in April 1961.

The Mack units had been designed, like the Leylands, on the principle of railway wagons. The whole road train now carried the equivalent of twelve K railway wagons and one KB wagon. Two of the remaining Leylands were converted to goods-carrying vehicles, which enabled the company to start transporting its own station stores and equipment. At first the Leylands were used with the stock crates removed to transport goods to the stations in the off-season. When it was realised that more regular trips were needed, two Mack trucks were purchased in 1962 to replace the goods-carrying Leylands.[24] The seventh truck, the Leyland Beaver, was also used for carrying stores to the stations until it was sold in 1964.

The introduction of the Mack trucks could have suffered an unfortunate delay but for good coordination by the Mack company between Brisbane and the United States. Cracks developed in identical spots in the connecting rods of two of the engines. Within three days, there were four new engines being shipped to Australia and one of the company's 'trouble-shooters', A. Halt, was sent from Mexico to Brisbane. His inspection revealed that two connecting rods with consecutive

21 Directors' Annual Report, Marion Downs Pty. Ltd., 30 June 1951.

22 Minutes of Directors' Meeting, Marion Downs Pty. Ltd., Brisbane, 15 December 1960.

23 W.F.Alexander, 'The Road Trains', in 'Ninety Years of the N.A.P.Co., 1878-1967'.

24 Minutes of Directors' Meeting, Marion Downs Pty. Ltd., Brisbane, 9 November 1961.

Ted Cottrell. Queenie Cottrell

numbers had been drilled too deep just at the end of a run before an inspection. Despite only two engines being affected, the Mack company insisted on completely dismantling the other two engines at no cost to the company.[25]

More transport firms were entering the industry during the 1960s and, as early as 1963, the directors discussed the possibility of disposing of the road trains. The units were idle for six months of the year, and being based at Mundoolun were not immediately available for jobs during that time. A concrete floor had been laid at Mundoolun in 1954 to help with the maintenance. Each May the trucks would move out in convoy to begin the season which, depending on the weather, would last until October or November.

In 1964 eight dog trailers, each 12.2 metres (40 feet) long, were purchased.[26] Despite this increased capacity, by the late 1960s, other companies were being hired to meet NAP's trucking requirements at the height of the season. A directors' meeting in December 1967 decided that instead of keeping the road trains at Mundoolun during the off-season they should be based in Brisbane.[27] The next year a block of 10 hectares (24 acres) of land was acquired at the corner of Blunder Road and Kippax Street, Oxley, for $60,000. By December 1968 a storage shed, garage and workshops were erected by the company, although there was already a residence on the property for the purchasing officer, Ted Cottrell.[28] The Oxley depot proved a successful base for the maintenance of the road trains and as an assembly point for goods before their transfer to the stations. The situation altered slightly in 1969 with the decision to leave all cattle trailers at Coorabulka at the end of each season, but the prime movers and other vehicles of NAP's Transport Section continued to return to Oxley.[29]

Two major developments followed in 1970 with the replacement of four of the Mack prime movers with four Leyland Buffaloes, and the purchase of three blocks of land in Cork Street, Winton. The dwelling on one of the blocks was used by the drivers for accommodation, and the land was useful for maintenance of the vehicles.[30] This attempt to increase the efficiency of the transport operation in the face of increasing competition from the growing number of private companies specialising in the field was not completely successful. NAP's Transport Section made losses in 1971 and 1972 and the decision was made to sell the road trains and the Winton land.[31] In 1974 the land was sold to Roy Shaw, a private operator from Winton. He leased the four Leyland prime movers and eight trailers from NAP until the following year, when he purchased the entire fleet. For NAP the return to Brisbane each season had proved costly, with key personnel having to be paid all year round to retain their services. Besides, the cartage of NAP's cattle was valued and sought after by contractors. In 1974 a Volvo truck and semitrailer were purchased to replace the Mack used for the general carrying, and in the same year a Leyland Super Hippo with two dog trailers was purchased for cattle cartage on Alexandria.

After Frank Russell's resignation in late 1949, the person in charge of the road trains was William Gibson, followed in 1952 by George Schultz who, like Gibson, had briefly been manager of Soudan. The drivers did their own servicing and minor repairs, although it soon became clear that a person

25 W.F. Alexander, 'The Road Trains', in 'Ninety Years of the N.A.P.Co., 1878-1967'.

26 Minutes of Directors' Meeting, Marion Downs Pty. ltd., Brisbane, 21 November 1963.

27 Minutes of Directors' Meeting, NAP, Brisbane, 13 December 1967.

28 Minutes of Directors' Meeting, NAP, Brisbane, 12 June 1968.

29 Minutes of Directors' Meeting, 20 August 1969.

30 Minutes of Directors' Meeting, NAP, Brisbane, 18 February 1970, 20 May 1970.

31 File, Road Train/Sale and Lease to Shaws Transport Pty. Ltd., 1972-75, NAPR.

with intimate knowledge of the Leylands would be particularly useful.

So began NAP's thirty-year association with Edward Charles (Ted) Cottrell,[32] a trained Leyland engineer, who, after the war, had the option of migrating to either South Africa or Australia. He chose the latter and worked for Leyland Motors in Brisbane before formally joining NAP in January 1952 at the age of 34. Overseeing all road train activities, he drove a support vehicle with the convoy, and spent the cattle turnoff season at the trucking camps and on the stations. In the off-season he overhauled the vehicles at Douglas Fraser's property at Mundoolun. When Richard Purvis-Smith, a grandson of Harry Bruxner and a truck driver with the road trains, took over supervision of the road trains in 1961, Ted became purchasing officer for the company and lived at the Oxley depot after it was established in 1968.[33] The late 1960s saw further change. Claude Patterson took charge of the road trains in April 1967 when Richard Purvis-Smith left, and when Patterson left in 1971 he was replaced by Bruce Cameron.

The many drivers during the road train era included Richard Purvis-Smith, Keith Milner, Dick Marshall, Reg Field, Norm Draper, Lawrie Stevenson, Gordon Rowe, Terry Mahoney, V.Houghton, B.Bartley, Ray Mantova and the cook, Keith Maclean. But the driver who became synonymous with these transports was Camillo Doglione who was with NAP for a total of thirty-four years, four with the road trains and the remainder in general carrying.

Born in Italy in 1919, Cam came to Australia when he was 14. He worked in the canefields of Ingham with his father and uncle where he gained experience driving trucks. After joining NAP in 1950, he spent the first eight months driving one of the Leylands at Alexandria, taking supplies to the bores and moving cattle. Manager Rowand Murray tried to persuade him to stay and work permanently on Alexandria, but Cam was more interested in life on the road with the trucks. Each trucking season, he would join the road trains and truck cattle from the stations to the railheads. His first years were filled with memorable experiences — the time the cattle at the Mount Isa railhead wandered into the lettuce patch of a nearby market garden; the rat plague at the Ranken when the cook would boil the corned meat in the same pot

Cam Doglione, NAP's longest serving truck driver, 1950-1987.

used to catch rats the previous night; and on one of the early trips to Dajarra, he recalled how the bulldust cleared long enough to reveal that the trailer had broken away from the prime mover some distance back.[34]

After four years Cam left the company to drive for an interstate carrier. He returned to NAP in 1957 and began driving a Leyland Comet 90 for general carrying. When two Macks were purchased for this purpose in 1962, he continued as one of the drivers, and remained in this job for the next 30 years until his retirement in 1987. At the time, his Volvo truck was still in excellent condition, despite having been driven by Cam for more than a million kilometres.

With his effervescent personality, his humour and above all his total commitment to his job, he will long be remembered in the company.[35] His loyalty was shown in the Transport Court Case of 1965 (referred to later) in his capacity as a witness. During cross examination, Arnold Bennett QC, senior counsel for the Crown, tried unsuccessfully to make him admit to certain allegations and in the process tried to prove that the neat handwriting on

32 Born 3 November 1917.

33 Ted Cottrell retired in on 31 December 1982 and died 28 January 1990.

34 Cam Doglione Interview, Brisbane, 17 October 1990.

35 Ross Brunckhorst, General Manager's Report, No.11/87, presented to Directors Meeting, 18 November 1987, NAPR.

Trucks from the road train fleet provide the fencing for tennis; on the clay pans next to Narrah Waterhole on Marion Downs, 1950s.

Richard Purvis-Smith

a certain document was his. It was not — and Cam added: 'Look, if I had handwriting like that, I wouldn't be driving a truck'.[36]

At first with the relatively small numbers trucked, Ted Cottrell liaised with William Doull, the secretary, and Bill Fraser, a director, on the running of the road trains. Later Cottrell and Ken Moore would plan the numbers to be sold, and where and when mobs were to be picked up and delivered, so that the road trains and the railway trains had full complements. Camps were established for the drivers at the trucking yards at the Ranken and Soudan, with a major camp on Marion at Narrah Waterhole. Suitable yards had to be built for loading. At Narrah Waterhole, for example, there were main yards, forcing-up yards, a long race and a loading ramp. A small holding paddock (Wanderer Paddock) held the mustered stock before loading. Loading was often at night, at first without lights. Bill Alexander recalls the expertise of head stockman Ernie George:

> He was the best man I have ever seen with cattle in a yard . . . He would not allow anyone make a sound . . . all you could hear were the bullocks breathing as they walked up the crush.[37]

To save droving from the Herbert Downs area, another set of yards, a paddock and watering facilities at Sandy Creek were completed in 1952. Since the Leylands running to Dajarra were not able to carry a full trainload in one journey, arrangements were made to hold stock in the Borthwicks-owned Flyveil Paddock just out of Dajarra. An unloading ramp was built, and as long as there were sufficient stock in the paddock, the road trains were not held to a tight schedule. Thus any badly bruised cattle could be spelled longer.

An array of tents, spare parts for the trucks and household chattels became home for the drivers. Of those bred in the city having their first taste of the bush, many stayed only one season. They were employed as stockmen, and the best drivers were not always available under these conditions of relatively low wages. But it did provide them with valuable experience which could lead to other trucking jobs in a burgeoning industry. The men provided their own entertainment, whether it was building makeshift boats from petrol drums to sail

36 Cam Doglione Interview, Brisbane, 17 October 1990.

37 Bill Alexander Interview, Marion Downs, 15 November 1990.

in a waterhole, or playing homemade musical instruments. They often played practical jokes on each other — like the time Ted Cottrell and George Schultz came out of a hotel to find the Bedford utility, which was used as a support vehicle, up on bricks. The culprits — the truck drivers — were across the road roaring with laughter. A driver once dressed up in a sailor's outfit and convinced the locals at one hotel that he was in the Navy and served on the 'Bore Drain Submarine'.[38]

While Dajarra was initially the railhead for Marion, Coorabulka and Monkira cattle, from 1956 Winton was used from Marion, with Coorabulka cattle being walked first to Marion. Winton was preferred because of the rough rail trip from Dajarra. As early as 1950 Alexandria cattle were trucked from Soudan to the railhead at Mount Isa using the NAP road trains. Later, Monkira cattle were trucked to Quilpie. The Mount Isa line connected with Cloncurry via Duchess and then through Julia Creek, Richmond and Hughenden to Townsville. From Winton, cattle could travel either north to the Townsville line connecting at Hughenden, or they could travel south to Longreach and then east to Rockhampton. The Quilpie line connected with Brisbane through Charleville, Roma, Miles and Toowoomba. Cattle from Quilpie were slaughtered at meatworks at Brisbane, and those from Winton or Mount Isa at Townsville, Cairns, Bowen or Rockhampton.

Road conditions in the 1950s were often atrocious. In 1953 heavy rain completely bogged the convoy in a narrow stretch along the Mitchell-Morven road. This section became so chopped up that it remained impassable to all traffic, even after the fleet had moved on.[39]

In November 1961 the Federal Government made a commitment which greatly assisted the pastoral industry in the Territory and the Channel country. This was the provision of £20,000,000 for the building of bitumen beef roads to rail and roadheads. The government recognised that such a development would lift Australia's beef cattle export earnings by at least £15,000,000 annually.[40]

For NAP, the roads provided improved links between Monkira and Quilpie, and between Marion Downs and Boulia, Dajarra and Winton. Indeed, one of the first beef roads completed in 1962 was between Boulia and Dajarra, which then became one of the largest trucking centres in Australia. Eventually the Mount Isa-Boulia and Windorah-Quilpie sections of the Diamantina Developmental Road and the Boulia-Winton section of the Kennedy Development Road serviced NAP properties.

By 1959 cattle numbers had increased by nearly one-third in two decades in the Northern Territory and the Channel country.[41] In that year, there was a record road movement of 10,000 cattle from Alexandria to Mount Isa. The seven Leyland vehicles with prime movers were used at that stage. It cost £3 a head to move stock by road transport compared with £1 to £1 5s to drove them. But depending on the size of the bullock, there was a saving in weight loss from 50 to 100 pounds and this meant money was gained at the meatworks.[42] Over the next few years, there were similar movements by road to Mount Isa.

One of the Alexandria jackaroos who worked with the cattle in Mount Isa during this trucking program was John Ohlsen. NAP employees were provided with meals by Mrs Dakin, wife of the slaughterman for Miners Meat Supply in Mount Isa, and it was here in 1962 that John met their daughter Wendy, later to become his wife. John became manager of Islay Plains (1970-80) and Alexandria (1981-91).

The company's Transport Court Case of 1965 is perhaps one of the best-remembered legacies of its road train era. The case had nothing to do with the transport of cattle; rather it concerned station supplies destined for NAP's Channel properties. Central to the issue was the Queensland Government's preference that goods travel by rail rather than road. To bypass the restriction in the Transport Act that legal road transport of goods intrastate be confined to a select area — with fees levied for greater distances — NAP in 1959 instituted a system of interstate transport to and from Tweed Heads. The company deemed this legal under section 92 of the Commonwealth Constitution which provided for freedom of interstate trade.

Between 1959 and 1962, NAP sub-contracted J.H.Nicholson Transport to transport station supplies from Brisbane to Nicholson's depot at

38 Cam Doglioni Interview.

39 W,F.Alexander, 'The Road Trains', in 'Ninety Years of the N.A.P.Co., 1878-1967'; Letter from E.C.Cottrell to W.Doull, Marion Downs, 22 April 1953, Correspondence file from Secretary's Room, A-L, 1951-1954, NAPR.

40 *Bulletin*, 12 April 1961, p.20.

41 *The Economics of Road Transport of Beef Cattle*, Canberra, Bureau of Agricultural Economics, 1959, p.1.

42 *Queensland Country Life*, 9 April 1959.

Tweed Heads. The goods had physically to touch the ground in New South Wales before being reloaded for the trip back to Brisbane by Nicholsons. NAP trucks then transported the supplies to the Channel properties. Alexandria's supplies went directly from Brisbane, already being interstate. On their return from the Tweed to Brisbane, Nicholsons also carried wool which NAP trucks had back-loaded to Tweed Heads from Eulolo. Wool was often carried from other stations, as well as tallow and hides from Mount Isa.

In August 1962, new arrangements were made as a further response to the Act. Nicholsons would take the goods destined for NAP properties to their depot at Tweed Heads, but instead of Nicholsons taking the goods back to Brisbane, NAP trucks would take the goods directly from Tweed Heads to the stations. Meanwhile, the NAP trucks continued to back-load with wool. The 'Interstate Carriage of Goods' books printed in 1962 were formulated by Fraser and Moore in order to satisfy administrative requirements, and the drivers were issued with strict instructions:

> Should you be stopped and questioned by any competent authority in this connection, present to them your letter of Authority, and proceed with your duties. Should they desire to detain you, please ring the office immediately. You are not to make any comment or expression of opinion beyond handing over the letter of Authority.[43]

The government, however, was increasingly concerned over such border-hopping activities. On 10 September 1964, six complaints were lodged by the Transport Department against NAP for alleged breaches during 1963. The ensuing court case began on 16 March 1965 and continued for four days. The decision went against NAP and the company was fined £1,302 15s 2d. In the judge's opinion, NAP and Nicholsons acted in collusion, with the essential commercial purpose being the carriage of goods from one part of Queensland to another.[44] The fine was minimal, considering there had been a total avoidance in fees over the eight years of £26,700.[45] Since the government was evidently serious in its levying of compulsory permits and fees, the company decided not to challenge their endeavours.[46] Permits were henceforth used for the goods trips and Nicholsons were no longer contracted to carry NAP supplies.

The Railways Department encouraged the transport of cattle by rail and loading facilities gradually improved. Increased numbers of cattle wagons were purchased by the railways and there was a gradual change from the slower steam locomotives, which required stopovers for water, to the faster diesel engines. NAP took part in trials in June 1963 in which 700 Monkira bullocks travelled in 40 railway wagons pulled by multiple diesel units. The trip from Quilpie direct to the meatworks at Brisbane took twenty-seven hours compared with the usual train trip of five days during which the cattle required frequent spelling.[47] So satisfied with the rail network were meatworks owners such as Thomas Borthwick and Sons (Australasia) Limited, that they saw no need to establish any new meatworks in the state.[48] Such improvements in transportation enhanced sales.

In June 1965, Thomas Borthwick and Sons bought 5,000 Shorthorn bullocks and 3,000 culled and spayed cows for £300,000 for their Merinda meatworks near Bowen. Believed at the time to be one of the largest single cattle purchases from one property ever finalised in Australia, all the stock came from Alexandria. Borthwick's general manager flew from Melbourne to finalise the deal on the station with Douglas Fraser, Ken Moore and Borthwick's resident Mount Isa livestock buyer. The cattle were road-trained to Mount Isa and spelled for 36 hours before the trip of 1,130 kilometres by diesel-powered cattle trains to the Bowen meatworks. Each train trip took nearly 50 hours, and it was a mammoth organisational task both for NAP and the railways. Each road train carried 250 stock, and with 600 to 1,000 head arriving at the meatworks each week, the Railways Department needed twenty-four trains and 417 wagons spread over two months.[49]

It is clear that NAP played a significant part in the establishment and progress of transport of cattle by road and rail. As experienced industry pioneers, the company was again in the vanguard of new developments.

---

43 Carting Instructions to C. Doglione, 1960, Transport File, NAPR.

44 Reserved Decision, Arnold James Horne, Sub-Inspector of Police versus Marion Downs Pty. Ltd., Breach of Section 49 of 'The State Transport Act of 1960', Magistrates Court, Brisbane, 9 April 1965, Transport File, NAPR.

45 K. Moore to NAP directors, 14 April 1965, Transport File, NAPR.

46 K.Moore, Aide Memoire of a conference without prejudice with Commissioner for Transport, Mr N. Kropp, 17 May 1965, Transport File, NAPR.

47 *Courier-Mail*, 17-18 June 1963.

48 Submission by Thomas Borthwick and Sons (Australasia) Limited to Meat Committee of Inquiry, *North Queensland Register*, 6 June 1964.

49 *Courier-Mail*, 24 June 1965, *Townsville Daily Bulletin*, 30 June 1965.

CHAPTER 11

The Modern Era: New Challenges and Expansion

The modern era for NAP began in the late 1960s and is marked by three major events. The first occurred in 1967 when the company's long-serving chairman, Douglas Fraser, relinquished the position and Michael Crouch began his own long tenure as chairman. The second major event occurred in 1968, when the company purchased both Glenormiston Station and freehold properties in Brisbane — head office premises in the city and an area of land in an outer suburb to be used for a depot. These were the first new properties purchased in nearly thirty years, and they marked the beginning of an expansionary phase that has seen six more cattle properties added since then (though of these, one, Islay Plains, has been sold). The third major event was the formulation in 1969 of the Federal Government's Brucellosis and Tuberculosis Eradication Campaign (BTEC) — to be discussed in Chapter 13 — which was to become a major preoccupation of the cattle industry, and which would have important and positive effects on property management practices.

NAP is different today from twenty years ago, having had an exciting but often anxious period of challenge and development. But in accounting for the successes of this era it should never be forgotten that the solid base established in earlier years by the efforts of the Forrests, the Collins', Fraser, Foster and many others both in Brisbane and on the properties, made it possible for those who followed to guide the company to the strong position it holds within the industry today.

The 1970s were challenging but successful years for NAP. Profits realised during the 1960s had allowed the company to expand, with the 1968 purchase of its head office premises, NAPCO House, at the corner of Edward and Margaret Streets, Brisbane. Glenormiston Station was purchased in 1968 and Islay Plains in 1970. Drought wreaked havoc on different properties at various times through the 1960s, but the drought of 1969-70, especially on Alexandria, was particularly devastating. By contrast, for the rest of the 1970s there were good seasons, with record floods in 1974 and 1977. But other factors such as the government campaign to eradicate tuberculosis and brucellosis, and the slump in beef prices in the mid-1970s, coupled with suddenly greater inflationary pressures on wages and improvement costs, imposed new financial pressures. By the end of the decade, beef prices had recovered and, with good seasons and BTEC on target, General Manager Ken Moore, due to retire in 1982, was generally pleased with the company's progress.

In retrospect, the turn of the decade was a relatively brief period of calm before new storms in the 1980s. Although significant progress had been made on many fronts during that decade, it has also been one of the most disruptive and challenging in the company's history. Severe drought conditions were experienced on Alexandria in the 1980s; the Channel properties also suffered from indifferent seasons; and the possibility of the company being taken over by outsiders (which will be discussed in Chapter 14) was a major distraction for directors and staff in 1984 and 1985.

For management, the focus in this period was very much on the well-being and productivity of the NAP herd — first in drought relief measures through agistment, supplementary feeding, and earlier weaning, and the purchase of five new properties (Connemara, Kynuna, Boomarra, Vergemont and a feedlot, Wainui). Second, and most important for the longer term, from 1982 the breeder groups within the herd were crossed progressively with Brahman bulls, a development which in 1988 was taken further with the introduction of genes from five different cattle breeds. Beef markets expanded in the 1980s, and more attention was given by the industry to marketing, especially in Japan and Korea.

Despite difficulties and rapid change, NAP has come out of the 1980s in good shape. Its profitability has increased, the breed is proving acceptable in the lucrative export market, and there is a new stability in the ownership of the company, following the takeover period, with the Foster family now the majority shareholders.

Just as in the past, the company's progress in the modern era can be attributed particularly to the efforts of a few key figures. During the 1970s, the dedication and application of two men was outstanding — Michael Crouch, the Managing and Executive Director and Chairman of Directors, and Ken Moore, the General Manager. The two had known each other since their rugby footballing days with the GPS club[1] in the 1930s. They served in the same unit in the army during World War II, and were firm friends long before Moore joined NAP. Crouch and Moore were also important influences through the 1980s, Crouch as Chairman and Moore as company consultant and board member. But other significant contributions were made in that decade, such as that of Henry Foster, the senior male member of the Foster family, who took a keen interest in the company, as did his brother William after he joined the NAP board in 1986. And since 1982, General Manager Ross Brunckhorst has played a crucial role in implementing company policy.

Edward Michael Crouch was born in Brisbane on 16 February 1916, and educated at The King's School, Parramatta. He was admitted as a solicitor in Queensland in 1940. Following war service in the Middle East and New Guinea between 1940 and 1945, he joined his father's legal firm in Brisbane where he gradually became involved in NAP's legal affairs. Following his father's retirement[2] from the NAP board in 1955, a year before Ken Moore joined the company, Michael

Kenneth Woodhead Moore: General Manager 1956-1982; Director 1980-1990.

Ron Spedding

took his father's place as alternate director for the English shareholders. In November 1972 he was appointed managing and executive director, a position he held until May 1981, having by then turned 65.[3] His 'hands-on' approach ensured that a good working relationship developed between Crouch, Ken Moore and Francis Foster. The bond of friendship and respect between the three was a significant factor in the company's stability and success. As Moore wrote of Crouch in a letter to Francis Foster in 1968:

> It is a pleasure to be able to have a good discussion on any matter and feel that the decision which is made has full backing and that prompt action will be taken.[4]

By the 1960s, the Crouch-Ingram association had existed for some thirty years and early in 1962, in recognition of the 'long intimate association and service of the Crouch family with the Company',[5] the Ingram family passed 250 shares to Michael Crouch, at a nominal price and with the consent of the board. In 1974 Moore and Crouch each took up a small parcel of shares following a proposal by other board members that they be given the

1 Greater Public Schools Club.

2 Edward Robert Crouch died in 1962.

3 Resolution, NAP Minutes, 15 May 1974; Minutes of Directors' Meeting, 13 May 1981.

4 K.Moore to F.Foster, 12 January 1968, File 4, Directors, 1968, NAPR.

5 Minutes of Directors' Meeting, 15 February 1962.

opportunity to do so, in view of their contribution to the company.[6]

Unlike his father who never visited the properties, Michael Crouch made visits soon after he became a board member and continued to do so regularly over the next thirty years. He was eager to learn about all the aspects of the industry and Bill Young, manager of Alexandria, taught him how to handle stock. Young relates one of the experiences:

> I let him have a go at branding and cutting calves and he wanted to have a go on a broncho horse. We gave him a go on a broncho horse and he was there for a long time and he never caught a bloody thing. We started to tease him. So he let fly and he caught a bloody steer. So he dug him out and didn't the blacks hoorah him. He took it all in good faith.[7]

He moved in quite different social circles in Brisbane, and for a period was the president of the Queensland Club. At the same time he never forgot the employees on the stations and for this he gained widespread respect and support. With staff often promoted by transfers around the properties, he and Moore helped to sustain the family atmosphere in the company. The loyalty this produced continues today and remains an underlying strength of NAP.

Crouch officially became chairman of directors after Douglas Fraser's death in 1968, although he had unofficially held the position since November 1967 when Fraser became too ill to continue. Crouch remained in the position for twenty-three years. The most immediate change Crouch made was to increase the detail given in the minutes of both directors' and shareholders' meetings. Shareholders became better informed about the company through his letters and the addresses he gave at the Annual General Meetings. He was a staunch supporter of Ken Moore, and he also provided support to Ross Brunckhorst when the cattle breed change was begun in the 1980s and more properties purchased. This was an era of drought and expansion, interrupted by the worry of takeover bids during the mid-1980s, and Crouch

6 Minutes of Directors' Meeting, 15 May 1974; Minutes of Shareholders' Meeting, 9 January 1974.

7 Bill Young Interview, Casino, 26 January 1986.

Chairman of Directors Michael Crouch and General Manager Ken Moore on a visit to Coorabulka, late 1970s. Steve Millard

Directors' Meeting, Alexandria, October 1987. Back, Left to Right: D.S.F. Alexander, R.A. Launder, C.B. Lyndon, W.M. Fraser, E.M. Crouch, J.W. Ohlsen (Manager, Alexandria), Dr L.G. McNicholl (Veterinarian). Front Left to Right: R.M. Brunckhorst (General Manager), K.W. Moore, W.F. Foster, H.F. Foster, D.H. Brown (Secretary).

was a key figure in all these matters, approaching each issue with an open, analytical mind.

Kenneth Woodhead Moore or 'KW' as he was known, was born in Brisbane on 13 April 1917. He joined the Queensland Meat Export Company in 1948 as a cattlebuyer. Promotion led to his becoming manager of the Ross River Meatworks at Townsville, which he left in 1956 to take up the position of assistant general manager at NAP. Douglas Fraser was still general manager at the time but because William Doull, the secretary, had resigned through ill health some months earlier, and because Fraser was at Mundoolun for part of the week, the company was without any full-time general management. A few months after he joined the company, therefore, Moore assumed the position of general manager, becoming at the same time general manager for the Collins White Company, a position Fraser had also relinquished. Fraser's influence continued, however, and he remained as Chairman of Directors of NAP for a further eleven years. Moore's Collins White position finished when that group disbanded, at the same time as Glenormiston, a Collins White property, was sold to NAP in 1968. When Jerry Chambers retired in 1975, Moore also took on the role of pastoral inspector, which meant frequent trips to the properties.

Just as Michael Crouch's time as chairman spanned the years of two general managers — Moore and Ross Brunckhorst — so Ken Moore's career as general manager spanned the eras of two chairmen — Fraser and Crouch. From 1967 Michael Crouch was Chairman of Directors for Moore's remaining fourteen years as general manager; like the preceding Moore-Fraser combination, Crouch and Moore virtually ran the company. Ken Moore joined the board in 1980 and for some years after his retirement in 1982 he was a consultant to the company. He remained a board member until his death on 21 September 1990 at the age of 73.

Moore's contribution during his twenty-six years as general manager was considerable. He guided the company through the road transport era when NAP was one of the pioneers in this field. He was also there when the company embarked on its own tuberculosis and brucellosis eradication campaign, before the program was made compulsory throughout Australia. His knowledge of the meat export industry was a great advantage in negotiations for the sale of cattle. Another asset was his ability to cooperate and liaise with government departments, which helped enormously when both the Northern Territory and Queensland leases were successfully extended.

He could be firm, even ruthless in his decision making but he is also remembered for his jovial attitude towards life and his ability to get along with others. In 1983, a cancer in his throat resulted in a total laryngectomy (removal of the larynx) but he could still communicate orally and contributed at board level remaining interested in NAP affairs and the people in the company until his death. Those on the stations particularly welcomed his support and encouragement, a contrast with the often abrupt ways of Douglas Fraser. The company's accountant Gordon Jamieson (1956-1985) and Nancye Bechtel, his private secretary (1958-1967), assisted Moore during his term as general manager. Moore was not afraid to launch into the new areas the industry offered — with one exception. Never completely convinced that a change of breed from the Shorthorn was in the company's interests, he left it to his successor, Ross Brunckhorst, who introduced Brahman into the herd.

Henry Francis Foster was born in Hobart on 19 January 1936, the elder son of Francis and Patricia Foster. Following his secondary education in Hobart, he attended Roseworthy Agricultural College, South Australia, where he graduated with a Diploma of Agriculture. He subsequently completed a degree in Agricultural Science at the University of Melbourne. Following employment with the Victorian Department of Agriculture for six months, he worked on farms in various capacities from overseer to temporary manager. Since 1965 he has been owner-manager of his own property, Fosterville, at Campbell Town in Tasmania, where he runs Merino and Cormo sheep and Hereford cattle. He has been a member and office-bearer of numerous rural organisations, and has always shown a keen interest in adding to his knowledge of the industry.

With his obvious interest in agriculture, particularly in scientific, land care and environmental issues, the period of his association with NAP has been one of genuine interest and significant contribution from the time he became his father's alternate director in 1968. His involvement increased when he replaced his father on the board in 1971. People who have known both father and son have commented on the similarities between them, in appearance and manner. As early as 1949, following a holiday at Mundoolun with the Frasers, Douglas Fraser wrote in a letter to Patricia Foster, the mother of 13-year-old Henry:

> There is one thing certain that while he [Henry] is alive, Francis will never be dead. He amused my wife and myself very much in his extraordinary likeness to his father in many ways.[8]

While following mainly agricultural and pastoral pursuits, he was closely involved as the spokesman for the Foster family in the takeover period of 1984-85. His calm disposition and measured judgment of both people and events was a key factor in the final outcome.

William Francis Foster was born in Hobart on 21 December 1948, the youngest of the six Foster children. He holds a Master of Science degree in statistics from the University of Melbourne, and during further study in London completed a Master of Science in economics. William now lives in Melbourne where he has worked in various consulting and research capacities in statistics, computing and economics in both the public and private sectors and at the University of Melbourne. With his background in finance and investment matters, he was well placed to assist Henry during the takeover period. He returned from overseas in late 1984 just as events were unfolding, and since then has been closely involved in the affairs of the company, especially following his appointment to the NAP board in 1986.

It could be said that he and Henry complement each other perfectly and provide a strong continuum to their father's involvement in the company. While Henry provides the agricultural expertise and has a quiet, serious manner, William contributes with economic/investment interests

8 D.Fraser to Patricia Foster, 25 May 1949, D.M.Fraser File, A to F, 1946-1949, NAPR.

General Manager since 1982, Ross Brunckhorst, receiving an award on the company's behalf from the Governor of Queensland, Sir Walter Campbell, following NAP's contribution to the Last Great Cattle Drive, a bicentennial project undertaken by the Stockman's Hall of Fame, 1988.

and has a more outgoing personality. William sums up the family's philosophy:

> There is no doubt that over the years we could have made a lot more from putting our money into blue chip shares or Commonwealth Bonds . . . The primary sector has not been particularly profitable; people tend to be in it for other reasons . . . We have a real feeling for this company and our family's involvement in it . . . just the whole concept . . .[9]

Ross Murray Brunckhorst was born in Warwick on 5 April 1935. Following a rural upbringing, he worked for five years for Grazcos, a sheep-shearing and wool-classing company. After two years overseas, he was employed by the United Graziers' Association in Brisbane where he was responsible for the formation of the Service Department. His association with Sir William Gunn during this period led to his employment in 1965 with Gunn Rural Management. For five years, from 1972, he worked for Unibeef Australia Pty. Limited, a company in which Gunn formerly had interests. The company owned pastoral properties in Queensland and the Northern Territory, its parent company being Israeli-owned. This background and experience was of great benefit to NAP when Brunckhorst joined the company on 1 February 1977 as assistant general manager. He was promoted to general manager on 1 July 1982, when Ken Moore retired.

Within days of Ross joining NAP, Ken Moore commented to the board of the 'wise choice' that had been made.[10] Especially after Ross assumed the general manager's position, his personable qualities ensured that good relations were maintained among staff at head office and between the stations and head office. Although he sets high standards for himself which are reflected in his expectations of others in the company, his fairness and thoroughness have been his trademark.

Negotiating the renewal of the property leases has been a most important aspect of the company's history and, in the modern era, success in this area

9 William Foster Interview, Brisbane, 11 June 1991.

10 Ken Moore to Directors, 4 February 1977, File 4, Directors, 1977, NAPR.

No. 49 Bore and Turkey's Nest, Alexandria, 1991. Christine Taylor

can be attributed to the careful work done by Crouch and Moore. With the Alexandria and Channel property leases due to expire in 1980 and 1982 respectively, preliminary consultation with the relevant authorities began during the 1970s. There was cause for concern with Alexandria because the total area of 16,359 square kilometres[11] exceeded the limit of 5,000 square miles (13,000 square kilometres) as specified under Section 38A of the Northern Territory Crown Lands Act of 1946. While with ministerial approval an upper limit of 20,000 square kilometres could be held, self-government for the Territory in 1978 had altered the position, in that all decision making was subsequently confined to the authorities in Darwin. With the Northern Territory Land Board fiercely patriotic to the Territory, some saw NAP as an outsider. A thorough three-day inspection of Alexandria was made in August 1979, and official hearings took place both on Alexandria and in Darwin. As a result of the disease eradication program, the property was well improved and the company was seen as a worthy lessee. There was much relief when the company was informed that new leases — with no loss of area — had been granted for fifty years.[12]

The Channel property leases were also extended — for thirty years from 1982. That the company was worthy of such an extension was revealed in the remarks of the Land Commissioner at Longreach concerning the renewal of the Marion Downs leases:

> Although the company seems to play a low profile in the cattle industry they are a proud company and are proud of their achievements and the overall management and condition of their stations . . . Although these properties are in a low rainfall area and are very remote, they are very well improved and all improvements are maintained to a very high standard . . . In these arid areas I feel that it is important to have responsible lessees and this company has proved itself.[13]

11 E.M.Crouch to Marshall Perron, Minister for Lands and Housing, Darwin, 22 November 1978, K.W.Moore and R.M.Brunckhorst File, 1.6.78-22.12.78, NAPR.

12 E.M.Crouch, Chairman's Address, Annual General Meeting, 15 April 1981.

13 V.W.Rosel, Land Commissioner, (Boulia and Jundah Districts), to Secretary, Land Administration Commission, (Brisbane), 30 November 1981, Marion Downs File, PH 04/5350, Lands Department, Brisbane.

1974 Floods, Alexandria. Ron Spedding

Hay making from Flinders grass near No. 36 Bore, Alexandria, 1983.

Monkira and Coorabulka were left intact and while small areas were excised from other NAP properties, it had a negligible effect on their performance. Glenormiston lost a small area (312 square kilometres) of the Toko ranges and Marion lost 117 square kilometres at the north-east boundary.[14] Because of drought conditions in the 1980s, the Land Act Amendment Act of 1986 granted a further extension of twenty years to the leases, which are now due to expire in 2032.

Seasonal conditions are always the most important factor in the fortunes of any pastoral company. The drought in 1969-70 particularly affected Alexandria; only 116.3 mm (465 points) of rain fell in 1970, the lowest ever recorded. Because of the need for more bores, the contractor then working on the Channel properties was sent to Alexandria in 1970 to sink five more.[15] The drought on that station was the worst since 1952. Although some agistment was available to the north and considerable sums of money were spent on fodder and bores, some 15,000 head were lost on all NAP properties. Included in this figure were unbranded calves lost through heat exhaustion and cattle lost through gidyea poisoning, especially on Glenormiston.[16]

The drought broke in March 1971 and a complete reversal of the seasonal conditions took place. All the properties benefited and there was good general flooding in the Channel country which allowed the company to sell 12,000 head in good condition instead of at 'give-away' prices. Good seasons followed and in March 1973, pastoral inspector Harold Chambers commented:

> This is about the best season I have seen in the Channel Country. They should have a record branding and a lot of choice young fats to turn off.[17]

A bigger season was to follow. During the first four months of 1974, up to three years' rainfall fell on the properties.

1974 RAINFALL

	millimetres	inches
Alexandria	879.5	35.2
Marion Downs	424.6	17.0
Glenormiston	679.2	27.2
Monkira	719.4	28.8
Coorabulka	535.4	21.4

It was the wettest season on record. Not only was water one-third of a metre deep through the homestead at Gallipoli but Ken Moore and office staff had a busy time rescuing files and records as the flooded Brisbane River swept through head office. Fortunately, the company's top stud bull for the AI program on Islay Plains, Weebollabolla Colonel, was safe above the flooding at the Artificial Insemination Centre at Wacol.[18]

About 6,000 head, including 2,000 calves, were lost on the properties in the floods.[19] Whole herds were swept over boundary fences on to adjoining properties. The clean-up was massive; kilometres of fencing, including most of the dingo netting fences, were lost and many of the earth tanks and other improvements were severely damaged.[20] The abundance of feed, however, outweighed any disadvantages. On properties such as Coorabulka, the Mitchell grass that had practically disappeared was now back in great quantity. Ken Moore enthused about Marion Downs:

> At Marion there is a terrific response to Flinders and Button grass. These grasses are over a foot high and extend right up to the Herbert Downs homestead. The hills at the back of Marion which are normally just bare stony hills, are all covered in grass. It was a pleasure to talk to Bill [Alexander (the manager)] and hear him speak with such enthusiasm.[21]

While the 1977 wet on the Channels was generally very good, on Alexandria it eclipsed even the 1974 season. There the rainfall was a record 1056.45 millimetres (42.25 inches). Staff were evacuated from Soudan and water entered four of the cottages at Alexandria.[22] The floods delayed the start to the season.

Productivity is affected not only by the amount of rain but also its timing; as Ken Moore noted in 1976:

14 E.M. Crouch, Chairman's Address, Annual General Meeting, 17 March 1982; Marion Downs File, PH 04/5350, Gregory North, Part 2, Lands Department, Brisbane.

15 E.M.Crouch, Chairman's Address, Annual General Meeting, 21 October 1970.

16 E.M.Crouch, Chairman's Address, Annual General Meeting, 20 October 1971.

17 H.W.Chambers to General Manager, 11 June 1973, File 14, Pastoral Inspector, 1973, NAPR.

18 Minutes of Directors' Meeting, 20 February 1974.

19 Ken Moore to Chairman of Directors, 18 July 1974, Security File, NAPR.

20 E.M.Crouch, Chairman's Address, Annual General Meeting, 23 October 1974.

21 Ken Moore to Directors, 25 January 1974, 'A' File, Greens, 1.11.1973-31.3.1974, NAPR.

> McGuire at Alex feels that we might now have had too much rain on that property. Before the February rain the grass was well grown with very good body and a lot of nutriment. He feels that this extra February rain will bring a big body of rank feed which will not be much good to the cattle at the end of the year, and also bush fire risks at that time could be great.[23]

He was correct about the bush fires. Major fires during 1977 resulted in 60 per cent of Alexandria being burnt out.[24] During 1974, serious bushfires had also occurred on Alexandria, Glenormiston and Herbert Downs with an unfortunate loss of 200 breeders in a fire at Soudan.[25] Overall, however, the 1970s were seasonally very good; in the Directors' Annual Report of 1978, Michael Crouch drew attention to the 'seven successive good seasons'.[26]

These good seasons prompted the company to experiment with haymaking on Alexandria in 1983. Using Flinders grass, this was so successful that the production level was lifted to 6,000 bales in 1984. A thousand of these bales were used on the Channel properties for feeding weaners, cattle held in yards, and horses. It was not intended as a supplementary feed but rather for convenience while stock were being held.[27] But drought conditions during the rest of the 1980s on Alexandria put an end to any further baling of hay except for a period in early 1987.[28]

When he took over as general manager in 1982 Ross Brunckhorst was faced with continuing dry conditions, especially on the Channel properties, and record numbers of stock on the company's properties. The 1982 season was indeed one of the driest on record on the Channel properties. Monkira, Marion Downs, Coorabulka and Glenormiston were all declared drought properties by the Department of Primary Industries, making them eligible for freight assistance. The average rainfall over these properties that year was 70.5 millimetres (2.8 inches), the lowest since 1965, when the average had been 67.8 millimetres (2.7 inches). The annual long-term average is 175-200 millimetres (7-8 inches). In 1965, however, there had been a total of only 6,636 head on the properties compared with 30,592 in late 1982.[29]

Alexandria had reasonable rain in the 1981-82 wet season, (494 millimetres or 19.8 inches), so in 1982 an unprecedented 8,000 head were moved from the Channel properties to Alexandria. With the congestion of stock on Alexandria some 2,000 were agisted to nearby properties. It made the control and implementation of the brucellosis and tuberculosis eradication program very difficult. The chairman of directors referred to it as a 'game of draughts, constantly shifting mobs almost on a daily basis and ensuring that dirty cattle were not mixed with those declared clean'.[30]

He also explained the situation in the subsequent months:

> Until approximately a month ago [March 1983], the whole position was so desperate that it may well have been that we had ended up with practically no cattle at all on the Channels and with very few left on Alexandria, whether they had been sold (for what we could get) or sent for agistment (if we could find it) and although some agistment was available, the costs and conditions were almost prohibitive. As it turned out, we were saved from these eventualities by a matter of days only. Alexandria received approximately 8" and a general overall fall of 2" on the Channels.[31]

The year 1984 was reasonably secure for Alexandria when 507 millimetres (20.3 inches) of rain fell. It was 1985 when drought really set in on the property, the average rainfall from 1985 to 1990 being only 272.6 mm (10.9 inches) in an area whose long-term average is 400 millimetres (16 inches). New drought management practices were needed, as General Manager Ross Brunckhorst explained in 1985:

> In bygone years the practices would have been to let the stock forage and take their chance of survival but now days this would be an unacceptable approach having spent several millions of dollars on disease eradication.[32]

With fences preventing natural distribution of stock over the property, and a shortage of watering

22 Ken Moore to Directors, 25 January 1977, File 3, Rainfall, 1977, NAPR.

23 Ken Moore to Directors, 13 February 1976, File 4, Directors, 1976, NAPR.

24 E.M.Crouch, Chairman's Address, Annual General Meeting, 9 November 1977.

25 Ken Moore to Chairman of Directors, 18 July 1974, Security File, NAPR.

26 Directors' Annual Report, 30 June 1978.

27 E.M.Crouch, Chairman's Address, Annual General Meeting, 18 April 1984.

28 E.M.Crouch, Chairman's Address, Annual General Meeting, April 1987.

29 Directors' Annual Report, 30 November 1982.

30 E.M.Crouch, Chairman's Address, Annual General Meeting, 13 April 1983.

31 *ibid.*

32 Ross Brunckhorst, Report to Directors, 5/85, 14 May 1985, NAPR.

points in areas where pasture was available, four new bores were drilled and equipped, turkey's nests constructed at these bores, polythene pipes laid from selected bores, and a watering lane constructed from East Ranken into No. 72 bore.[33]

Nevertheless, in 1986, the company was forced to sell 10,000 of its breeder herd in order to reduce stocking to safer levels. With the exception of Monkira, dry conditions on the Channel properties meant that they could not accept any store stock from Alexandria. Furthermore, the disease regulations concerning the sale or transfer of stock from the Northern Territory into Queensland made it difficult to sell into store markets. It was a case of the company having too many cattle for the amount of available pasture. On Soudan in particular there was a build-up of over 20,000 head at a time when rain was needed urgently to provide sufficient feed. The company responded quickly with the purchases of Connemara in January 1986 and Kynuna two months later. A feedlot on the Darling Downs, Wainui, had been purchased in 1985. This enabled the substantial build-up to be dispersed with a minimum of loss:[34]

7,242 to Kynuna
2,086 to Connemara
5,942 to Monkira
5,337 to Wainui Feedlot

Despite the high turnoff from Alexandria in 1986 and despite its suffering one of its driest periods, 18,572 calves were branded during the year which increased the Alexandria herd to 66,364 by the end of 1986.[35] While the 1987 season showed some improvement, the seasons from 1988 to 1990 were very poor and by November 1989 stock numbers had been reduced on Alexandria to 33,815 head compared with a normal capacity of 60,000.[36] In 1989, in the biggest ever single sale from Alexandria, pastoralist and cattletrader John Dunnicliff purchased over 14,000 head.

Not until early 1991 did the property receive sufficient rain to recover and allow the return of stock from agistment. The Channel properties also benefited from the good season. 'The overall result is very satisfying',[37] wrote Ross Brunckhorst in February 1991, the first time in nearly a decade that there could be such optimism.

It is very reassuring for the company that the management practices undertaken during the drought, in particular destocking, have not gone unnoticed by the authorities. A 1991 report commented on Alexandria's low stock numbers as 'indicative of proficient management' and of 'benefit to range land condition'.[38] Such a run of poor seasons, however, resulted in the cattle eating poor quality and old feed which was lacking in nutrients.[39] Consequently, in 1983, feed supplements in the form of lick blocks were introduced on Alexandria for the first time since the 1960s, when molasses and urea were provided in licker drums. The use of the lick blocks increased in the late 1980s and helped not only in stock survival, but also in calving rates as the condition of the breeders was maintained. This contributed to record brandings on Alexandria in 1988 and 1989 — 25,383 and 24,227 head respectively.

Agistment was also arranged for Alexandria cattle in 1989[40] although it was more expensive than providing feed supplements.[41] Another response to drought was to wean calves very early, at weights as low as 50 kilograms liveweight. Calves as young as one month were weaned and placed in a calf feedlot where they remained until they reached 100 kilograms. This practice, together with the use of growth promotants, ensured a rapid increase in weight and an early turnoff.

New developments to assist management were also important in handling the drought conditions of the 1980s. Improvements associated with the BTEC program (discussed in Chapter 13), have been very important, while polythene pipes have been used extensively and successfully on most NAP properties since the mid-1980s to transfer water away from bores and other watering points to open up new grazing areas. The station managers have been helped by an increased number of bores, earth dams and desilted waterholes; better communication with more vehicles with two-way

33 *ibid.*
34 Directors' Annual Report, 30 November 1986.
35 Stock Returns, Alexandria; E.M.Crouch, Chairman's Address, Annual General Meeting, April 1987; The calving improved to a record branding in 1988 of 25,171.
36 Directors' Annual Report, 30 November 1989.
37 R.Brunckhorst, Report to Directors, 2/91, 12 February 1991, NAPR.
38 Inspection Report, Alexandria Station, 17-18 July 1991, Department of Lands and Housing, Darwin, Alexandria Lease File, NAPR.
39 Minutes of Directors' Meeting, 2 August 1985.
40 R.Brunckhorst, Report to Directors, 9/89, 5 October 1989, NAPR.
41 R.Brunckhorst, Report to Directors, 12/88, 7 December 1988, NAPR.

Fred Shephard (Manager of Boomarra) and Nick Murray (Manager of Kynuna) inspecting carcasses during an AUSMEAT Workshop, Brisbane, 1991.

radios; and mustering aids such as motorcycles, fixed wing aircraft, ultra-light aircraft and helicopters. From the late 1980s, there was a change in the configuration of station staff with the employment of jillaroos.

An important concern in cattle management continued to be disease prevention and cure. The recent major issue here has been the brucellosis and tuberculosis eradication program (BTEC). Among other diseases, the only significant recent outbreak of consequence of redwater from ticks for NAP occurred at Alexandria in the years 1974 to 1979 when there were stock losses. The 1974 wet was so big that it was not surprising there would be such an outbreak, but the company was not prepared for the extent of the tick infestation and the five years it continued. Ticks were detected initially at Gallipoli where a spray dip was used first and then a permanent dip constructed. A plunge dip was built at No.36 bore.[42] Although Gallipoli cattle were inoculated, by 1978 ticks had spread to the main lease areas of Alexandria. Plunge dips were installed in the home yards and at No.16 bore but, as reported in the Directors' Annual Report for that year, the mustering, dipping and inoculation of 22,000 head against redwater was a costly exercise.[43] A dip needed to be built at Soudan for cattle travelling interstate to the Channel properties.[44] While preventative measures helped, a series of dry seasons subsequently reduced the tick population.

Jack Maunder from Brisbane was the first veterinarian to work for NAP. Records show that he was at Alexandria in 1951, inoculating stud bulls against pleuro-pneumonia.[45] Previously, the pastoral inspectors — Bill Frith, followed by Harold Chambers — in conjunction with DPI officers, attended to any problems with the stock. Pleuro-pneumonia had been evident since the Alexandria leases were first taken up, but a concerted Australia-wide attack on the disease in the 1950s and 1960s resulted in Alexandria being given a 'pleuro-free' certificate for moving stock to the Channel properties — a movement vitally important to NAP's viability.

---

42 Minutes of Directors' Meetings, 15 May 1974, 20 November 1974, 19 February 1975, 20 October 1975, 26 November 1975.

43 Directors' Annual Report, 30 June 1978.

44 Directors' Annual Report, 30 June 1979.

45 D.M.Fraser to R.H.Murray, 1 May 1951, R.H.Murray File, 1951-1955, NAPR.

Botulism was also a recurring problem at Alexandria and increased to worrying proportions in the 1960s during Bill Young's time. Caused by a toxin produced by *Clostridium botulinium* growing in decomposed organic matter such as mouldy hay, decomposing grass and dead carcasses,[46] it was also thought on Alexandria to be linked to a deficiency of phosphorus and protein. Cattle which suffered from a phosphorus deficiency had a craving for bones and in an attempt to take in this mineral, they were poisoned by the toxin when chewing the bones of infected cattle carcasses. The appropriate management strategy involved burning carcasses and feeding animals a dry lick mixture of approximately equal proportions of bone meal, ground rock phosphate (from Christmas Island) and salt.[47] There was a period in 1967 when the lush growth of grass around the bores made it unsafe to burn carcasses — surely the only time a good season would prove to be a disadvantage.[48] Treatment was by vaccination with a serum which was considered to be effective for two years. At Alexandria there was a marked decrease in the incidence of botulism, due to the untiring efforts by Bill Young to inoculate against the disease. By 1969, 90 per cent of the herd had been vaccinated, and Young proudly stated in September of that year: 'If we had not jumped into it 13 years ago we might have been in the same position [as others] today'.[49] Botulism was not a problem on the Channel properties, so vaccination was confined to Alexandria.

Other less important factors also threatened the well-being of the cattle. Dingoes were a pest until baiting with 1080 poison was allowed in the Northern Territory from the early 1970s.[50] NAP had entered into an agreement in 1969 with the NT authorities for 1080 to be used on selected parts of Alexandria as part of an experimental program. In other areas of the NT, strychnine-based baiting had to be used even though 1080 had already been used successfully in Queensland, but fortunately the Territory ban on 1080 did not last for long.[51] Pigweed *(Portulaca filifolia)* caused deaths from time to time on Alexandria and in 1978 the DPI was concerned at a possible outbreak of bluetongue disease. The Alexandria herd tested negative on this score, however, which ensured that cattle movements were not upset.[52] Three-day sickness sometimes affected the herd, especially after a wet, but provided they were not moved, the stock generally overcame the disease with few losses. St George's disease caused some loss on Marion Downs in the 1950s, and on Coorabulka in the 1960s, while gidyea poisoning has always been a problem on Glenormiston and Herbert Downs.[53] The shrub, Parkinsonia *(Parkinsonia aculeata)*, which grows around the edges of turkey's nests, is also an ongoing problem.

In the mid-1970s, a crisis developed in beef prices and markets. Domestically, a heavy market influx of local cattle following an increase in Australia's herd size caused prices to drop sharply. Secondly, following high prices in the early 1970s, demand in two of Australia's major export beef markets, Japan and the United States, fell dramatically during 1974 after massive increases in the world oil price. Increased fuel costs were felt in Australia too, and compensation prices for tuberculosis and brucellosis-infected cattle in fact represented a better return for the local producer than domestic and export prices. As NAP's veterinarian Lee McNicholl commented, 'the best market was shooting them on Alexandria and putting them in a hole in the ground'.[54] Beyond that, relief was offered by the Department of Primary Industry (DPI), which ruled that if it were too expensive to test a whole herd, destocking orders would enable stock to be sold to meatworks and the company compensated for any loss.

In 1973 prices for NAP cattle averaged out at $48 per 100 pounds dressed weight, but in 1974 this dropped to $18 per 100 pounds, at a time when the break-even price was $27 per 100 pounds. With the prolonged meat strike of 1974, three-quarters of the killing season was lost[55] and it was only because of the extremely good relations that Ken Moore had built up with Borthwicks and Swifts over the years, that large numbers of NAP stock could be slaughtered during the shortened season. The sale of beef to Russia in February 1975 at unprofitable prices was accepted just to absorb surplus meat and

46 T.G.Hungerford, *Diseases of Livestock*, 8th revised edition, Sydney, McGraw Hill, 1975, p.248.

47 H.W.Chambers to Bill Young, 13 January 1964, File 1, Alexandria, 1967, NAPR.

48 Bill Young to Ken Moore, 22 January 1968, File 1, NAP Stations, 1968, NAPR.

49 Bill Young to Ken Moore, 22 September 1969, Correspondence File 1, NAP stations, 1969, NAPR.

50 Ken Moore to Manager, Glenormiston, 5 January 1971, File 1, NAP Stations, 1971, NAPR.

51 Minutes of Directors' Meeting, 19 July 1972.

52 Minutes of Directors' Meeting, 15 February 1978, 19 April 1978.

53 General Manager to Manager Coorabulka, 15 November 1967, File 1, Correspondence, Coorabulka, 1967, NAPR.

54 Lee McNicholl Interview, Dulacca, 16 July 1991.

55 E.M.Crouch, Chairman's Address, Annual General Meeting, 23 October 1974.

not until 1978 were directors encouraged by a slight upturn in cattle prices.[56]

Although NAP took steps to reduce costs, some factors were beyond the company's control — increases in land rental and local authority rates, increases in fuel prices because of the international oil crisis, and increased costs through inflation which was nearly 18 per cent in 1976-7.[57] Rises in labour costs were also spectacular, with the Station Hands' Award increasing 75 per cent through 1973 and 1974.[58] There were also improvements in award conditions for employees with a 44-hour hour week spread over five and a half days, overtime in excess of this, and substantial camping out allowances. Michael Crouch summarised for shareholders the downward movement in beef prices compared with the increase in wages between 1968 and 1975:[59]

Year to 30th June	Gross Sales $	Per Head $	Weekly Wages $
1968	1,789,554	104.81	37.00
1973	1,703,686	156.00	56.55
1974	2,969,645	206.74	68.15
1975 (a)	1,395,266	93.54	92.35
1975 (b)	1,029,730	61.63	102.47

(a) Twelve months to 30th June
(b) Six months to 3rd December

In 1974, despite a reduction in staff from the previous year, there was an increase in the total wage bill.[60] Total staff numbers fell from 157 in June 1974 to 108 in June 1976. With improved efficiency this trend has continued to the extent that in 1990, with a total staff of 149, there were only twenty-three more persons employed than in 1980 despite the company's owning five more properties.[61]

In rationalising staff numbers, the first to be put off were domestic staff and cowboys who worked around the station. Aborigines were no longer a cheap source of labour and mostly whites were employed. More was demanded of the station manager and his wife and their social life tended to decrease — they were simply too tired. Wives were required to cook for the staff and in 1972, for the first time, they received some payment.[62] Gradually more recognition was given to their role but their contribution was generally underestimated. It took one of the English shareholders in 1978 to highlight this aspect:

> I think that we place a very considerable, largely unpaid, burden on the wives of Station managers who are apt to have to act as nurse, deputy manager and general confidant when they may be qualified in none of these respects. They also have the problem of educating their children and being separated from them later on for long periods. I believe that we should make special efforts to keep them happy and show them that the company is aware of their importance.[63]

Indeed, as the correspondence reveals, wives were significant when station manager positions were being filled:

> (1947)
> I don't think he drinks, and he has been very satisfactory at Herbert, and I like his wife from what little I have seen of her.[64]
> (1970)
> His wife is a fairly good cook, is a bush girl and never seems to complain about what she has got or is given.[65]

But the 1980s have brought different demands. While some station wives are quite content with station life, difficulties can arise if they want to pursue a career of their own. The company has improved the standard of living, with the provision in the late 1980s of television and airconditioning in the homesteads. The facilities of STD telephone and facsimile machines are now available; yet the factor of isolation remains.

The 1970s were proving to be financially tough times for the company. Despite economies such as reductions in staff numbers, the overdraft with the bank had reached a point where it became necessary to sell part of the company's share portfolio.[66] In 1975, for the first time since 1952, no dividend was paid to shareholders. Although there had been years

56 Directors' Annual Report, 30 June 1978.
57 E.M.Crouch, Chairman's Address, Annual General Meeting, 9 November 1977.
58 E.M.Crouch, Report to Shareholders, 24 June 1975, E.M.Crouch Papers.
59 E.M.Crouch to Shareholders, 16 June 1976, E.M.Crouch Papers.
60 E.M.Crouch, Chairman's Address, Annual General Meeting, 23 October 1974.
61 E.M.Crouch, Chairman's Address, Annual General Meeting, 26 April 1990.
62 Ken Moore to all station managers, 15 May 1972, File 1, Glenormiston, 1972, NAPR.
63 Michael Ingram to E.M.Crouch, 22 May 1978, E.M.Crouch Papers.
64 D.M.Fraser to R.Murray, 16 January 1947, Alexandria File, 1944-1947, NAPR.
65 Ken McGuire to General Manager, 14 October 1970, File 1, Correspondence, Marion Downs, 1970, NAPR.
66 E.M.Crouch to Shareholders, 16 June 1977, E.M.Crouch Papers.

previously in which dividends had not been paid, these were confined essentially to years prior to 1940 and there was some sense of shock now. Losses continued to 1978, but apart from 1975, dividends were paid and a total of $588,600 was distributed to shareholders, albeit at lower rates than previously.[67] Dividends had been particularly high in the 1960s, however, and Michael Crouch reminded shareholders in 1975 that between the years 1939 and 1974 a total amount of nearly $4 million had been paid back to shareholders on an original paid up capital of $324,000.[68]

In August 1974, however, the situation was so serious that a meeting was held at Alexandria between the board, station managers, the company's bankers and a member of the Meat Exporters Board. They resolved to do what had, in fact, always been done in similar tight situations — breed the company out of trouble. They discussed ways of increasing the turnoff, of improving the quality of the herd and of the tuberculosis and brucellosis eradication program.

Labour-saving herd management practices associated with the eradication program helped enormously. Portable steel yards, road transportation of livestock within the property, the use of light vehicles by fewer pumpers to service the bores, and increased numbers of yards and loading facilities, all helped not only to keep staff numbers down but to increase efficiency.

Accounting procedures were rationalised in 1979 with the formation of the NAPCO cattle partnership, so that tax losses and profits could be averaged over all the subsidiary companies. There was also a change in the company's accounting year so it would finish in November of each year rather than June. This resulted in annual general meetings being held in the first half of the year — usually April — rather than after June.[69] In 1991 there was a further change in the accounting year and it now ends on 31st December.

Computerisation followed in the 1980s. In December 1983 the board authorised the general manager to investigate the use of a computerised accounting system.[70] The person who was primarily involved in this task was chartered accountant David Brown, who had joined NAP in 1969. He left the company in 1975 but rejoined two years later as the company's secretary. David has been assisted in the development and implementation of new computer systems by computer programmer Ken Hearder, who joined the company on a full-time basis in 1989.

Following the financial pressures of the mid-1970s, cattle prices markedly improved late in that decade, a time which also saw Korea emerging as an importer of Australian beef. Such improvements were destabilised in 1981, however, when the beef substitution scandal (in which kangaroo meat was sold as Australian export beef) created an atmosphere of doubt and mistrust in the minds of overseas buyers. Beef prices fell in the early 1980s, but by the mid-decade with the help of a drought, Australia's cattle numbers had dropped to more manageable levels — from around 34,000,000 in 1976 to 23,000,000 — and prices for NAP cattle gradually increased as the following table shows:

Sale Price: NAP cattle: Cents per kilogram[71]

	1980	1981	1982	1983	1984	1985	1986	1987	1988	1989	1990
Steers	141	119	117	143	164	170	174	205	226	233	221
Cows	119	110	95	121	131	133	108	168	182	196	193

Marked increases in meat consumption in Japan and Korea have effectively forced prices upwards. With Australia's home consumption continuing to fall, these expanding overseas markets have been very important. As Australia's external markets have increased and diversified, NAP has obviously benefited along with the rest of the industry.

Effective and innovative marketing has been the key to success in the new Asian markets where a premium price will be paid for a better product. While Australia's beef producers have been slow to respond to specific consumer demands, the Australian Meat and Livestock Corporation has introduced a 'uniform national description language' through the Aus-Meat system to help pursue these markets. In practical terms, this provides a set of broad stock descriptions: carcass weight ranges, fat scores, standard definitions and, where relevant, a system of bruising and conformation scoring. The livestock producer is paid according to the yield and quality of the meat

67 E.M.Crouch, Chairman's Address, Annual General Meeting, 18 October 1978.

68 E.M.Crouch, Chairman's Address, Annual General Meeting, 21 October 1975.

69 Thus there was no annual meeting held between October 1979 and April 1981.

70 Minutes of Director's Meeting, 14 December 1983.

71 Directors' Annual Report, 30 November 1990.

HEAD OFFICE STAFF, BRISBANE – AUGUST 1991

Left to Right: Ken Hearder (Data Processing Manager); Joan Heymer (Chief Accounts Clerk); Ron Spedding (Purchasing Officer); David Brown (Company Secretary); Lisa Luppi (Office Secretary); Steve Millard (Projects Officer); Val Martin (Accounts); Joan Weller (Accounts); Ross Brunckhorst (General Manager); Barbara Beswick (GM's Secretary).

produced, and the stock is classified in a particular category for a particular market.[72]

General Manager Ross Brunckhorst, who is directly involved with the sale of the company's stock, summarises the company's position:

> NAP has already recognised the importance of producing lean beef with the gradual infusion of the Alexandria breeding herd with Bos indicus blood . . . The current breeding programme is heading in the right direction to produce slaughter animals ideally suited to the Aus-Meat carcase requirements. In recent years half of the red meat produced in Australia has been sold on the local market, but the most lucrative market for producers is the chilled beef trade with Japan. NAP has the ability to produce slaughter animals to meet both the local trade and export markets, particularly the chilled ox trade with Japan. Every effort should be made to ensure that the maximum number of turnoff cattle meet the Aus-Meat high yield and high quality indicators.[73]

Despite the difficult seasonal conditions, branding and turnoff rates have generally increased in recent years. Losses associated with the BTEC program at one point caused a drop in herd numbers as breeders, often pregnant, had to be killed; at best, they were spayed and sold. Branding rates, however, continued to increase. The spaying of poor quality heifer calves at the time of weaning ensured that only superior animals were left for breeding. Furthermore, the disposal of older breeders, the introduction of maiden heifers into the herd, earlier weaning and supplementary feeding all helped to create an environment where calving rates increased. By the

72 R.Brunckhorst, Report to Directors, 7/87, 15 July 1987, NAPR.

73 *ibid.*

late 1980s, calving rates were over 80 per cent in some groups and it could be said that the company was again breeding its way out of its problems. Another factor, a change of breed, was also becoming significant.

In mid-1976, when ticks had been a real problem particularly in the Gallipoli area, Lee McNicholl suggested that Sahiwal bulls, because of their tick resistance, be introduced in the East Gallipoli area. His reasoning for this was the gradual development of a tick-resistant herd.[74] The board did not take up his suggestion and continued with the exclusively Shorthorn policy. The only concession, passed at a board meeting twelve months later in 1977, was that

> every effort should be made to select sleek coated Shorthorn bulls from within the Company's herd or by acquisition from other breeders in order to breed animals more suited to tick infested areas.[75]

Since the time Alexandria was declared stocked in 1884, the station had always run a pure Shorthorn herd. Almost 100 years later, in December 1982, this changed when the first bulls of the *Bos indicus* breed, Brahman, were introduced into the herd. Such an introduction did not occur without debate. Older board members and managers felt that the Shorthorn had served the company so well for so long there was no reason to change, but younger members did not have the same loyalties. The introduction of *Bos indicus* breeds was known to produce animals which were more resistant to heat and ticks, which had improved calving rates, and which possessed an ability to survive the stresses of drought. In addition, the progeny produced lean meat which was more acceptable to market demands.

Other Barkly stations had made the change to *Bos indicus* breeds as early as the 1960s. For instance, Santa Gertrudis cattle had been introduced at Brunette, Avon and Alroy. The first official evidence of NAP's change of policy was in the minutes of a board meeting on 23 September 1982 in which it was unanimously agreed that *Bos indicus* bulls be introduced into a segregated portion of the Alexandria herd. Sufficient Brahman bulls were purchased to effect a 3.5 per cent bull ratio in the East Gallipoli breeding herd of approximately 3,000 breeders.[76] The results were encouraging, so more Brahman bulls were purchased. The progeny were referred to as the F1 group. Santa Gertrudis bulls, also a *Bos indicus* genotype, were purchased in 1983-84 to be mated with the Brahman cross heifer progeny.[77]

In 1984, the first full year for Brahman cross calves, the branding percentage in this group was an excellent 77.9 per cent.[78] The company could clearly see the benefits of cross breeding, and continued with the program. By 1987, a total of 20,430 Shorthorn cows had been joined to Brahman bulls.[79] The abattoir feedback sheets gave encouraging slaughter results for the F1 bullocks, and animals suitable for the Japanese market could be turned off grass younger at between 2.5 and 3.5 years of age.[80]

While the cross breeding of *Bos taurus* (Shorthorn) and *Bos indicus* (Brahman) breeds increased productivity markedly, the company was aware that subsequent generations of the crossbreed would lose the benefit of the hybrid vigour of the initial cross. To address this problem, NAP sought veterinary advice from the DPI, the CSIRO, the Gatton Agricultural College and other producers. As a result, in 1988 a new cattle-breeding program was implemented to upgrade the herd to an eventual composite breed involving Brahman, Africander, Shorthorn, Hereford and Charolais genes[81] where many of the initial advantages of the changes would be retained. Just as it will be some time before the initial Brahman cross is complete, so it will be many more years before the composite breed replaces the entire Alexandria herd. This new program has required close attention by Alexandria staff in the selection and segregation of stock, with Max Ferris particularly involved with the program on Alexandria until his transfer in 1991 to manage Connemara.

The Channel properties maintained an all-Shorthorn breeding herd until 1989, when Brahman bulls were introduced. The impetus came during the previous year when consignments from Coorabulka had been downgraded because the butts of the Shorthorn carcasses had become

74 Minutes of Directors' Meeting, 21 July 1976.

75 Minutes of Directors' Meeting, 15 June 1977.

76 Minutes of Directors' Meeting, 23 September 1982.

77 E.M.Crouch, Chairman's Address, Annual General Meeting, 18 April 1984.

78 E.M.Crouch, Chairman's Address, Annual General Meeting, 17 April 1985.

79 Directors' Annual Report, 30 November 1987.

80 Directors' Annual Report, 30 November 1990.

81 Minutes of Directors' Meeting, 23 March 1988.

very uneven. Meanwhile, the *Bos indicus* cross cattle were fulfilling market requirements.[82]

In 1988, in keeping with the company's support of genetic developments in the industry, NAP became a member of the Boran and Tuli Producer Consortium, which combined with CSIRO to bring two breeds of East African cattle into Australia — the Boran and Tuli.[83] The venture is designed to inject the benefits of the first direct African imports of tropically adapted cattle into Australia to provide a previously untapped gene pool. As part of the evaluation of these breeds, 206 of NAP's F1 heifers were inseminated with semen from the Boran and Tuli breeds in 1990. A control group of F1 heifers were mated naturally with Belmont Red bulls and paddocked with the inseminated heifers at Wainui.[84]

NAP's position as a producer improved in April 1984 when the company was advised of its successful application to the Land Board of the Northern Territory to convert the Alexandria leases to Perpetual Pastoral Leases. This was the first successful application in the Northern Territory. One emerging factor, however, was clouding the future tenure of Alexandria — Aboriginal Land Rights Claims.

The Aboriginal Land Rights (NT) Act of 1976 gave Aborigines an avenue to make traditional land claims. In 1983, prior to the granting of the perpetual leases, the company received a demand from the Federal Minister for Aboriginal Affairs that an area be excised for an Aboriginal community on Alexandria.[85] Land offered by the company was rejected — the area between River Paddock and the Mittiebah boundary and also land in the vicinity of No.47 bore, north-west of Alexandria homestead.[86] The issue lay dormant for some years until 1987 when there was a claim for 500 square kilometres around Lorne Creek in addition to stock routes and reserves.[87] This time, the issue did not go away. By 1990, four scheduled areas could pass legally to the Aborigines under the Aboriginal Land Rights Act (NT) Amendment Act 1989:

1. Lorne Creek Locality
2. No 47 Bore Locality
3. Ranken River Locality
4. Soudan Locality

The company's offer to negotiate alternative living areas was ignored.[88] Discussions are continuing, on issues such as continued use of stock watering facilities within the areas and mustering rights to retrieve stock.

The composition of the board has changed considerably during the modern era. At the beginning of the 1970s the directors were:

Michael Crouch (Chairman)
Francis Foster
Charles Bruxner
Bill Alexander
Bill Fraser
Sir Herbert Ingram

Francis Foster resigned in 1971 because of advancing years, his position being taken by his elder son Henry. He remained as Henry's alternate. Harold Chambers joined the board in 1971 making seven directors at that time. Charles Bruxner, a director since 1957, died in 1976. He and his father had served a total of forty years on the board. He was replaced by Bill Norton, a Brisbane branch manager of Borthwicks, Chairman of Queensland Meat Exporters Association and later a director of Stanbroke Pastoral Company. Bill Alexander, a McIlwraith descendant, who had been a board member since 1934, retired in 1976 after a record forty-two years as director, and was replaced by his son David. With such a breadth of knowledge of the company, Bill Alexander had researched and written an unpublished history of the company from its inception to 1967. As the chairman of directors said at the annual general meeting following Alexander's retirement in 1976:

> His contribution to the success of the Company has been significantly important and of great value. He loved and still loves this Company and gave it his all.[89]

He remained an alternate director for his son until his death on 22 September 1978. Further changes to the NAP board followed in the 1980s, particularly after the takeover period.

82 Minutes of Directors' Meeting, 28 September 1988.

83 Minutes of Directors' Meeting, 17 September 1987; Directors' Annual Report, 30 November 1988.

84 Minutes of Directors' Meeting, 17 August 1988, 28 September 1988; R.Brunckhorst, Report to the board, 2/91, 12 February 1991, NAPR.

85 Directors' Annual Report, 30 November 1983.

86 Minutes of Directors' Meetings, 21 September 1983; 23 November 1983; 29 February 1984.

87 Minutes of Directors' Meeting, 17 September 1987.

88 Minutes of Directors' Meeting, 7 August 1990.

89 E.M.Crouch, Chairman's Address, Annual General Meeting, 27 October 1976.

In 1982 Brahman bulls were introduced into the Shorthorn herd on Alexandria.

The young bull in the foreground is a one-half Brahman, one-quarter Charolais and one-quarter Shorthorn crossbred photographed at Kynuna Station, 1991.

CHAPTER 12

The Properties: Old and New

And they'll just keep going westward; Till they think they can't go on
But their pioneering spirits; Will keep 'em coming on [1]

Alexandria Station, 1991. Philip Myers

After three decades in which NAP held only four properties — Alexandria and the three Channel properties which had been acquired in the 1930s — since 1968 a further seven have been purchased (and one, Islay Plains, sold). These properties will be discussed in order of their acquisition by NAP.

ALEXANDRIA, the largest and oldest property, has always been the subject of special attention within the company. Certainly this was the case in the early 1970s with the rebuilding of the station complex and the introduction of the disease eradication program. A southern Shorthorn expert said of the Alexandria herd in 1968:

> The cattle we saw on Alexandria, in regard to bone, length, depth, thickness and conformation, were

1 From the poem 'Birdsville '89' by the Boulia Balladeer, Jamie MacFarlane.

> amongst the best cows I have seen. Their tail settings were well nigh perfect, beautiful mossy coats with a length of hair that could not be bettered.[2]

The same glowing compliments could not be paid to the buildings. The homestead was nearly sixty years old, and some of the station buildings were beyond repair. With the good season in 1971, 'the best season for fifteen years', said Bill Young,[3] the company was ready to spend money on improvements. The result was the pulling down of many buildings at Alexandria and the complete rebuilding of the station complex. This activity coincided with a change of manager when Bill Young left at the beginning of 1972. Ken McGuire took up duty on 1 January 1972 and construction work began in that year.

McGuire had been at Marion Downs for sixteen years and his elevation was in line with company policy of promoting from within the company wherever possible. He enthusiastically supervised the rebuilding program, checking regularly the contractor's progress. McGuire's wife Lorna went to Brisbane to select furniture and furnishings and by October 1972 the family had taken up residence in the new homestead. By the end of 1972 also, men's and jackaroos' quarters had been completed, together with secondary buildings such as the ablution areas, the canteen and office building, the medical centre, the married men's cottages, the store, the kitchen and dining room, the workshop and the lighting plant. A tennis court was completed in 1973. Over three years, the program cost $275,000. This was in addition to the extensive improvements from the mid-1970s associated with the brucellosis and tuberculosis eradication campaign (BTEC), including fences, crushes, bores, holding yards and portable yards, as well as improved amenities in the stock camps such as mobile kitchens.

While new buildings are a definite aid to management, suitable station personnel hold the key to the smooth running of the property. One vital assistant to the manager is the bookkeeper. James Broadbridge had been invaluable during Holt's and Johnston's time; as were Jack Owen, and Jack Palmer during Murray's time. Prince was employed as bookkeeper in 1947 but his strong Yorkshire accent was quite unintelligible over the radio transceiver. 'An important thing in a bookkeeper here is one with a good voice able to use the Transceiver', said Harold Chambers.[4]

McGuire was assisted by a capable bookkeeper Ron Spedding, or 'Bookie', who was appointed in 1972. Ken Moore wrote to Ken McGuire, noting that Spedding was

> combining many duties beyond that of bookkeeper. He fits in well with all concerned and you consider him a key man, with which I whole-heartedly agree.[5]

Spedding left NAP in 1977, but rejoined the company in 1982 after Ted Cottrell's retirement, as purchasing office at the Brisbane depot at Oxley. The depot[6] was transferred to leased premises at Rocklea in 1988 and Spedding retired in 1991.

Peter Lowe commenced as a bookkeeper on Alexandria in 1977 and stayed until 1989. He was one of the more successful of the long run of office workers in the field whose role has been to keep local management running smoothly. Former manager Bill Young had the final word on bookkeepers: 'I have given up as to what is the best, Single, Married or Female', he said in 1966.[7] Such a statement can only be eclipsed by Betty Murray's minimal requirements for a governess at Alexandria in 1947:

> I have a new governess - morals A1 - looks fair - intelligence keen - sense of humour good - age 22 - so wish me luck. . . She seems to be managing the boys and that is all that worries me.[8]

One of the memorable characters on Alexandria during the 1970s was James Robert Flood — 'Floodie'.[9] He had been a camp cook on Marion before being appointed pumper at No.36 Bore. He is remembered as a great storyteller and a rascal; all those who knew him can remember at least one instance where they had a 'falling out' with him. Staff used to receive poison-pen letters from him, and on at least one occasion he went to the union over a wage dispute. He referred to No.36 bore as the 'LR2 Ranch in the Heart of the Adder

2 W.Stuart, quoted by E.M.Crouch, Chairman's Address, Annual General Meeting, 14 November 1968.

3 W.E.Young to K.W.Moore, 20 August 1971, File 1, Correspondence, Alexandria, 1971, NAPR.

4 H.W.Chambers to D.M.Fraser, 5 October 1947, Fraser and Chambers File, April 1945-October 1949, NAPR.

5 K. Moore to Ken McGuire, 4 June 1973, File 1, Correspondence, Alexandria and Glenormiston, 1973, NAPR.

6 The Oxley Depot was sold in 1988 and the land used for further expansion of the Forest Place Retirement Village.

7 W.E.Young to Ken Moore, 7 October 1966, File 1, Correspondence, Alexandria, 1966, NAPR.

8 Betty Murray, Alexandria Station to Harold Chambers, 8 February 1947, Alexandria Correspondence File, 1947, NAPR.

9 He died August 1987 and is buried at Mount Isa.

Country', LR2 being the model of the Lister engines used on the bores. He was very proud of his vegetable garden and his goats which he called his 'dancing girls'.

There were many excellent Aboriginal stockmen in the industry, but from the 1970s with the award of equal wages, Aborigines played a progressively smaller part in pastoral operations. The influence of alcohol, as well as what some saw as interference from politicians and unions, caused an exodus of Aborigines from the stations to the towns, reserves and missions. The attraction of the cattle station as a place for food, clothing and shelter, and for medical attention, was no longer present, so by the 1980s Alexandria had few Aboriginal employees.

After eighteen years of discussion, delay and frustration, an important public amenity — a school — was provided by the Northern Territory Administration[10] in 1972 on Alexandria. Northern Territory station-owners had resisted earlier moves to establish schools for Aboriginal children arguing that it was not in the Aborigines' best interests. Douglas Fraser commented in 1954 that he was

> only too willing to educate the boys to become good stockmen, pumpers and motor drivers, and the girls to become good cooks, seamstresses, and versed in domestic duties, but to teach them to read when they will then become experienced in everything that is undesirable, is the last thing that we should do for them in my opinion.[11]

Alexandria's manager Rowand Murray cited the example of Yuendumu, a government mission in the Territory's centre where Aborigines were being educated. When employed on Soudan and Gallipoli he found that Aborigines there gave trouble and did little work:

> I have decided to get rid of them and will in future employ blacks who are amenable to reason and who have not been unbalanced by a system of education.[12]

The years dragged on but the Administration persisted and gradually NAP acceded to the idea of having a school. In December 1963, there was a partial surrender of Pastoral Lease 427 to provide an appropriate area,[13] the company declaring their willingness to allow the NT Administration 'to build whatever school and accommodation they desired'.[14] But Douglas Fraser added:

> Other than providing an area of land which you and I have both inspected for the purpose of this school, we are not prepared to offer any other co-operation.[15]

A school at the Ranken, for Alexandria as well as other surrounding properties, was considered by the Administration in 1964 but nothing eventuated.[16] Finally, eight years later, a mobile school with accommodation for a teacher was established at Alexandria. Up to twenty-three children attended, including children of white and black employees. The immediate success of the school was due in no small way to the efforts of the first teacher, Kath Featherstone, and the company saw increasingly the benefits of having a school for the children of white as well as black employees. Its presence was one way to encourage married employees to remain on the property without having to worry about their children's education. As Michael Crouch explained to shareholders in 1978:

> The problem of education is paramount. Most parents will place the future of their children above their own welfare and advancement in the Company.[17]

To help cover the cost of secondary education at boarding schools, a bursary scheme of financial support was established by the company in 1972, and many station families continue to benefit from it.

Communication gradually improved on Alexandria with the purchase of four portable wirelesses in 1969,[18] which enabled staff, particularly the manager, to keep in contact while out on the run. Although cumbersome, they were also purchased for the Channel properties. From 1973 there were three radio bands available for communication on Alexandria, with a further two available in the mid-1970s. In 1985, a digital radio

---

10 Until self-government in 1978, the Northern Territory was administered by the Federal Government under the auspices of the Northern Territory Administration.

11 D.M.Fraser to R.H.Murray, 26 August 1954, R.H.Murray File, 1954, NAPR.

12 R.H.Murray to D.M.Fraser, 4 September 1954, R.H.Murray File, 1951-1955, NAPR.

13 Meeting of Directors, 19 December 1963.

14 Meeting of Directors, 22 August 1963.

15 D.M.Fraser to H.C.Geise, Director of Social Welfare, 1 February 1966, Green Copies, Letter File, November 1965-February 1966, NAPR.

16 Ken Moore to Bill Young, 19 October 1964, Green Copies, Marion Downs Pty.Ltd., October-December 1964.

17 E.M.Crouch, Chairman's Address, Annual General Meeting, 18 October 1978.

18 Meeting of Directors, 19 February 1969; Bill Young to Ken Moore, 28 January 1969, Correspondence File 1, NAP stations, 1969, NAPR.

Soudan out-station, Alexandria, with frontage to the Barkly Highway.

Gallipoli out-station, Alexandria.

Alexandria School, 1990. Standing Left to Right: Anne-Maree Lonsdale, Fiona Ferris, Wendy Ferris, Marcella Rankine. Sitting Left to Right: Waylon Birt, Damian Rankine, Heather Birt, Rowan Rankine, Scott Birt, Waylon Rankine. Teacher: Darryl Orr.

Mount Isa Staff: Chris Lilford and Sandra Garland with 'EFI'. Mount Isa is central to the company's western activities with company transport and personnel constantly passing through the city.

concentration system was connected at Alexandria which provided an STD telephone service.[19]

The transport of goods and personnel from Brisbane, and between stations, was adversely affected in 1973 when the Federal Labor Government cut existing subsidies on far-western routes for mail services and air freight. A year later the fuel subsidy was withdrawn and regular air services to the Channel properties were severely curtailed. The resulting disruption of links with head office prompted the company to purchase its own aircraft, a Cessna 206, in April 1974. Based in Mount Isa, with the call sign VH-EFI, and affectionately known as 'Effie', it is still in use.

The purchase was wise. The aircraft was immediately used as a courier service for mail, spare parts and perishable foods. It was also used for spotting cattle during mustering and bush fires, and it provided an impetus to improve communications, with the result that two-way radios were more widely used in station vehicles. The company has had a number of pilots, including Phillip Packer (1974-75), Peter Lehmann (1975-78), Greg Blackmore (1976), John Murphy (1978), Noel Stanley (1978-79), Casey Herman (1981-83), Ian Bean (1979-83) and Chris Lilford from 1983. In 1976 a pilot's residence/depot was purchased at Beta Street, Mount Isa. As the workload with the company's Mount Isa operations increased, Sandra Garland was employed in December 1980 and a centralised office was leased. In 1982 a second aircraft, a Cessna 182, was purchased for use on Alexandria. Company pilot Chris Lilford also manages the Mount Isa operations, a most important role within NAP. The company purchased new premises for a permanent office in Mount Isa in 1991.

Ken McGuire resigned from Alexandria in 1976 because of ill health and was replaced by George MacQueen from Monkira. MacQueen's five years as manager coincided with a continuation of the disease eradication program. Although he started work in a good season, he had to cope with bushfire problems in mid-1976, and severe flooding early the following year, which caused damage to fencing and the boxing of clean and dirty cattle from the disease program. MacQueen left in 1981, during a dry period on Alexandria when only 165.6 millimetres (6.6 inches) of rain fell in 1980 — but worse seasons were to follow in the 1980s.

The difficulties of the next decade were negotiated with skill and dedication by John William Ohlsen, 'Long John', Alexandria's manager from 1981 to 1991. Born in Brisbane on 10 February 1940 and educated at The Slade School, Warwick, Ohlsen joined NAP as a jackaroo at Gallipoli in 1957. His potential was noticed, and Jerry Chambers commented in 1961 that Ohlsen was a 'good worker and reliable'.[20] After two years working with the studs at Alexandria head station (1959 to 1961), he was appointed to Gallipoli as manager. His time on Gallipoli and Alexandria between 1957 and 1970 almost mirrored Bill Young's time as manager on Alexandria (1956 to 1971). It was no coincidence, therefore, that Ohlsen's style, when he became a manager, was similar to Bill Young's. Ohlsen left Gallipoli in 1971 to manage Islay Plains after it was acquired by NAP, returning to Alexandria ten years later.

Managing such a huge property as Alexandria, even in reasonable seasons, was a challenge, but the drought years of the 1980s created enormous logistical problems. To John Ohlsen's credit, during his period as manager, the calving rates increased, there was a successful outcome to the eradication of tuberculosis and brucellosis, and the company introduced Brahman bulls into the Shorthorn herd. He retired in 1991 after thirty-four years with the company. His replacement in early 1991 was 40-year-old Ross Peatling who had previously managed Wondoola Station (1980-87), Mount Bundy and Elsey stations (1988), Wallaroo Station, Injune (1990) and had been a livestock manager for Twynan Pastoral Company in 1989.

The management of MARION DOWNS has been stable throughout NAP ownership with only five managers in 57 years. William Gordon Shiels Alexander, the manager since 1971, was born in Scotland on 18 June 1941 and came to Australia as a child. He was educated at the Church of England Grammar School, Brisbane, and started work in 1958 as a jackaroo on Eulolo Station. He subsequently worked on Strathfield Station, another Collins White property, and Eulolo again before taking up the management of Coorabulka for NAP in 1964. Not only was Bill Alexander one of the youngest managers appointed by NAP, but he was the first NAP manager to obtain a pilot's licence. It was Alexander who introduced the use of fixed wing aircraft in cattle work. He purchased his own aircraft, a Cessna 150, and as manager of

19 Minutes of Directors' Meeting, 15 May 1985.

20 H.W. Chambers to Chairman of Directors, 4 January 1961, Green Copies, Marion Downs Pty. Ltd., December 1960-February 1961, NAPR.

Marion Downs, has successfully used it there. The company purchased the aircraft from him in 1978.[21] Like other NAP managers before him, Bill has served a term (1976-79) on the Boulia Shire Council.

He married Rhondda, Sam Hill's daughter, in 1965. Of their first home, Coorabulka, Rhondda described the homestead:

> There were holes in the walls . . . Tommy Groom didn't want anyone to look into the house so he painted the louvres dark, dark green and by the time I got there the paint had started to flake off. . . The floors had warped . . . The kitchen had a wood stove sitting against one wall; the sink was sitting against the other wall and the outlet pipe went through the wall . . . into a flour drum outside . . . The floor boards were never together . . . [they] were like a pallet.[22]

The Marion Downs homestead, however, had all the amenities of a comfortable station house at that time.

Marion Downs had generally suffered poor seasons in the 1960s. As Ken McGuire said in 1971 as he was leaving:

> I trust my successor has many more good years on Marion than I have had, three out of sixteen is a poor average in anyones language.[23]

Seasonal conditions certainly improved in the 1970s. The 1974 flood at a height of 7.42 metres was the highest on record.[24]

While the early management of Marion's out-station, Herbert Downs, was synonymous with the name of former manager Sam Hill, there have been several managers there in recent years. Don Smith was manager from 1980 to 1982; Peter McLaren for the following seven years; and the present manager, Bob Kirk, from late 1990. Peter McLaren commenced with NAP in 1974 at Alexandria, and following experience as a grader/truck driver and a short period away from the company, he returned in 1982 to manage Herbert Downs. His stay at Coorabulka in 1990 was cut short by the untimely death of Jamie MacFarlane, whom he succeeded as manager of Monkira.

MONKIRA has had only six managers since it merged with NAP in 1939. George MacQueen followed Bob Gunther as manager in 1961. MacQueen was born on 16 February 1931 at Casino. He was known in the company as 'Slippers' after his feet were badly burnt in his first year at Monkira when the lighting plant exploded. MacQueen had worked in the Ord River district of Western Australia, and also on Strathmore, Iffley, Wondoola and Glenore stations, before joining NAP in 1956 at the time Alexandria was managed by his stepfather, Bill Young. He was appointed manager at Gallipoli in 1958 and in 1961 became manager of Monkira where he stayed for fifteen years before returning to Alexandria as manager. He retired in 1981 after twenty-five years with the company.

He was at Monkira during the dry seasons of the 1960s when dust storms were not unusual. For instance, a letter he wrote to head office in 1967 was interrupted:

> There is a beautiful dust storm raging in from the west so I will finish this letter when it passes . . . The dust storm lasted for two days and the worst I have seen since I have been here.[25]

He was at Monkira also during the record floods of 1971 and 1974. In 1971 the river rose to 5 metres, the highest since 1950 when it had risen to 5.7 metres. MacQueen described the situation in March 1971:

> There are very few Monkira cattle in the channels. The cattle are very good in this respect. As soon as it starts raining they move out and the Alex steers follow them. The water spread for 15 miles. The flood went halfway out to Yenda Tank on the western side and out to Rusty's Tank on the eastern side above the station, a distance of 15 miles . . . Below the station the water was from the Bullock Paddock fence at Appermilkie out through the sandhills to Milkinbrinna.[26]

John Crane followed MacQueen as Monkira's manager in 1976. Before joining NAP in 1972 Crane had jackarooed in the Moree district of New South Wales, had managed some small Queensland and New South Wales properties and had been a livestock manager on Brunette Downs.[27] In 1977,

21 Minutes of Directors' Meeting, 19 July 1978.

22 Rhondda Alexander Interview, Marion Downs, 15 November 1990.

23 Ken McGuire to General Manager, 26 Oct 1971, Correspondence, Marion Downs, 1971, NAPR.

24 It was considered that a flood level of 5.57 metres (18 feet 2 inches) gave a good flood out which covered everything.

25 George MacQueen to General Manager, 27 December 1967, File 1, Correspondence, Monkira, 1967, NAPR.

26 George MacQueen to General Manager, 17 March 1971, Correspondence File 1, Monkira, 1971, NAPR.

27 Ken Moore to station managers, 8 November 1971, File 1, Correspondence, Alexandria, 1971, NAPR.

Wendy and John Ohlsen. John was Manager of Gallipoli 1961-1970: Islay Plains 1970-1980; Alexandria 1981-1991.

Ross Peatling, Manager of Alexandria since 1991.

Manager of Marion Downs since 1971, Bill Alexander, at the Old Stake Yards in Hamilton Paddock. The yards were built from coolibah trees during the Mackinnon's time.

Rhondda Alexander, Marion Downs, 1990, conducting one of the two daily radio sessions (7am and 1pm) between NAP's Channel properties. In 1988, Rhondda was awarded Citizen of the Year from the Boulia Shire Council for her community service with the Isolated Children's and Parents' Association, the Rodeo Association and The Flying Doctor Service.

June and Jamie MacFarlane: Jamie was manager of Gallipoli 1982-1984; Coorabulka 1985-1989; Monkira 1990. He was killed in an aircraft accident on Coorabulka 22 July 1990. Steve Millard

George MacQueen: Manager, Gallipoli 1958-1961; Monkira 1961-1976; Alexandria 1976-1981.

George MacQueen

while he was manager, fire almost totally destroyed the Monkira homestead. His successor in 1985 was Bob Wales who had joined NAP as a jackaroo and had been manager of Soudan for five years (1980-85). Bob Wales' twin brother Aidan, who had been head stockman on Monkira, followed him to Soudan as manager from 1985 to 1989.

On 22 July 1990, while checking cattle from a light aircraft over Coorabulka, Bob Wales' successor, Jamie MacFarlane, was tragically killed when the aircraft crashed. The young pilot, Steven Litherland, who had been with the company for only a short time, was also killed. The aircraft had been purchased with Vergemont Station in early 1990.

Jamie MacFarlane commenced work with NAP in 1981 as head stockman on Soudan and was soon promoted to the position of manager on Gallipoli, where he remained until July 1984 when he resigned to move his family closer to schooling. He returned in November 1985 to manage Coorabulka, where he remained until he was transferred to Monkira in January 1990. The minutes of a board meeting after the tragedy paid tribute to him:

> Jamie's untimely death is mourned by his family and a wide circle of friends who knew him as a person of great intelligence, personable character, a sporting achiever, tireless community worker and one of nature's gentleman.[28]

Typical of the family-oriented nature of the company was the immediate offer by Bob Wales to return to Monkira to take over the management temporarily during the busy turnoff period. Peter McLaren, who had been on Herbert Downs, was appointed manager of Monkira in late 1990, after a short period managing Coorabulka.

A list of COORABULKA managers completes the jigsaw puzzle as managers have moved from one NAP property to another.

1964 - 1971	Bill Alexander
1972 - 1976	John Crane
1976 - 1985	Steve Millard
1985 - 1989	Jamie MacFarlane
1990	Peter McLaren
1990 - 1991	Dieter Jordan
1992 -	John Rickertt

All managers have been mentioned except Dieter Jordan and John Rickertt who were new to NAP, and Stephen Millard who has been with the company since 1973. Following an education at the Church of England Grammar School, Brisbane, Millard completed a Diploma in Animal Husbandry at the Queensland Agricultural College, Gatton, and two years of veterinary science at The University of Queensland. He worked on Avon Downs on the Barkly in 1969 and 1970 and from 1971 to 1972 was employed by Pioneer Sugar on their cattle properties at Bowen and Ayr. He was appointed stud overseer at Alexandria early in 1973, managed Gallipoli from 1973 to 1974 and Soudan from 1975 to 1976. For nine years he was manager at Coorabulka (1976-85). He is the only NAP station manager to have

28 Minutes of Director's Meeting, 7 August 1990.

made the transition to head office, becoming Projects Officer at the Brisbane office in 1986. With more stations being purchased in the 1980s, the extra workload warranted more staff and his appointment, as well as that of Research Officer Colleen Dixon in 1991, has taken some of the pressures off General Manager Ross Brunckhorst, particularly with respect to the properties purchased since 1985: Wainui Feedlot, Connemara, Kynuna, Boomarra and Vergemont.

While the company was essentially preoccupied with disease eradication and the cattle market slump through the 1970s, there had been some expansion before then with the purchase of GLENORMISTON Station in 1968 and Islay Plains Station in 1970. In 1967, with large numbers of stock on the Channel properties requiring feed and with little prospect of selling them, NAP took the opportunity to purchase Glenormiston Station situated on the Georgina River in the Channel country. Drought-stricken at the time, it was owned the Collins White company which was then being wound up.

Collins White had owned Glenormiston since 1899, when William Collins took it up on the company's behalf as relief country for cattle from drought-affected Beaudesert Station in the McKinlay district. The Collins brothers had also considered purchasing Brighton Downs and Davenport Downs, certain that whatever western property was purchased it was sure to 'acquire greatly increased value' on account of impending Federation. 'Barring ticks I think we have a good prospect before us generally', said Robert Collins.[29] The drought of 1898 to 1902 was too extensive, however, and three-quarters of the 12,500 cattle delivered to Glenormiston to provide relief perished.[30]

Following unsuccessful attempts to sell Glenormiston during the 1940s,[31] the property was not offered for sale again until the mid-1960s. The sale to NAP eventuated in 1968, although since 1964 there had been a steady purchase by NAP of Collins White shares as they became available, to the extent that by 1967 NAP owned some 30 per cent of the Collins White Company.[32] When the agreed purchase price of Glenormiston was given as $232,306.50, therefore, only 70 per cent of that figure was actually paid by NAP.[33] Within a year, good rains had fallen and the station was carrying 5,000 head. 'The gamble paid off', Michael Crouch told shareholders at the annual general meeting in 1968.[34]

GLENORMISTON

Glenormiston has brolgas and big reds.
Land turns red as the sunset spreads
Early morning rises in the cattle season.
Never say die when the weather is freezin'.
On the hot plain the cattle fight for shade.
River runs long in the channels it has made.
Monstrous bullocks we sell each year.
Infected cattle we find none here.
Sizzling steak on the barby at night.
Toko Ranges are a wonderful sight.
On the big flat we see a drover.
NAPCO has stations like this all over.

Matthew Debney, aged 12,
Glenormiston Station

Glenormiston takes its name from Glenormiston No.1 run which, in 1876, was the first run in the area taken up by Colclough Kirwan, the lessee of Herbert Downs Station. By 1878, the station was known as 'Glenormiston, Kelvin Grove and Idamea Lakes'; the last two names were eventually dropped. When Kirwan died in 1878, the leases passed to Walter Douglas of Inverlocky, near Goulburn. Henry Skuthorp was his manager on Glenormiston, with Skuthorp's name appearing as lessee until 1882. Two years later, the property was acquired by James Tyson with William Collins

29 R.M.Collins, England, to W.Collins, 7 September 1899, Fraser Family Papers, Mundoolun.

30 W.Collins to Minister for Lands, 26 March 1904, LAN N43, Glenormiston Consolidation, QSA.

31 Minutes of Directors' Meetings, Collins White and Company, 1941 and 1947, NAPR.

32 Collins White and Company Pty. Ltd., Sale of Shares to NAP, 1964; Ken Moore to Francis Foster, 28 February 1964, Green Copies, Letter File, Marion Downs Pty. Ltd., February-May 1964, NAPR.

33 Minutes of Directors' Meeting, NAP, 8 January 1968.

34 E.M.Crouch, Chairman's Address, Annual General Meeting, 14 November 1968.

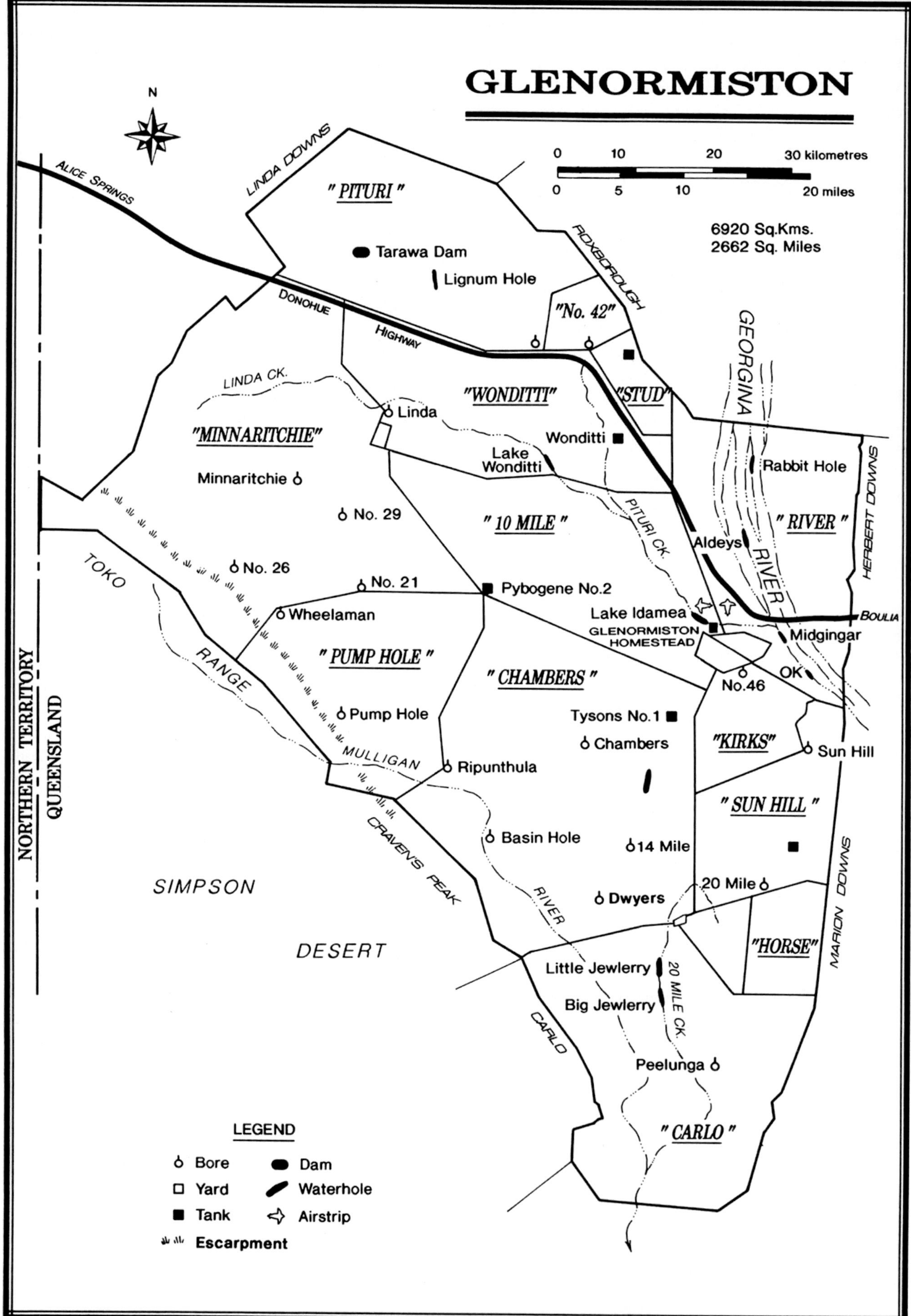

Map of Glenormiston.

Glenormiston with Lake Idamea in the background.

taking up the leases on behalf of Collins White from Tyson's estate in 1899.[35]

Glenormiston, with 6,920 square kilometres (2,662 square miles), is the most varied of NAP's Channel properties. Francis Foster described it in 1913 as consisting of saltbush, thick gidyea, Mitchell downs, limestone ridges, barren stony ground, sandhills covered with fine spinifex and cane grass, and flooded bluebush country.[36] On the east, the property adjoins Marion Downs while the Toko Ranges form a natural western boundary bordering the Simpson Desert. To the north are Roxborough Downs and Linda Downs stations. Situated in a dry, 200-millimetre (eight-inch) rainbelt, the country is watered primarily by the Georgina and Mulligan rivers, and the Pituri and Linda creeks. There are large permanent waterholes such as Lake Idamea on which the homestead is situated and which was made permanent in the early 1970s by the repair of an old overshot dam.[37] Lake Wonditta in Pituri Creek is rarely dry and good waterholes on the Georgina River include Rabbit, Aldey's, Midginjar and O'K holes.[38] The cleaning out of three natural springs in the vicinity of the Mulligan River in Carlo Paddock in 1989 has enhanced the water supply in that area. In appraising the property, a Lands Department report in 1981 stated that 30 per cent was inferior; 5 per cent was scalded claypan and saltpan areas along the Mulligan; 7 per cent was flooded channel areas; and the balance was mixed country of mainly open plains and gidyea flats.[39]

Glenormiston has always suffered from water problems. Before NAP ownership, up to eighteen bores had been abandoned, after either finding no water, salt water, or an insufficient supply. By 1981, records revealed a total of thirty-six unsatisfactory

35 LAN N43, Glenormiston Consolidation, QSA.

36 Francis Foster, Record of trip to Queensland, 1913, Foster Family Papers.

37 There are also remnants of stones walls for dams built by the Chinese in the 1800s.

38 Also spelt Wonditti, Aldeys, Midgingar, OK.

39 V.W.Rosel, Land Commissioner, Report for Appraisement of Rent, Boulia District, 10 December 1981, 04/364, Glenormiston, Part 3, Lands Department, Brisbane.

(dud) bores of which twenty-six were salty.[40] It has always been difficult, therefore, to service all parts of the property. In the first year of NAP ownership, the company put down eight successful bores,[41] but not without problems. It was extremely difficult to find reliable and experienced drillers who would work in such remote areas for extended periods of time. New roads had to be established through difficult country to the sites; boring equipment and iron tanks had to be hauled over many hundreds of kilometres. One can well appreciate the enormity of the task in earlier times, of hauling steam engines and associated equipment over such terrain. The situation has gradually improved to the extent that there are now approximately twenty working bores which have either windmills or pumps attached. Water is mostly stored in steel or concrete tanks, the earth being too porous for turkey's nests.

The problem of gidyea poisoning on Glenormiston was mentioned in correspondence as early as 1903.[42] Interestingly, although gidyea grows throughout the property, the gidyea in the limestone country in the north is poisonous, while the gidyea in the southern sandhill country does not adversely affect stock.[43] The poisoning happens mostly when feed is scarce and cattle are forced to eat gidyea leaves. The cattle die at bores soon after they drink water. One of the worst times for this problem was in 1971, when 878 cattle perished.[44] In Minnaritchie paddock, one of the best fattening paddocks on the property, there were no more than 150 on the bore, but 40 of them died.[45] Another difficult period was in 1980-81 when there was a combined loss of 1,000 head on Herbert Downs and Glenormiston.[46] With no economic control measures known, the company closely monitors the situation and has contributed financially to scientific research into the problem of gidyea poisoning. As recently as 1988 during drought conditions, the General Manager Ross Brunckhorst wrote that with the scarcity of feed 'risks will have to be taken with the gidyea poison areas'.[47]

Herd numbers on Glenormiston, as elsewhere, have fluctuated with the seasons. Despite limited watering points, the property carried more than 13,000 head in the good seasons of the early 1920s, but later in the 1920s drought caused the numbers to drop to 1,000. Cattle have been agisted at various times to other Collins White or NAP properties. The 1960s were especially dry when numbers dropped to 400 in 1964[48] and the average annual rainfall for the decade was only 88.3 mm (3.5 inches). Ken Moore sympathised with Glenormiston manager Peter Donkin in 1969:

> You sound, in this letter, to be a trifle downcast. However, we appreciate that these conditions are quite worrying. . . Unfortunately you cannot control the weather, but you can only do as you are doing, and that is make the best of the situation, that exists, and keep us informed as you have done.[49]

Conditions could change dramatically. In 1971 Glenormiston experienced its third highest flood on record.[50] The flood of 1974 eclipsed all other floods with the Georgina River reaching a peak of 8 metres, approximately half-a-metre higher than the record 1953 flood.[51] Seasonal fluctuations are a continuing problem, however, with statistics such as the following making special demands of management:

Glenormiston: long-term annual average 175-200 millimetres or 7-8 inches.

Years	Average Rainfall	
1956-1965	93.3mm	3.7inches
1965-1973	171.3mm	6.9inches
1973-1979	395.5mm	15.8inches
1979-1983	159.8mm	6.4inches

Glenormiston has had three NAP managers:

Peter Donkin	1968 - 1971
Jim Dwyer	1971 - 1986
Malcolm Debney	1986 -

When NAP took over in 1968, Martin Hayward, who had been at Glenormiston for a record thirty-three years, retired. His successor, Peter Donkin,

40 *ibid.*

41 Minutes of Directors' Meeting, 18 September 1968.

42 F.H.Story, Manager of Glenormiston, to W.C.Frith, Pastoral Inspector, 10 September 1903, Letterbook, 1903-1906, held at Glenormiston Station.

43 Malcolm Debney Interview, Bedourie, 17 November 1990.

44 P.Donkin to Ken Moore, 22 April 1971, File 1, Correspondence, Glenormiston, 1971, NAPR.

45 P.Donkin to K.Moore, 10 February 1971, 31 March 1971, File 1, Correspondence, Glenormiston, 1971, NAPR.

46 E.M.Crouch, Chairman's Address, Annual General Meeting, 15 April 1981.

47 R.M.Brunckhorst, Report to Directors, 3/88, 23 March 1988, NAPR.

48 Jim Dwyer Interview, Mount Isa, 12 November 1990.

49 Ken Moore to Peter Donkin, Manager of Glenormiston, 8 July 1969, Correspondence File 1, NAP Stations, 1969, NAPR.

50 Peter Donkin to Ken Moore, 22 April 1971, File 1, Correspondence, Glenormiston, 1971.

51 Glenormiston Station Report, January 1974, Records held at Glenormiston Station.

Station hands, Glenormiston 1913. Henry Foster

had started as a jackaroo on Marion Downs in 1960. He showed promise from the start. Harold Chambers stated in 1961 that two jackaroos — John Ohlsen and Peter Donkin — were likely to become managers and he proved correct.[52] Donkin managed Coorabulka in 1963-64 and, after a trip overseas, relieved as manager on NAP properties before taking up the Glenormiston management. He became restless, however, and resigned in 1971 after some agonising over the loyalty he felt towards the company. In his letter of resignation to Ken Moore he stated:

> I regret that by doing this, I'll probably be letting both you and Mr. Crouch down, but this is the decision I've made.[53]

The wet of 1971 prevented his departure until May of that year, when Jim Dwyer took over. An experienced man, Dwyer had already worked on Glenormiston and knew it well.[54] During his management, he was a warden under the Aboriginal Relics Preservation Act and took a keen interest in this position.[55] Bob Kirk was Dwyer's head stockman while Kirk's wife, Val, a daughter of Sam Hill[56], was elected as councillor to the Boulia Shire Council for three years from 1973.[57]

Malcolm Debney started work with NAP as a jackaroo at Marion Downs in 1971. He was 18 and had worked for a short time for his father on Arrabury Station, near Betoota, after leaving Toowoomba Grammar School. His great-grandfather's brother, George Leonard Debney, had been a part-owner and manager of Monkira in the 1880s and 1890s.

Debney was at Alexandria and Coorabulka in 1972, and from 1973 to 1977 he worked as a

52 H.W. Chambers to Chairman of Directors, 4 January 1961, Green Copies, Marion Downs Pty.Ltd., December 1960-February 1961, NAPR.

53 Peter Donkin to General Manager, 21 October 1970, File 1, Glenormiston, 1970, NAPR.

54 Jim Dwyer has already been discussed.

55 K.Moore to Department of Aboriginal and Islander Advancement, 6 April 1979, File 04/364, Glenormiston, Part 3, Lands Department, Brisbane.

56 Sam Hill was manager on Herbert Downs 1947-1980.

57 K.Moore to J.Dwyer, 18 April 1973, File 1, Glenormiston Station, 1973, NAPR.

Malcolm and Elizabeth Debney, Glenormiston. Mal has been manager of Glenormiston since 1986.

jackaroo, then a grader and dozer driver on Islay Plains, where he was also involved in the artificial insemination program. He was head stockman at Monkira in 1977 and 1978, and following time away from the company returned to Alexandria as overseer in 1981, a position which became known as Alexandria out-station manager. He has been manager at Glenormiston since 1986 and is a councillor on the Boulia Shire Council. He is one of the younger NAP managers and one of the few to hold a pilot's licence. His skills are used to the company's advantage when he flies the Cessna 172F, which was purchased in 1989 and is used primarily for mustering and spotting cattle. He met his wife, Elizabeth, at Coorabulka, and like all station families, the Debneys have had to cope with educating their two children by correspondence lessons and the School of the Air. Secondary education has meant boarding school.

While an 1878 exploration map[58] locates the Glenormiston homestead in its present position on Lake Idamea, the present building of stone and pise was not erected until 1898. Ruins in the south of the property are probably the remains of dwellings built by Afghan traders or Chinese. Dams built by Chinese on the property suggest they were in the area from the late 1860s to early 1870s, at the time of gold discoveries in the Northern Territory, when Glenormiston was on the route to the Territory via the Georgina River, then called the Herbert River. The old station store next to the homestead remains while other outbuildings such as the overseer's cottage, the office, the men's quarters and the workshop have been replaced. The kitchen complex, including the cold room, was accidentally burnt down in 1985 and was replaced by a modern structure.[59]

While the manager's role is still essentially concerned with the survival of cattle, the pace of station management has accelerated considerably. This is exemplified by a selection of entries over a number of days taken from the station diaries of two managers of Glenormiston — William Barnett in 1912 and Malcolm Debney in 1988.[60]

1912: William Barnett manager
Entries during 1912

> McCraw in with mail by packhorse; McCraw Jnr swam river with mail, river too high for crossing; Piccaninii born in camp; Making soap; Cut up soap (next day); made buggy shed good with boughs; Water baling for garden; had blacks clearing up wood heap; Making hobbles; Milker brought in with killer; Dray in for rations and left for Carlo; Mr Chapman left by coach; carting wood.

1988: Malcolm Debney manager
Entries during 1988

> Self for fly around Carlo and over river country; Working on Toyota - fitted new engine mounts, made spare tyre carrier in back; On the mail plane was grader radiator and water pump; Repaired pipe on septic tank at house; Pruning trees and put drip line in; On mail plane to start holidays; Picked up new motorbikes; Serviced bore motors; To Boulia for Rodeo Association Meeting; Set up wireless in new Toyota; Roads graded; Two DPI blokes arrived and overnight to inspect for drought relief; Australian Pacific Tours coach rolled trailer and blew up the diff just past the river; Self and family to Bedourie for ICPA [Isolated Children's and Parents' Association] Meeting; TV Satellite dish man arrived and laid cables; Self and family to Mt

58 Frank Scarr Exploration Map, 1878, Mortlock Library, Adelaide.

59 E.M.Crouch, Chairman's Address, Annual General Meeting, 16 April 1986.

60 Both diaries are held at Glenormiston Station.

School of the air lesson for Matthew Debney at Glenormiston, 1990.

Isa and then onto Brisbane for Conference and Expo; Self mustering with plane; PCAP [Priority Country Area Programme] Teachers arrived; Surveyors arrived to look for Surveyor Bedford's survey line from Boulia to Tobermorey - located exact spot of 65 Mile post at Mark Tree; Did a bore run in the plane and then over to Marion Downs to assist with ICPA catering for Endeavour Car Rally; Self to Mt Isa for Gidyea Poison Meeting.

The purchase of NAP's next property, ISLAY PLAINS, was essentially as a result of the 1969-70 drought on Alexandria. The Alexandria rainfall in 1969 was 233 mm (9.3 inches) while in 1970 it was lower, at 116.3 mm (4.7 inches). During the 1969 season, it was found necessary, for the first time since Marion Downs was purchased in 1934, to sell store steers (2,000) to outside parties. As Ken Moore explained:

> A further $65 per head gross profit on these cattle could have been realised to produce a further $130,000 clear profit, had we been able to carry them for another twelve months.[61]

61 K.W.Moore to ANZ Bank, 29 May 1970, Islay Plains Initial Reports, 1970s, NAPR.

So, with the unpredictability of seasons in the Channel country, the board looked towards purchasing a property in a higher rainfall belt. In August 1970, Islay Plains was bought for $410,000 (plus $107,670 for the 1,600 head on the property) from James Douglas Goodwin who had run the property since 1923. Situated 96 kilometres from Clermont and 72 kilometres from Alpha, it was in an official rain belt of 650 mm (26 inches) and consisted of three leases — Beelarong and Terrara Holdings, and a smaller special lease. Originally it was three runs — Islay Plains, Berri Downs and Horbury Grove — taken up in 1885 by Walter Henry Holt. Transferred to The Bank of Australasia during the drought at the turn of the century, William Cope took up the leases in 1907 and ran them from his adjoining property, Banchory. Thomas Borthwick and Sons (Australasia) acquired the leases in 1914, and after the government resumed Islay Plains in 1922 it was acquired by James Goodwin the following year. Borthwicks kept Banchory and not until after the purchase by Goodwin was a homestead built and the property run separately.[62]

62 Islay Plains Consolidation, LAN AF975, QSA.

The Belyando River crosses the property, which at the time of the NAP purchase was 36,879 hectares (91,092 acres)[63] of mixed gidyea scrub and open coolibah with plains intersected by flood channels. Waterholes in the Belyando and other creeks lasted up to six months and there was a permanent hole at the homestead. Because the property was in a higher rainfall area, twelve strategically placed dams meant that only one bore was needed. At the time of purchase some 12,100 hectares of gidyea and brigalow scrub had been cleared and large areas sown with buffel and green panic grasses.

The property was put to immediate use with 1,200 steers transferred from Alexandria as a drought-relief measure.[64] It was estimated that with further scrub clearing, the property had a capacity for 5,000 to 5,500 head. The first season was excellent and 1,200 head were turned off in 1971. NAP set about further improving the property with the repair of river crossings, the replacement of existing fencing and the re-equipping of Duncan's Dam.[65] About 2,400 hectares were successfully burnt and seeded in 1971.

Islay Plains was not as isolated as the other NAP properties and enjoyed facilities such as a twice-weekly local mail service and a telephone connection to Alpha. It was conveniently situated for the railway and markets and so could be used to fatten cattle brought up from the Channel properties, breaking the journey to the meatworks.

From 1972 a problem appeared with the turnoff weights. Whereas 3.5-year-old Islay Plains cattle weighed 254 kg (560 pounds), *Bos indicus* infused cattle of the same age from a neighbouring unimproved property averaged 313 kg (690 pounds).[66] The Alexandria-bred cattle apparently needed to acclimatise, a process which effectively set them back a full season. Lee McNicholl reported a similar situation when the King Ranch company tried to fatten their stock at Tully in north Queensland. With little evidence available for the Islay Plains area, McNicholl advised the board that they should either send weaner steers to the property or cross breed with *Bos indicus* lines. The latter suggestion was implemented with the cross-breeding of 300 Shorthorn heifers using Sahiwal semen in an artificial insemination program. By 1973, this program was proceeding satisfactorily with Malcolm Debney becoming an experienced inseminator.[67] But manager John Ohlsen advised that, unfortunately, the heifers used in the program were 'knocked back . . . at a time they should have been going ahead fast'. He suggested that red Brahman bulls be purchased and allowed to run with the heifers because this would be simpler and cheaper, and it would also give the highest hybrid vigour to the progeny.[68] Although the company persevered with the AI program, by 1978 it had not determined the best breeding and fattening policy for the property. Islay Plains was proving an unexpected challenge. In 1974, following a huge wet, there were also losses from tick fever and part of the property had to be spelled to alleviate the problem.[69] Meanwhile, TB and brucellosis testing had commenced which further interrupted the stockwork. From 1978, the property operated independently, with its own breeders.

John Ohlsen was NAP's manager at Islay Plains. He and his wife Wendy arrived in early 1971 from Gallipoli and with their two young sons enjoyed the benefits which a more closely settled area afforded. By the time the property was sold ten years later, Ohlsen had proved to be one of NAP's most capable managers. It was logical, therefore, that he was promoted to the management of Alexandria when the position became vacant in 1981.

The bank loan to purchase Islay Plains was paid off in 1979, but since it was not properly fulfilling its planned fattening requirements, the property was sold in December 1980 to the Stanbroke Pastoral Company, for around $1,250,000.[70] This was the first NAP property to be sold since 1910, when the north Queensland properties, Woodstock and Inkerman, were sold. The decision to purchase Islay Plains had not necessarily been an incorrect

63 This is equivalent to 369 square kilometres (142 square miles).

64 Minutes of Directors' Meeting, 22 April 1970.

65 Minutes of Directors' Meeting, 17 March 1971.

66 Ken Moore to Directors, File 2, Correspondence, Islay Plains, 1972, NAPR.

67 L.McNicholl to General Manager, 28 December 1972, Vet's File 1973; Meeting of Directors, 22 March 1973; John Ohlsen to General Manager, 14 April 1973, File 2, Correspondence, Islay Plains, 1973, NAPR.

68 J.Ohlsen to General Manager, 26 October 1973, File 2, Islay Plains, 1972, NAPR. The suggestion was not carried out.

69 Minutes of Directors' Meeting, 22 September 1976.

70 Minutes of Directors' Meeting, 15 December 1980.

one. Rather, as the directors reported in December 1980,

> whilst this property had made profits, it did not fit geographically into our type of operation as well as could be wished. . . A substantial profit resulted from this sale.[71]

The WAINUI Farm and Feedlot on the Darling Downs has proved a definite aid in marketing the company's product. Purchased in mid-1985 from Tancreds for $2.1 million, it was immediately used to improve the condition of drought-affected stock. In the first five months, 6,162 head of NAP cattle were fed and marketed through the feedlot, helping the turnoff in 1985 to total 27,900 head.[72] Again in 1986, it was of great benefit because with the drought conditions further west, 17,635 head were transferred there for fattening before slaughter.[73] In 1987, with sufficient grass on the Channel properties, there was no need to send any store cattle to Wainui and the feedlot successfully custom-fed cattle for outside owners.

Situated midway between Toowoomba and Dalby, with frontage to Oakey Creek, Wainui allows an extensive area of the farm to be flood irrigated from a ring tank. The farm is capable of producing 25 per cent of its own grain requirements — from sorghum, barley and wheat; the feedlot has a capacity for 3,000 head. The aggregation of 1,286 hectares (3,176 acres) was increased in 1988 with the purchase of adjoining properties: The Willows, 250 hectares (618 acres) and Rangeview, 264 hectares (652 acres). The former property allowed relocation of the feedlot yards and both helped overcome concerns of the Jondaryan Shire Council about the drainage of the feedlot pens and the ingress and egress by road trains from the Warrego Highway.[74]

While Wainui has been crucial to the survival of NAP's cattle during drought, it has also provided a cash flow throughout the whole year instead of income being confined to the period of the cattle season. The property has also been used as a 'finishing off' facility to ready stock at a certain time for particular markets. By forward selling, cattle can be at their prime when a slaughter date is known. The marketing results have been most encouraging. Steers fed through the feedlot have been reaching Jap Ox[75] weights at 2 years of age[76] compared with 3 to 3.5 years for purely grass-fed stock.

There has been some upgrading at Wainui with the installation of more storage silos and a more sophisticated grain handling arrangement. NAP managers there have been Peter Gilbert, followed in 1989 by Philip Myers. While the directors of the company and General Manager Ross Brunckhorst make the major policy decisions, projects officer Steve Millard has an important supervisory role in the administration of Wainui. The feedlot industry has become a recognised part of the pastoral industry and with plans for expansion of NAP's Wainui property to ultimately accommodate 7,200 head, the company is well placed to react to market demands.

The purchase of CONNEMARA Station unstocked (except for some 2,300 head of bush cattle),[77] from Tancreds for $800,000 in January 1986 provided immediate relief for drought-stricken cattle from other company properties. Manager Arthur Pryce and his staff gave special attention to fencing, road improvements and the repair of windmills, while a major destocking of bush cattle from the station in 1987 qualified the property for disease-free status the following year. The property played a major role in helping the company negotiate the drought of the late 1980s. In 1989, following excellent rain, Connemara began receiving steers from Alexandria. As Ross Brunckhorst pointed out in October 1989,

> When the season finally turns around on the Company's northern properties everyone should pause for a moment and consider the contribution that Connemara has made during a most difficult period of extended drought.[78]

Connemara, in the Jundah district, is 5,025 square kilometres (1,933 square miles) in area and comprises open Mitchell grass downs with open plain and coolibah frontage to Farrar's and Portia creeks. There are also extensive areas of Flinders

71 Directors' Annual Report, July 1979-Dec 1980.

72 Directors' Annual Report, 30 November 1985.

73 Directors' Annual Report, 30 November 1986.

74 E.M.Crouch, Chairman's Address, Annual General Meeting, 19 April 1989.

75 Bullocks (or Oxen) which meet carcass specifications required by the Japanese market.

76 Directors' Annual Report, 30 November 1990.

77 Directors' Annual Report, 30 November 1985; Report by S. Millard, 3 December 1987, Finance Committee File, 1987, NAPR.

78 R.Brunckhorst, Report to Directors, 9/89, 5 October 1989, NAPR.

Wainui Feedlot, 1991. Steve Millard and Phil Myers (Manager, Wainui) with Boomarra cattle newly arrived at the feedlot.

Connemara, 1986.

Arthur Pryce, Manager of Connemara, 1981-1991.

Michael Ingram. Director representing the English shareholders, 1980-1991.

Crossbred heifers at Alexandria, 1991.

Chris Lyndon

grass, gidyea, ironstone ridges and edible mulga country. Although the property was purchased only recently by NAP, its connections with the company can be traced back to 1875 when it was taken up by De Burgh Persse, whose descendants married into the Collins family.[79] He named the property after the Connemara district in Ireland whence he came. Thomas Harding was the first manager of Connemara until 1882, when he became the first manager of Alexandria.

Other owners of Connemara[80] have included:

1877 Bentinck Yelverton Bingham (Tabragalba)
1881 George Henry Cox and Septimus Alfred Stephen (Sydney)
1898 Queensland Trustees Limited
1899 V.J. Dowling, S.A. Stephen, George H. Cox, Sir Edward Ward
1905 Lease forfeited
1906 James Rutherford
1913 James McIntosh Allison and George Frederick Wigan
1938 Burns Kidman Reid (grandson of Sidney Kidman)
1972 Australian Continental Resources Ltd. (Australian Pastoral Company)
1980 Townsville Cold Stores Pty. Ltd. (Tancreds).

Only two months after the purchase of Connemara, KYNUNA was purchased unstocked in March 1986 for $1,600,000 from the Australasian Grazing Company (Tancreds). This was the third Tancreds property purchased by NAP within twelve months, and it was immediately stocked with cattle from Alexandria.

Kynuna has a higher annual average rainfall (375-400 millimetres or 15-16 inches) than Connemara (250-275 millimetres or 10-11 inches) while both are higher than Marion Downs, Coorabulka, Monkira and Glenormiston (average 175-200 millimetres or 7-8 inches).

It was once a much larger property of 2,587 square kilometres, which took in not only Kynuna township but also the Combo waterhole which, it is said, inspired Banjo Paterson to write *Waltzing Matilda*. By 1913 the property was only 1,737 square kilometres, and Francis Foster noted during his western trip of 1913 that Kynuna carried 120,000 sheep and was relatively close to the sheep property Eulolo.[81] Following further resumption, the property is now 881 square kilometres (339 square miles) with the headwaters of the Diamantina River forming part of its northern boundary. While the property has run cattle for some years, some of its neighbours still carry sheep.[82]

In 1911 Edmond Jowett sold Kynuna to the Scottish Australian Investment Company. At the time Jowett also owned Vergemont, later an NAP property.[83] Two owners followed: Ted Naughton and then the Australasian Grazing Company (Tancreds). Nick Murray has been manager since 1979 when Kynuna was purchased by Tancreds. The property was well developed at the time of the NAP purchase although more management aids have been added, especially fencing and earth dams. While it did not have the excellent seasons that Connemara enjoyed in the late 1980s — Kynuna was declared drought stricken in 1988 — the property was still capable of holding over 2,000 head in such difficult conditions.

The long 1980s drought on Alexandria reminded the company of its dependence on that property for breeding. This concern resulted in the purchase of BOOMARRA Station prior to auction in April 1989 for $5,790,000. Situated 140 kilometres north of Cloncurry in the Gulf region, at the time of purchase it was an established Brahman breeding property carrying 10,000 head. Boomarra had been owned[84] by:

1878 Edward Palmer and John Stevenson
1887 Scottish Australian Investment Company/The Australian and New Zealand Mortgage Company/ Australian Joint Stock Bank
1897 Reginald Edward Rowe Hillcoat
1900 Bank of New South Wales
1912 Reginald Edward Rowe Hillcoat
1920 Sydney Malcolm Chaplain
1921 Chaplain Brothers Pty. Limited
1979 Australian Grazing Pty. Ltd. (Tancreds)
1986 Angus and Christine McClymont

The 1,080-square-kilometre (415 square miles) property is in a high rainfall belt, 500-750 millimetres (20-30 inches), and has a carrying capacity of approximately 13,000 head. The country is flat to undulating, with low spinifex hills. Dismal Creek, and its flooded channels, runs along the eastern boundary, while the balance of the country is grassed with Mitchell, Flinders and summer grasses, and lightly shaded with whitewood, mimosa, coolibah and vinetree. There are fifteen sub-artesian bores, with polythene pipe

79 Persse also acquired Palparara, Buckingham Downs, Hawkwood and Rocky Springs.

80 Connemara Consolidation File, LAN AF265, QSA; Connemara Lease File, NAPR; Connemara File 28/348, Jundah, Lands Department, Brisbane.

81 F.H.Foster, Record of Trip to Queensland, 1913, Foster Family Papers.

82 Kynuna Consolidation, LAN AF284, QSA.

83 *Brisbane Courier*, 14 March 1911.

84 LAN AF6 (Boomarra), LAN AF13 (Canobie), LAN AF14 (Canobie North), LAN S5, S6 (Occupation Licences), QSA.

connected to two bores to reticulate water to required areas. The purchase not only provided another breeding herd but also increased the company's potential herd to 110,000 head.[85] Boomarra's manager is Fred Shephard, who remained in the position after the company's purchase from the McClymont family.

The purchase of 3,516-square-kilometre (1,352 square miles) VERGEMONT Station, unstocked, at auction in March 1990 for $2.25 million increased the total number of properties owned by NAP to ten, the total size of the NAP holdings to 56,462 square kilometres (21,717 square miles) and the total herd potential to 120,000. Situated in the Longreach district and surrounded by properties running sheep — Vergemont also ran sheep in former times — the property was purchased essentially by NAP for its feed. With an average rainfall of 350mm (14 inches), Vergemont has extensive channels and creek flats in the Vergemont and Central Creek systems, while opal miners have leases on the station in areas not so suitable for cattle fattening. Owned by P.C.Abbott Pty. Ltd since 1987, Vergemont had previously been owned by Frederiks Realty Pty.Ltd. between 1977 and 1987 and prior to 1977, by W.E.and H.Cameron.[86]

When purchased, the buildings were in poor condition and NAP has since made extensive improvements, including the building of a new homestead, repairing bores, windmills and fences, and constructing earth tanks and dams. Following the clearing of 3,240 hectares of gidyea scrub, the area was aerially seeded with buffel grass.[87]

The manager, Rod Murphy, commenced with NAP in 1981 as head stockman of Alexandria's No.2 camp. After four years, he was promoted to manager of Gallipoli and in 1990 was offered the position of manager of Vergemont. It has been an exacting task in view of the run-down condition of the improvements but the company has been prepared to outlay the necessary funds for upgrading.

The strategic placement of NAP's properties for breeding and fattening has ensured that the total herd is accommodated even in drought times. The spread of properties over several river systems, the inclusion of a feedlot to facilitate marketing, and a meticulous approach to improvements, has meant that the company has great flexibility to cope with whatever the future might bring.

---

85 Boomarra File, NAPR.

86 Vergemont Lease File, PH 29/1201, Part 3, Lands Department, Brisbane.

87 R.Brunckhorst, Report to Directors, 12/90, 12 December 1990; Meeting of Directors, 7 August 1990, NAPR.

Typical of the Brahman-based Boomarra herd at Bog Hole Dam, Boomarra, 1991. Richard Launder

Kynuna Station, flood rains, Dec/Jan 1990-91. View from the Kynuna Townside; the rain caused extensive flooding in the channels of the Diamantina River.
Nick Murray

NAP Management, 1991. Standing: Left to Right: Nick Murray (Kynuna), Peter McLaren (Monkira), Steve Millard (Projects Officer, Brisbane), Dieter Jordan (Coorabulka), Chris Lyndon (Chairman of Directors), Bill Alexander (Marion Downs), Mal Debney (Glenormiston), Fred Shephard (Boomarra), Max Ferris (Alexandria out-station), Ross Brunckhorst (General Manager) Seated: Left to Right: Glenn Page (Gallipoli), Rod Murphy (Vergemont), Philip Miller (Soudan), Phil Myers (Wainui).

Vergemont Station, in a west, north-west direction from Mount Haystack, 1991.

Boomarra Station. Corb's Dam on the division fence between Rainers Paddock and West Pelican Paddock, 1991. The windmill pumps water from the dam to a turkey's nest incorporated in the dam wall. Water then flows by gravity from the turkey's nest into two troughs.

CHAPTER 13

Brucellosis and Tuberculosis Eradication Campaign (BTEC)

While the tapping of water in the artesian and sub-artesian basins was the most crucial development for Australia's cattle industry in the nineteenth century, the Brucellosis and Tuberculosis Eradication Campaign (BTEC) is probably the most significant development in the twentieth century. Not only were two major diseases eradicated, but other advantages also followed: calving rates and survival rates improved, resulting in increased turnoffs, sales and dividends; botulism vaccination was incorporated into BTEC; more efficient mustering and trucking procedures were developed; much better data became available on herd composition; range management and land care policies began to be accommodated; and labour costs fell as properties could be run more efficiently. The campaign also profoundly changed herd management practices in the industry. While other factors such as seasons and markets are of utmost importance in the overall performance of a pastoral company, the spin-offs from BTEC played a major role in increasing profits for cattle companies. The following scenario can be attributed not only to NAP, but to the industry as a whole.

> Overall the BTEC campaign not only made dollars for NAPCO but it brought their management into the twentieth century . . . The TB scheme jerked them into gear. They had to do it, and it introduced modern management techniques.[1]

The government-sponsored Brucellosis and Tuberculosis Eradication Campaign (BTEC) was begun in 1969. Although by 1989 the campaign had cost the Federal and State Governments and the industry $543,800,000,[2] it was money well-spent. Australia was declared free of brucellosis in July 1989, and although the scheduled date for eradication of tuberculosis is 1992, all NAP herds were confirmed free from both diseases by the end of 1988; Alexandria was cleared in 1987 and the Channel properties were cleared at various times between 1984 and 1988.

Tuberculosis (TB) was suspected in the Alexandria herd in the mid-1950s. It became a huge problem to the industry for the next three decades and, indeed, the Northern Territory is still trying to eradicate it — mainly from buffalo herds. Testing for the disease began in the Territory in 1934 but only limited eradication schemes had been attempted.[3] The impetus for the complete eradication of the disease in Australia came in the 1960s when export markets were threatened by countries concerned that meat from diseased cattle was being exported. It was thought that bulls introduced into the Alexandria herd had first carried the disease which then spread to the rest of the herd through respiratory transmission. Alexandria's manager Bill Young also suspected that TB was transmitted when cattle licked on molasses in licker drums.[4] The first glands affected

1 L.McNicholl Interview, Dulacca, 16 July 1991.

2 Strategic Plan for BTEC in the Northern Territory 1989-1992, BTEC Management Committee of the Northern Territory, NAPR.

3 BTEC Committee of the NT, 'Application for an extension of the provisionally free and the impending free areas for Tuberculosis in the Northern Territory', 1990, p.1, NAPR.

4 Bill Young to Ken Moore, 18 July 1969, Correspondence File 1, NAP Stations, 1969.

were those around the throat and neck, with the infection subsequently spreading down to the lungs.

In 1956 Jack Maunder, the Brisbane veterinarian, began testing the Alexandria herd for TB. This involved injecting cattle with a small dose of tuberculin into the skin under the butt of the tail. A lump developing after three or four days indicated the presence of TB. Certain factors, however, lessened the reliability of detection; a lag of two to three months between infection and the point when sensitivity was sufficient for a positive reaction; the fact that advanced cases lost sensitivity to the test; and the problem of incomplete musters when cows with newly born calves and unmusterable rogue animals, could not be brought in.[5] There was no treatment for the disease. Destroying the animals was the only way to eradicate it — a procedure Maunder followed after the first test on the stud in 1956 at Alexandria when, with 8.5 per cent of aged cattle showing a positive reaction, 17 out of a draft of 204 were shot.[6] He continued to test and slaughter any reactor cattle after mustering and branding each year, although the November weather was very hot and stressful for the animals. In 1957 Maunder tested the Coorabulka stud and found 10 reactors in the 155 tested, indicating that TB was more widespread in the NAP herds than hitherto believed.[7] Since Alexandria was considered the most important property on which to begin a program, all efforts at eradication were directed there.

By 1968 the company was pleased when the percentage of TB reactors in the stud was down to 0.4 per cent; but complete eradication seemed impossible, for as Maunder said: 'This could be as low as we are likely to get'.[8] By 1970, the figure had moved up to 0.66 per cent,[9] and reactors were being detected at meatworks from cattle in the main herd. After fourteen years of annual testing, eradication still had not been completely effected in the stud, which suggested that whole herd testing was not likely to achieve eradication. A new approach was needed.

Tuberculosis was not the only disease requiring eradication. By the late 1960s NAP's attention was drawn to another disease — brucellosis. While brucellosis, leptospirosis, vibriosis, trichomoniasis and infectious bovine rhinotracheitis all contributed to low calving percentages, it was brucellosis (contagious abortion) which gave most concern. The disease is transmitted through contact with an affected or diseased part of an animal, such as an aborted foetus, and with an average of 1,800 head congregated daily at bores during the hot summer months, it was inevitable that the disease would spread. Alexandria's calving percentages in the 1960s dropped sharply to what at one point was thought to be as low as 40 per cent, although following the bangtail muster in 1972, a more accurate estimate put the figure rather higher at around 60 per cent.

The test for brucellosis involves collecting blood from under the tail of each female beast and sending the samples to laboratories for analysis. Bulls were also sometimes tested, in order to contain the spread of the disease. Those animals which tested positive were either spayed or sold for slaughter (early in the campaign) or, when compensation became available in July 1976, they were shot.[10] There was considerable calf loss, however, when pregnant cows were slaughtered. Two clean blood tests indicated that a beast was disease-free. Unlike TB for which there was no vaccination, brucellosis could be prevented. Initially, in 1969, a vaccine called Strain 19 was used in inoculation at Alexandria, but this was compulsorily replaced in the Northern Territory after 1 January 1973 by Strain 45/20, because of difficulties created by the former in laboratory testing. Two vaccinations were required — the second to be no less than six weeks and no more than six months from the first. With meatworks beginning to reject brucellosis reactor cattle in the 1970s, a massive input of Federal funds was needed to accelerate such eradication — hence the implementing of the national campaign on the Barkly in the early 1970s. The situation was not as far advanced in western Queensland, with the DPI doing only monitoring tests for brucellosis in 1973.[11] The United States authorities anticipated the eradication of the disease in their herds by 1975, but although this was not achieved, it was

5 J.Maunder, 24 June 1960, File 15, J.Maunder Reports, 1960, NAPR.

6 W.E.Young to D.M.Fraser, 5 December 1956, Alexandria Correspondence, 1956, NAPR.

7 J.Maunder, Report, May 1957, J.C.Maunder File, 1957, NAPR.

8 J.Maunder, Report, May 1968, J.C.Maunder File, 1968, NAPR.

9 Minutes of Directors' Meeting, 21 October 1970.

10 Minutes of Directors' Meeting, 21 July 1976.

11 National Brucellosis Eradication Campaign Report, Australian Meat Export Federal Council, 29 May 1973, Correspondence File 1, Alexandria and Glenormiston, 1973, NAPR.

Dr Lee McNicholl TB testing at Monkira. Lee McNicholl

imperative that Australia should attempt to do likewise as soon as possible.[12]

In 1972 NAP instituted a program to simultaneously eradicate both diseases on Alexandria. NAP's veterinarian Lee McNicholl was adamant that if a successful campaign was to be mounted to eradicate rather than merely control TB and brucellosis, a weaner segregation system was essential. Property improvements in the form of new paddocks, yards, crushes and bores were necessary to carry out weaner segregation successfully. This approach represented a radical break from past cattle management systems, when the whole herd ran on an open-range system. McNicholl advocated that the best method of managing the herd was to wean at five to six months; to test the weaners for both diseases; to segregate clean heifer calves into a clean paddock; and then to proceed with a planned program of vaccination and blood testing for these heifers to ensure they achieved 'confirmed free' status. At the same time, aged animals were to be culled and sold. To begin NAP's program, some 3,000 weaned heifers were sent to Gallipoli in 1971, although they were already over twelve months of age. Not until 1972 did the full program of earlier weaning begin. During this introductory period, in order to build up Alexandria's disease-free herd, some weaners and maiden heifers were sent up from the Channel properties.[13]

The testing had to be thorough and the cattle passed through the four stages: 'infected', 'restricted', 'provisionally clear' and 'confirmed clear'. To be confirmed free of the disease, weaners were tested at sixty-day intervals until two clean successive tests were obtained. In the early days when the progeny from heavily infected cows were being tested, six tests on each group were often required to give confirmed-free status. Later in the campaign, the interval between tests was lengthened — up to 360 days for the final two clean tests before the cattle could leave Alexandria for the Channel properties. Such a change lengthened the process and where the DPI agreed it was too expensive to test a whole herd, destocking orders were available whereby stock could be sold to meatworks and the company compensated for any loss incurred.

12 Ken Moore, 2 June 1973, Vet's File, 1973, NAPR,

13 Ken Moore to all station managers, 9 March 1973, File 1, Glenormiston Station File, 1973, NAPR.

Bronchoing, Glenormiston, 1964. W.F. Alexander

Drought on Monkira, 1980. E.M. Crouch

Although not compelled by regulations to do so, NAP thought it economically prudent also to begin eradication in Queensland as soon as possible. Since an approach there could work in conjunction with weaner segregation programs at Alexandria, McNicholl was confident of achieving the eradication requirements.

In 1972, a massive improvement program was begun on Alexandria. Such an upgrading was timely since many earlier improvements were in need of attention. The homestead complex was being built at the same time, and overseeing both projects placed great strain on manager Ken McGuire. Some neighbouring properties such as Alroy, Avon and Brunette had already commenced TB testing. These were pioneering days, however, the Barkly being one of the first areas in Australia to commence large-scale testing. Fences, particularly on the boundaries, were of utmost importance considering the ease with which such diseases spread. With the resumption of Mittiebah and Mount Drummond in 1965, NAP was anxious to build a boundary fence, even though the new owners were having cattle tested prior to arrival. While the fences were being built, holding paddocks, crushes and watering facilities such as bores were also constructed. The first of many sections of double fencing was erected along the entire boundary between Gallipoli and the rest of Alexandria (35 kilometres) to ensure that tested groups were kept apart. From the early 1970s, portable yards were used, which were a tremendous advantage when feed was scarce and cattle were too weak to walk vast distances.

The program worked southwards from Gallipoli; after paddocks were mustered clean, necessary improvements were built and the paddocks were restocked with clean cattle. Lee McNicholl explains the operation:

> Each year you had to develop an extra area to take the heifers . . . It was like a giant chess game . . . you had to draw up the paddocks where you were going to put the heifers . . . and work out how many holding paddocks, what areas were needed . . . and then what waters would be required.[14]

The scarcity of appropriate facilities on Alexandria had become apparent during pleuro testing in the 1950s but the eradication of TB and brucellosis certainly required a more sophisticated approach. Each improvement was established on the basis that, while it was a direct aid to the eradication campaign, it also had to be useful in future normal running operations. This policy has proved to be of enormous long-term benefit. While the government provided taxation incentives and met some of the testing expenses of the disease eradication program, no direct government assistance was available for capital improvements. By 1983, NAP had spent over $1 million on improvements associated with BTEC, and to this can be added unseen costs — mustering and holding of stock, the return of stock to paddocks, associated fodder costs, and comparatively lower brandings because of the delayed matings of heifers.[15] Indeed, towards the end of the program when reactors were few and thus compensation was less, there was an increased financial burden on the company, because mustering and testing had to continue. This was offset to some extent by increased turnoffs and sales; and of course there were compulsory sales under destocking orders. The benefit of the campaign, however, was not apparent in monetary terms until the higher beef prices of the late 1970s.

The frequent handling and holding of stock for testing and inoculation caused the animals to become stressed. In 1972, in response to the animals' stress, bulls were taken out of the commercial and top studs for a time in an attempt to control the breeding period, so that calves would be born in mid-January to early March instead of in the hottest period from October to December.[16] This strategy was carried out intermittently over the next few years.

It took only one stray beast to reinfect a clean herd, which meant a constant planning process for managers to keep clean and non-tested animals separate, and also to separate different classes of stock. This was especially difficult during the drought conditions of the 1980s when feed was scarce and the testing program still required cattle to be segregated. Accuracy was essential and the paddocks had to be mustered clean. As Bill Alexander, manager of Marion Downs, commented: 'It would have broken my heart to get everything clean and then just because you had been lazy [when mustering] to get one reactor'.[17]

14 Lee McNicholl Interview, Dulacca, 16 July 1991.

15 K.Moore, 'Managerial constraints in Program Planning; NAP in Q'ld/NT', Bovine Brucellosis and Tuberculosis Eradication Campaign Property Assessment Workshop, Toowoomba, 18-22 April 1983, NAPR.

16 Ken McGuire, Stud Report, 31 December 1972, Correspondence File 1, Alexandria and Glenormiston, 1973, NAPR.

17 Bill Alexander Interview, Marion Downs, 15 November 1990.

Private veterinary practitioners were contracted by the government to do the TB testing. Station labour collected the blood samples for brucellosis under stock-inspector supervision, because the Northern Territory Administration realised there were not enough government vets and stock inspectors to collect the samples. Property owners were subsequently reimbursed for each sample when station staff carried out the work.

In the early stages, the key to the whole operation was careful planning, and this is where Ken Moore, the general manager, Ken McGuire, the manager of Alexandria, and Dr. Lee McNicholl, the company's veterinarian, played a vital role. The company had previously advertised for a vet in 1969 but decided against proceeding with an appointment then because of the drought on Alexandria. Dr Lee McNicholl commenced work with NAP on a full-time basis on 2 June 1972. He was good-humouredly known as 'Leethal Lee' or 'Doc' and had participated in testing programs on neighbouring properties Avon Downs and Alroy in 1971 and 1972. During this time he also tested the stud on Alexandria for TB. From 1974 McNicholl was based at Mount Isa in order to be strategically close to both Alexandria and the Channel properties. During the late 1970s he spent some time at James Cook University in Townsville researching brucella problems. He resigned from full-time work with NAP in 1981 but continued on contract as a consultant until 1988. His period of employment with the company mirrored the era of brucellosis and TB eradication on the company's properties. During that period, approximately 800,000 TB tests and 500,000 brucellosis tests were carried out under his supervision[18] — a monumental task. Lee sums up his feelings:

> It was demanding and physically exhausting . . . It was boring and repetitive. The only thing that kept me going really was trying to achieve a result . . . The prime motivation in the end was the challenge.[19]

To his credit, there have been no breakdowns within any of the NAP herds since the completion of testing. As Ross Brunckhorst confirmed, 'He was an invaluable member of the team and his thorough methodical approach was vital to the success of the program'.[20]

18 K.Moore, 9 February 1981, TB and Brucella Report, 1980, NAPR.

19 Lee McNicholl Interview, Dulacca, 16 July 1991.

20 R.Brunckhorst Interview, 26 July 1991.

Cooperation between McNicholl and other contracted vets — Jim Dawes, Gordon Barnett, Lindsay Allan, Murray Cameron and Ian King — was important. They in turn had to cooperate with station staff: Ken McGuire; out-station managers Graham Stewart (Soudan 1972 to 1975), Syd Galvin (Gallipoli 1971 to 1973), and Steve Millard (Gallipoli 1973 to 1974); and overseer Rodger Johnston. The role of Ken Moore at head office was also crucial in those early stages of planning. The very first instructions of the campaign issued in July 1971 to Bill Young in his last year at Alexandria were daunting:

> Clean out Gallipoli by pulling out all the stock left there and putting some in Bullock Paddock . . . Clean up the Adder Block and Soudan and put into Gallipoli all selected heifers. These must bear a Brucellosis earmark and preferably a Botulism earmark . . . You should finish Gallipoli by end of August; Adder and Soudan by end of September and then be available for Maunder.[21]

Herd figures for Alexandria in the 1950s and 1960s were not always accurate. In 1968, Alexandria had 66,500 cattle on the books[22] and despite a huge subsequent turnoff, and a severe drought in 1969-70, the book figure in 1972 was still a high 51,141. A bangtail muster in 1972, however, showed around only 38,000 head on Alexandria, some 13,000 short of the book figure.[23] The severe drought of 1952 and losses from botulism had obviously affected the figures more than realised, and standard mortality write-offs had also been underestimated. For the disease program to be successful, however, numbers had to be as accurate as possible. On the smaller Channel properties, it was not so difficult to estimate the size of the herds.

The program caused management practices to change dramatically. Until the late 1960s, cattle had been managed under an open-range system where fences were few, and paddocks were mainly small holding paddocks. It had been considered essential to run steers with the breeders who knew the property, especially the watering points. But with more fences, closer watering points and facilities such as crushes for testing cattle, this policy changed. Different classes of stock— weaners, spayed cows,

21 Ken Moore, (General Manager) to Bill Young, (Manager Alexandria), 26 July 1971, Correspondence File 1, Alexandria, 1971, NAPR.

22 Discussions held at Alexandria Station between K.W.Moore, E.M.Crouch, F.H.Foster and W.E.Young, 23 July 1968, File 1, NAP Stations, 1968, NAPR.

23 Minutes of Directors' Meeting, 18 October 1972.

Brucella testing at Monkira. Lee McNicholl

breeders and steers — could be segregated and areas spelled and allowed to recover. The company was therefore in a better position to handle efficiently any range management, land care and environmental issues.

The pace of station work picked up. Stockmen became experienced in the segregation and handling of weaners which were without the guidance of older stock. There was an increased use of road trains within Alexandria as weaners were transported to various paddocks on the property. Eventually, with smaller paddocks and segregated areas, mustering could be done more efficiently with fewer men — a pertinent factor during the cattle slump of 1973-74 when men had to be put off. At the same time, award conditions for the men improved, resulting not only in wage increases but also improved facilities, such as mobile kitchens. Managers were forced to upgrade their bookkeeping procedures, because they were dealing more specifically with cattle numbers within paddocks rather than overall numbers on the property. They had to be very precise in their mustering program. They had to check on the contractors working on the improvements, and had to liaise closely with the veterinarians and with the DPI over the planning and implementation of the testing program. It was difficult for some older men who had not been schooled in such ways; some younger managers who grew up with the program readily adjusted.

The program gradually achieved results and by September 1976 the northern part of Alexandria virtually had all clean stock.[24] Testing facilities had been set up around the property — in yards at the homestead, Ranken, Soudan, Gallipoli and at Bores 3, 17, 24, 36 and 83. For testing the blood samples, the company converted a caravan at the homestead into a laboratory. Later, the government built a centralised permanent laboratory at Avon Downs, although at times blood samples were also sent to Alice Springs or to privately owned facilities at Brunette Downs. In order for the brucella testing to coincide with the TB testing, the results of the blood tests had to be back within three days. At first mobility was a problem, until aircraft were used both to transport staff to the testing points, and also to enable spotting for any stock that might have missed the muster. Motorcycles were also used in the musters. Lee McNicholl obtained a pilot's licence and put it to good use after 1974 when the board agreed that chartered air transport could be used, with costs recouped from government BTEC funds.[25] That same year the company purchased its own aircraft for general company use between the properties and for use in the disease eradication program.

24 Minutes of Directors' Meeting, 22 September 1976.

25 Minutes of Directors' Meeting, 15 May 1974.

Jackaroo School, Alexandria, 1982.

Drilling a bore which proved to be a dud at Sun Hill Paddock, Glenormiston, 1980. E.M. Crouch

Given the traditional movement of stock from Alexandria to the Channel properties, it was not surprising to find brucellosis and TB in the Channel herds. Reactor percentages were generally lower than at Alexandria, however, because of greater stock dispersal around natural waters, instead of vast numbers congregating at bores. Furthermore, stock were tested prior to leaving Alexandria. While NAP was not the first to implement the campaign in the Northern Territory, it was one of the first companies to begin simultaneous testing of TB and brucellosis in Queensland. The first full-scale testing on NAP's Channel properties took place at Coorabulka and Monkira in 1974. These two were chosen because they were sufficiently improved to provide the stock control required; they were the most manageable from a topographical and mustering point of view; and they provided an appropriate infrastructure in which the problems of disease control could be met and overcome. Nevertheless, until the introduction of BTEC, there were relatively few improvements on the Channel properties. Like Alexandria, they had been managed on an open-range system where the whole herd ran together.

The Monkira plan, a pilot for the area, was based again on the weaner segregation approach. Because the property was reasonably well sub-divided into paddocks such as Neuragully, Yenda, River, Bullock, Sallen and Bush, each year's weaners went into a 'clean' area, with the 'dirty' herd isolated in an area which was gradually destocked.[26] The Coorabulka plan partly destocked the station through the sale of fat bullocks and breeders. The steers and heifers were kept and tested and any reactors destroyed. Valroy, Grasstree, Bush, Bindy, Spell Horse, Bull, and No. 2 and No. 9 Paddocks were used, although further subdivisional fencing was required. Two of the larger paddocks so created were named after two of the most significant men in the company's history, Douglas Fraser and Francis Foster.

The company set about its task of disease eradication on the Channel properties with the same determination it had displayed on Alexandria. As a result, the DPI was soon offering congratulations because of the 'valuable stimulus' NAP provided 'to other owners with comparable disease control problems in remote areas'.[27]

Gradually results showed through. By May 1977 the heifers on Monkira and Coorabulka were confirmed TB free.[28] Islay Plains, where the program was also under way, and Monkira were both confirmed TB and brucellosis free within twelve months.[29]

Attention was also given to NAP's other Channel properties. In September 1976 Ken Moore could state that fencing was sufficient on Marion Downs and Glenormiston for testing to begin the following year.[30] As on Alexandria, the building of associated improvements followed. It was especially difficult on Glenormiston, however, because of the prevalence of gidyea poisoning, the high incidence of salty bores, and the difficulty of getting complete musters in its heavily timbered bush areas. The western side where the Mulligan River ran through the Toko Ranges was very difficult, because it was virtually impossible to erect and maintain fencing. The testing was carried out in the paddocks — Pituri, Wonditti, River, No.4 and No.42 — but because of the problems mentioned, Glenormiston was one of the last of NAP's Channel properties to be given confirmed-free status.

Similarly, at Marion Downs it was important to make the best use of the paddocked areas such as Herbert Downs, 8 Mile, Hamilton, Sandy, Sam's, Narrahs and Wanderer Paddocks, although with some caution on the fencing of Herbert Downs because of the threat of gidyea poisoning. Bill Alexander's aircraft greatly assisted in the complete muster of the station. By 1980, the property was divided into twelve main paddocks together with smaller homestead paddocks. Eight drafting/loading yards and thirty-five broncho yards existed.[31] Other improvements included twenty-four bores (nine flowing and fifteen equipped), one equipped well, nine earth tanks and six desilted waterholes.[32] Such aids to management did not go unnoticed during an inspection by Lands Department officers at the time of the application for the renewal of leases in 1981. As the Land Commissioner at Cloncurry commented:

26 L.McNicholl to Director, AIB Branch, DPI, Brisbane, 7 December 1973, Vet's File, 1973, NAPR.

27 Letter from DPI quoted in letter from Ken Moore to NAP Directors, 21 December 1976, File 4, Directors File, 1976.

28 Minutes of Directors' Meeting, 25 May 1977.29 K.W. Moore, NAP Disease Control Report for 1978, NAPR.

30 Ken Moore to DPI, 30 September 1976, File 4, Directors, 1976, NAPR.

31 Marion Downs, Gregory North, PH 380, Part 2, Lands Department, Brisbane.

32 V.W.Rosel, Land Commissioner, Cloncurry, to Land Administration Commission, 24 November 1981, Marion Downs, PH 380, Part 2, Lands Department, Brisbane.

The Company has been deeply involved for many years in disease testing on all their properties and DPI officers have informed me that they are leaders in this field. To assist this disease testing programme they have built testing paddocks and yards of good design. All boundary fencing has easily accessible graded tracks along and this is an achievement when the rough nature of some of the country is noted. In comparison, I could not help but notice that there was practically no tracks along neighbouring property boundaries.[33]

In 1983 a breakdown occurred in the disease program when TB reactors were detected in two groups of cattle at Alexandria. It was a severe blow to the morale of those involved and destocking had to take place in 1984 which affected the turnoff rates in future years.[34] Such a ruthless approach to disease eradication was imperative but a successful conclusion was at hand. At the Annual General Meeting in April 1985, Chairman Michael Crouch was able to tell shareholders:

Over 73,000 TB and Brucellosis tests have been carried out in the breeding herds showing no evidence of Brucellosis and only one TB lesion. This is by far the best annual result since these programmes commenced.[35]

Drought on Alexandria prevented any breeder herd testing there in 1986. Testing on the Channel properties was also limited, although no TB or brucellosis was detected in the 27,898 NAP cattle sent to slaughter or the several thousand which were required to undergo tests when moving.[36] Monkira was the first property to be completely free of both diseases — in 1984 — while tests on Alexandria in 1987, and on some remaining Channel properties a year later, finally gave the confirmed-free status the company had worked so hard to achieve for nearly twenty years.

To celebrate the successful eradication of TB and brucellosis on Alexandria, a cricket match was held at the station on Saturday 31 October 1987. As most of the Channel station managers had previously worked on Alexandria, Bob Wales (Monkira), Jamie MacFarlane (Coorabulka), Bill Alexander (Marion Downs) and Mal Debney (Glenormiston) were invited. To provide a spirit of competition Mittiebah (MCC) was invited to provide the opposition with the help of officers from the DPI. Ross Brunckhorst later reported:

Local rules applied to the game. No LBW decisions were allowed and nobody went out for a duck unless they were dismissed three times without scoring. Batsmen were required to retire after scoring 20 runs.[37]

Bob Wales summarised the success of the day — and of the campaign — when he said:'I thought it was a great morale booster for all concerned particularly the NAP employees'.[38]

It was still early days for the confirmed-free status — interrupted at Alexandria in August 1988 when one young bull was identified as a TB suspect. The twenty-four hour wait for the laboratory result placed the company in a precarious position as arrangements to load heifers at Alexandria for transfer to the Wainui Feedlot were delayed while waiting for the result. Fortunately the suspicious lesions were found to be consistent with a diagnosis of Actinomycosis and the favourable laboratory result relieved a great deal of tension. Had the animal tested positive to TB, Alexandria would have been considered to be an infected property and no further transfers of store steers to the Channel properties would have been permitted under the movement regulations. As from July 1988, stock from infected properties could only be sent to meatworks or to approved (infected) feedlots. As General Manager Ross Brunckhorst assessed the situation,

For this reason it would be injudicious to dispense with the breeder herds on the channel properties. It is too soon for the Company to be highly confident that TB has been eradicated from the Alexandria breeder herd even though a confirmed free status has been achieved through a long and rigorous testing programme.[39]

There have also been anxious moments on the Channel properties. In July 1987 there was an unfortunate test result for Coorabulka which had previously been given confirmed-free status. A

33 V.W.Rosel, Land Commissioner, Cloncurry, to Secretary, Land Administration Commission, Brisbane, 30 November 1981, Marion Downs, PH 04/5350, Lands Department, Brisbane.

34 Directors' Annual Report, 30 November 1983.

35 E.M.Crouch, Chairman's Address, Annual General Meeting, 17 April 1985.

36 Directors' Annual Report, 30 November 1986.

37 R.Brunckhorst, Report to Directors, 11/87, 18 November 1987, NAPR.

38 Bob Wales quoted in R.Brunckhorst, Report to Directors, 11/87, 18 November 1987, NAPR.

39 R.Brunckhorst, Report to Directors, 8/88, 17 August 1988, NAPR.

BOARD OF DIRECTORS
The North Australian Pastoral Company Pty. Limited, October 1991. Standing – Left to Right: Andrew Ingram, Henry Foster, Richard Launder, William Foster. Seated: David Alexander, Michael Crouch (Deputy Chairman), Christopher Lyndon (Chairman), William Fraser.

lesion (TB) was detected in a cow being sent to slaughter from the station, despite the fact that the animal had successfully passed several on-property tests.[40] The newly purchased properties also had to be tested free of disease. Connemara, purchased in 1985, required destocking and only after Kynuna (purchased 1986) and Boomarra (purchased 1989) had been given confirmed-free status could the Directors' Annual Report of 30 November 1989 confidently state that: 'All the Company's herd are confirmed free of Tuberculosis and Brucellosis'.[41]

The most telling results which clearly show the advantages of the BTEC program have been in the company's branding and turnoff statistics. For the six years prior to the campaign (1967-72), the average yearly branding on Alexandria was 13,272. For the six years following (1973-78), the average had risen to 15,401, an increase of 14 per cent.[42] This increase was in fact more significant because of the loss of breeders which reacted positively in the brucellosis campaign. But other factors associated with BTEC contributed to the increase: the quality of the herd gradually improved through spaying of poor quality

40 Directors' Annual Report, 30 November 1987.
41 Directors' Annual Report, 30 November 1989.
42 Alexandria Stock Returns, 1968-78, NAPR.

A manager's work is never done. John Ohlsen in his office at Alexandria, early 1980s.

heifers; the losses from botulism were minimised; there were better management practices through the availability of more improvements, which meant there were fewer mortalities; and earlier weaning allowed breeders to come into calf earlier.

The 1980s have seen branding increase to an average for the period 1981-90 on Alexandria of 18,979, an increase of 30 per cent over the 1967-72 figure.[43] Brandings for the total NAP herd over the same periods have also increased — from an average of 19,223 to 25,091, an increase of 23.4 per cent. Notwithstanding the introduction of both supplementary feeding and very early weaning on Alexandria as drought measures in the 1980s, both of which have boosted brandings, the figures are still a clear indication of the success of the BTEC program.

The results are also impressive when turnoff statistics are compared. The average annual turnoff (sales and stores) from Alexandria increased by 39.5 per cent in the seven-year period following the commencement of BTEC.[44] Subsequently, the total NAP turnoffs to meatworks increased by 19.2 per cent when comparing the 1983-90 and 1978-82 periods.[46]

Financial advantages resulted from these extra sales; dividends to shareholders increased and more funds became available for the purchase of other properties. Indeed, more properties were required to accommodate the increase in the total NAP herd.

Not only has NAP carried a major disease eradication program to a successful conclusion but valuable long-term economic advantages have also resulted. In so doing, the company has assumed a leadership role for which it has gained recognition and respect within the industry. Indeed, there is continuing involvement for General Manager Ross Brunckhorst, who was appointed in 1989 to the Northern Territory BTEC Management Committee.[47]

43 Statistics in this section are taken from NAP records.

44 This figure is compared with the 1958-72 period.

45 This figure is compared with the 1972-78 period.

46 NAP Stock Returns, 1978-90, NAPR.

47 Minutes of Directors' Meeting, 15 August 1989.

Chapter 14

The Takeover and Beyond?

After Francis Foster purchased Philip Forrest's shares in 1937, the ownership of NAP remained in mostly the same hands for nearly fifty years. The descendants of the Foster, Collins, Ingram, Alexander and Warner families owned the company and the distribution of shares between the various groups in 1984 was very similar to the pattern of the late 1960s. The Foster family stopped short of becoming majority shareholders in their own right, believing that about 40 per cent was a workable proportion. The idea was to ensure, as Henry Foster subsequently stated, that 'a spirit of co-operation, mutual trust and respect' between the partners was maintained.[1] The takeover period of 1984-85 was a dramatic testing time for the relationship, with the final outcome — formal control for the Fosters — decidedly changing the configuration of ownership.

The first sign of any outside interest in purchasing a substantial part of the company came in September 1984, when the board discussed an informal approach made two days earlier by a former Federal Minister for Primary Industry, Peter Nixon, to Michael Crouch, the company's chairman.[2] Nixon's initial offer price per share was soon raised to $4.40.[3] While the board rejected the proposal, considering the amount much too low, the company was nevertheless taken by surprise, never before having been confronted with such a sudden approach from outside. But corporate aggression was part of the Australian business scene in the 1980s and, as Henry Foster soon determined, NAP was a prime target because 'it has shown itself to be successful, has cash on hand and has readily realizable assets'.[4]

No more was heard of Nixon's offer in the next few months. By December the office staff were busy relocating, following the sale in October of the company's Edward Street property, NAPCO House, for $1,537,750. A new strata-titled office[5] at 25 Mary Street had been purchased for $310,000 and the shareholders were pleased with a special dividend of 10 per cent ($327,000) paid in November, largely as a result of profits realised from the NAPCO House sale. The last mail before Christmas, however, altered the focus of their attention.

Far North Pastoral Holdings Pty. Ltd., a Melbourne based-company, wrote to all NAP shareholders offering to purchase their shares for $4 each. Under NAP's articles, any share transactions had to be approved by the board, and any shares for sale had to be offered first to existing shareholders. The company's auditors, Coopers and Lybrand, had recently valued NAP shares for one such internal sale at around $1.50 per share on an earnings basis, the traditional method of valuing relatively small, non-controlling share parcels in a private company. The price of the Far North offer thus undoubtedly appeared attractive to some

1 Henry Foster to Shareholders, 26 March 1985, Takeover File, NAPR.

2 Minutes of Directors' Meeting, 19 September 1984.

3 Michael Crouch Interview, Brisbane, 12 June 1991.

4 Henry Foster to E.M.Crouch, 22 September 1984, Takeover File, NAPR.

5 An adjoining strata titled office (Unit 22) was purchased in January 1986 for $325,000 and after leasing to tenants until the end of 1988, the company increased its office space by expanding into this unit.

NAP shareholders, being considerably more than double the amount per share they would expect to realise under normal circumstances.

But the board recognised that the appropriate method of valuing shares when purchase of the whole company was contemplated — as opposed to the transfer of a minor, non-controlling interest — was on a liquidation basis, since a successful bidder would have access to company assets for subsequent sale if desired. The nature of primary industry is such that the two methods of valuation typically give very different results for rural enterprises, and NAP was no exception. Far North's offer price of $4 was judged by the NAP board to be much too low on a liquidation basis; if the company were to pass to Far North as the latter intended, then shareholders were entitled to a much better return for their shares. The board's general view was that at least $10 was a more realistic figure at that time.

Far North consisted of former Prime Minister Malcolm Fraser, who was a former NAP shareholder and a nephew of Douglas Fraser; Peter Nixon; Queensland pastoralist Don McDonald; Sir Robert Sparkes, the then president of the National Party of Queensland; and Ronald Melgaard, a management consultant.[6] In his letter to shareholders, Fraser spoke of his previous involvement in the company, saying that when he was Prime Minister he had sold his shares only because of a conflict of interest arising from government assistance to the pastoral industry during drought. He stated that, since leaving Parliament, he again wanted to involve himself in the pastoral industry.[7]

According to the tactics of takeovers, the timing of the offer was perfect. An unsettling impression of haste and urgency was generated because vital days for considering the issue were lost through the holiday period, and discussion among shareholders was hindered by the absence of advisers. Indeed, the NAP office was supposed to close for the Christmas break, but staff had to be recalled from holidays to deal with events. Shareholder groups became polarised, and there was enormous pressure on the Collins and Fraser families who had the largest number of shareholders: 25 of the 40. Some shareholders saw the takeover offer as an uninvited and unwanted intrusion into their lives, while others saw it as a financially opportune time to leave the company.

6 Statement A, Offer Document, Far North Pastoral Holdings Pty. Ltd., 19 December 1984, NAPR.

The Far North offer was quickly followed by further bids and counterbids. Such bids involved Collins group shareholders, Margaret Hockey and Des Persse, which caused further dislocation within the Collins and Fraser families.

3 Jan 1985	Azentaur Pty Limited (Margaret and Patrick Hockey) in association with Consolidated Press Holdings Limited (Kerry Packer) $5 per share Valued NAP at $16.35 million
21 Jan 1985	Far North Pastoral Holdings Pty. Limited Increased offer from $4 to $5.01
13 Feb 1985	Australian Agricultural Company Limited (AACo) through Des Persse's company, Bun Bun Holdings $6.50 per share Valued NAP at $21.26 million

On 19 February the chairman was advised that well-known pastoral landholder Peter Sherwin, with Elders IXL Ltd., would offer $7.50 per share,[8] but nothing more formal eventuated. The paperwork involved in responding to the offers by the required dates fully occupied the board, and the takeover proceedings took precedence over other matters. Liaison with Alexandria, the registered office of NAP in the Northern Territory, was important because formal documents were delivered there — once even by helicopter, as with material sent by AACo from its neighbouring Brunette Downs Station.

Each offer assumed that people wanted to sell, but this was not necessarily so, especially in the case of the Foster family. Their 43.2 per cent holding held the key to the outcome. With such a large amount of capital invested in the company, they had the most to lose by selling their shares at too low a price. As William Foster pointed out later, the rights of the larger shareholders had also to be recognised.[9] None of the offers in their view represented full value for the company, so the Fosters saw their own moving to majority control as the best avenue to settle the issue as quickly as possible. They therefore sought sufficient NAP

7 Malcolm Fraser to Shareholders, 21 December 1984, Takeover File, NAPR.

8 McIntosh, Hamson, Hoare, Govett Ltd. to E.M.Crouch, 19 February 1985, Takeover File, NAPR.

9 William Foster Interview, 11 June 1991.

shares to bring their holding to a clear 51 per cent. Prevailing Northern Territory legislation permitted partial takeovers and allowed a takeover to be effected through the purchase of shares from only three shareholders.[10] In fairness to all NAP investors, however, the Fosters made their offer to every other NAP shareholder.

On 8 March, Wivenhoe Pty. Limited, the Foster family's company, formally offered to purchase 15 per cent of other shareholders' shares for $8 each, in order to make up the difference they needed to gain majority control. Any shortfall of the required 283,000 shares (8.65 per cent) of the company[11] would be met by further shares, beyond 15 per cent, that people wanted to sell, these extra being selected from shareholders on a pro rata basis. In the event, the target was easily reached, with eight shareholders choosing to sell just 15 per cent of their shares, nineteen others selling up to 20.4 per cent to make up the shortfall, and the remaining shareholders not selling their holdings.[12]

Although for the Fosters the final outcome was successful, various pressures were brought to bear with the aim of thwarting their intentions. Six days after the Foster offer of 8 March, Far North increased its bid to $6.75 for 90 per cent acceptances.[13] Thus by the time of the board meeting of 27 March, although the Azentaur bid ($5) had lapsed and AACo had withdrawn its offer ($6.50), considering that the Foster plan would make it impossible for any outside organisation to make a successful takeover, the Far North offer ($6.75) remained. The board considered the two remaining offers — from Wivenhoe and from Far North — and recommended both to shareholders, depending on their circumstances. For those who wanted to sell all their holdings, Far North was recommended and for those who wanted to sell part of their holdings, Wivenhoe was recommended. The other option available to shareholders was, of course, to take up neither offer.

Before coming to a decision, shareholders were reminded to consider seriously the relative chances of success of the two offers.[14] There was indeed a fundamental problem for those shareholders who wanted to sell their holdings for the highest return. The $6.75 offered for all their shares was no doubt a better deal than $8 for 15 per cent of their shares — but only if that offer succeeded. The Far North documents stated that without acceptances for 90 per cent of shares their offer would not proceed, and so no funds would be forthcoming to accepting shareholders. Given the position of the Fosters, the Far North offer was bound to fail.

Nevertheless, within days of Far North's 14 March offer, Malcolm Fraser was claiming that Far North had acceptances for 40 per cent of shares.[15] In his communication of 26 March he said this had risen to 48 per cent of the company's shares.[16] It remained impossible for him to reach the 90 per cent acceptances his offer required, because the Fosters did not wish to sell; but the evidence from the time also suggests that it was impossible for him even to reach a majority holding in excess of 50 per cent of shares. Directors Crouch, Moore and Alexander publicly stated in the board's Part B Statement, responding to the original Far North offer, that they did not intend selling. The Fosters had also received verbal support from other shareholders to bring the combined total of shares that would not be sold to at least 50 per cent. Fraser nevertheless laboured the points that minority shareholders would suffer financially, and that such a partial takeover in any other State would require an independent expert's report. He criticised the NAP board for apparently supporting the interests of one dominant shareholder above the interests of all shareholders.[17]

At 5pm on 9 April 1985, when their offer period closed, the Fosters had won control despite an increased offer one week earlier from the combined Sherwin/Elders/Far North group (in reality Sherwin and Elders were using Far North as a vehicle for eventual Sherwin control) of $7.50 per share. But shareholders continued to be lobbied and the Far North offer was extended until 9 August. Inevitably the offer failed and Peter Sherwin was especially disappointed, because at

10 Companies Act of 1978 of the Northern Territory, Section 180C, p.215, NAPR.

11 Wivenhoe Pty. Limited required more than 7.8 per cent because the shares in the B.L.Foster Trust were administered independently by Perpetual Trustees and National Executors (Tasmania) Limited. During the transactions, 20.4 per cent of this holding was sold.

12 Takeover File, Foster Family Papers.

13 R.Gregory Melgaard, Director, Far North, to NAP Shareholders, 14 March 1985, NAPR.

14 The chairman, E.M.Crouch, in correspondence to shareholders of 23 January 1985, 27 February 1985 and 22 March 1985, had advised them to consider the possible success of any of the offers.

15 Malcolm Fraser to E.M. Crouch, 19 March 1985, Takeover File, NAPR.

16 Malcolm Fraser to E.M.Crouch, 26 March 1985, Takeover File, NAPR.

17 Malcolm Fraser to E.M.Crouch, 19 March 1985, 26 March 1985, NAPR.

the time he was anxious to acquire more cattle properties and he hated losing.

There is no doubt that the Fosters had significant points in their favour — the extent of their holding; the unanimous decision making within the Foster family; and their strong financial resources. But it was the simplicity and decisiveness of their offer which won through. With advice from Val Smith, the Fosters' Hobart solicitor, Henry and William Foster devised their strategy by late January — and did not waver from it. In contrast, the Far North offer changed three times. Henry Foster met with shareholders in Brisbane on 19 February 1985, so that everyone 'heard the same story'.[18] He patiently and systematically explained the family's position and reinforced the points he had made to them in a letter of 31 January, wherein the Fosters made certain commitments to other shareholders, to the effect that if they succeeded:

1. The current direction and management of the company would be maintained;
2. Family group representation on the board would continue. E.M. Crouch would be asked to continue as chairman and William Foster would be appointed as an additional director;
3. Share transfers would be registered after the 1985 Annual General Meeting so that shareholders would receive any dividend declared out of the 1984 profits;
4. In the event of a subsequent takeover offer acceptable to the Fosters being made before 1 January 1987, shareholders who had accepted the Wivenhoe offer in the meantime would be paid the difference between the higher price and $8 in respect of those shares purchased under the Wivenhoe offer;
5. Should an offer be made for the Foster majority shareholding a sale would only be made if all shareholders were made the same offer;
6. If a shareholder wished to subsequently sell all or part of their holding, the company would endeavour to find a buyer and a share price acceptable both to the company and the seller.[19]

The Fosters conscientiously adhered to these undertakings, although in the outcome points four and five were not applicable.

The rash of offers created difficulties for shareholders. In families whose forebears had worked long and hard to keep the company together, the situation created confusion and distrust and inflicted wounds which seemed impossible to heal. As Henry Foster stated at the time, the influence of the uninvited take-overs and media speculation was 'bewildering to shareholders, demoralising to staff and extremely damaging to the company'.[20]

Margaret Hockey (Azentaur Pty Limited), who had already combined with Consolidated Press Holdings in a failed bid, made one final effort to influence the outcome by calling a meeting on 6 April just before the Foster deadline.[21] At that meeting feelings were expressed and suggestions made, but nothing happened to alter the final result. The matter was finally settled some months later, when the Northern Territory Government confirmed that no breaches of the N.T. Companies Act 1978 had occurred during the takeover period.[22]

One of the undertakings made by the Fosters concerned creating an opportunity for shareholders who still wished to sell all or part of their holdings to do so at a price acceptable both to the company and the seller.[23] Throughout the remainder of 1985 and early 1986, the board endeavoured to find a suitable person or company who would be interested in purchasing such shares. Richard Launder, a part-owner in his family's costume jewellery company, was known to William Foster in Melbourne and emerged as that person. He was essentially a businessman but had run for some time a rural property in Victoria, and his wife was a member of the Angliss family which in former times had strong links with the pastoral industry.

At a board meeting in December 1986, the chairman tabled a proposal by Launder Nominees Ltd., Richard Launder's company, to purchase any number of shares in NAP up to 19 per cent of the issued capital at a price of $3 per share. The proposal also provided an attractive opportunity for NAP to purchase a 25 per cent equity in H.P.Launder (Aust) Pty. Ltd., a costume jewellery wholesaler and retailer, at an outlay of $2 million.[24]

18 Henry Foster, Transcript of speech to shareholders, Brisbane, 19 February 1985, Foster Family Papers.

19 Henry Foster to Shareholders, 31 January 1985, Takeover File, NAPR.

20 *ibid.*

21 Notes taken by Ken Moore, Meeting of Shareholders, 6 April 1985, Takeover File, NAPR.

22 Ian Tuxworth to R.M.Brunckhorst, 30 September 1985, Takeover File, NAPR.

23 Henry Foster to Shareholders, 31 January 1985, Takeover File, NAPR.

24 Minutes of Directors' Meeting, 17 December 1986.

The board approved both aspects of the proposal and the whole exercise proved very successful, although some shareholders could not suppress a quiet chuckle at the Annual Report of 30 November 1987 which listed among NAP's principal activities 'Manufacturers Costume Jewellery'.[25] Their amusement notwithstanding, a respected shareholder had entered the company and Launder's expertise on the board as a director since 1987 has brought a valuable new perspective. There were also financial gains when, just over twelve months after purchasing, NAP sold its 25 per cent equity in H.P.Launder (Aust) Pty. Ltd. at a profit of $1 million.

The proceedings leading up to the Launder purchase were not completely smooth. At the general meeting of shareholders on 27 February 1987 which was held to approve the Launder transactions, a last-minute attempt by Azentaur Pty. Limited to purchase 'call options' for 19 per cent of the company's shares at a price of $3.50 per share was foreshadowed.[26] The Launder proposal was successful, however, and a total of 18.9 per cent of NAP shares changed hands in 1987 as a result. Because there were both buyers and sellers from among NAP shareholders, Launder needed to purchase only 13.1 per cent of these. Shares amounting to 5.8 per cent were transacted between existing NAP shareholders — also at the price of $3 per share. It is interesting that among the eight buyers, there were three shareholders who had previously sold shares during the Foster offer. Indeed, it made economic sense to sell for $8 per share and purchase shares back at $3. A total of eleven shareholders (or family companies) sold some or all of their NAP shares.[27]

PRINCIPAL SELLERS	PRINCIPAL BUYERS
Fraser family (most)	Launder Nominees Limited
Bruxner family (all)	Wivenhoe Pty. Limited
Warner family (all)	Ingram family

Those who sold 20 per cent of their shares to the Fosters at $8 per share and the remainder during the Launder purchase averaged $4 for their total holding, the same as Far North originally offered.

25 Directors' Annual Report, 30 November 1987, p.21.

26 Minutes of General Meeting, Parkroyal Hotel, cnr. Alice and Albert Streets, 27 February 1987, NAPR.

27 Appendix to the Minutes of the Minutes of the Directors' Meeting, 15 April 1987.

By 1991, the pattern of ownership in NAP clearly reflected the increased influence of the Fosters and the decrease in the Collins/Fraser holdings. The Ingram and Alexander holdings remained approximately the same as before the takeover, while the complete sell-off of the Warner shares ended over 100 years of involvement in the company by that family. At the same time, Launder's entry into the company has been obviously significant. Subsequent to the takeover, the board also approved the purchase of small parcels of NAP shares by two other new shareholders — Doma Pty. Ltd. (a company associated with Hobart solicitor Val Smith) and solicitor Christopher Lyndon, Michael Crouch's son-in-law and his partner in the law firm, Crouch and Lyndon. The full pattern of changes of ownership of NAP since its founding can now be appreciated:

PERCENTAGE HOLDINGS IN NAP

Family	1880	1910	1937	1966	1984	1991
Forrest	25.0	31.3	—	—	—	
Foster	—	—	18.5	43.2	43.2	57.8
Collins	25.0	31.3	37.1	33.5	32.3	9.2
Alexander(a)	25.0	6.2	7.4	5.5	6.0	5.3
Ingram	12.5	15.6	18.5	13.5	14.0	12.5
Warner	12.5	15.6	18.5	4.1	3.2	—
Launder	—	—	—	—	—	13.5
Other (b)				0.2	1.3	1.7
	100.0	100.0	100.0	100.0	100.0	100.0

(a) Thomas McIlwraith's daughter, Blanche Alexander took up shares from the McIlwraith Estate in 1910.

(b)			
Crouch	1966 0.2%	1984 0.7%	1991 0.7%
Moore		1984 0.6%	1991 0.6%
Doma Pty.Ltd.			1991 0.3%
Lyndon			1991 0.04%

While the takeover bidders, the board and the shareholders directly affected the final outcome, the company's staff at head office and on the stations closely watched the takeover proceedings. Their livelihoods were at stake, and they had no control over the course of events. The station staff wondered whether their future would be as secure under different management, and head office staff wondered whether their jobs would be there at all. Would the head office remain in Brisbane? Would all the properties remain within the company with the same administrative staff? The atmosphere was tense and feelings ran high.

Michael Crouch's roles and responsibilities were the broadest and most difficult at this time. He was torn between his loyalties as chairman of the board, where he was trying to shape a future direction for the company; his responsibilities to shareholders as a board member, endeavouring to provide them with the best possible return on their investment; his position as an alternate director for the Ingram family in England, representing all the English shareholders, keeping them informed and seeking their opinions; his role as the company solicitor, giving legal advice; his responsibilities to some shareholders from the Collins and Fraser families who were private clients in his legal firm; and finally his own position as a shareholder in NAP.

It was an intolerable situation for him and certainly it was one of the most stressful periods of his life. Although painstaking in separating his roles and their implications, he was inevitably criticised by some, though never with substance. While as chairman and board member he would have recommended selling to outsiders if the price had been right, at the personal level his deep affection and respect for the company had developed over some forty years of involvement, and he was particularly concerned that in some hands properties would be individually sold and The North Australian Pastoral Company Pty. Limited would cease to exist as a significant entity. With the company in a healthy economic position and owning sound, viable properties, he saw no reason for that situation to change.[28]

The involvement of the media was significant — and damaging. Newspaper reports were often inaccurate and had sensationalised headlines, such as 'Hectic bidding duel for cattle run jewel'[29], 'Fraser wins fight for cattle giant'[30], 'Fraser set to lock horns in battle for NAPCO'[31], 'Fraser camp draws wider support'[32], and 'Territory investigates the fight for NAPCO'[33]. Press releases were fed to reporters from some of the takeover groups, to the extent that news of further bids was often first read in the papers. The result for shareholders trying to make an informed judgment was confusion.

The most significant advantage of the takeover period has been the stability that has followed. No longer is the company under threat of takeover. The pattern is clear; the Fosters have the majority holding and an increased influence on the board, but this does not mean that the company is now managed in an overbearing fashion. It would not be in the Fosters' interests for this to happen. Henry and William Foster have numerous other interests to attend to; they do not reside in Brisbane, nor do they wish to. They need the company to be managed capably and while they contribute to, and take a keen interest in, its progress and development, they also want consensus between board members and between themselves and the minority shareholders.

Another result of the takeover has been a keener awareness of the interests of minority shareholders, in particular the Collins group (including the Frasers). Their shares were eagerly sought by those trying to take control of the company at a time when the influence of the Collins family was declining. The Anna Bertha Collins line was the only one to produce males in any number — in the Fraser family[34] — and the females of the Robert Martin and William Collins families had virtually no say in such a male-dominated industry. Although they held strong loyalties to the family — hence the strong family argument Malcolm Fraser could put to them — their feelings towards the company had been influenced by years of frustration and neglect. During the takeover period, such grievances were aired. The period since, however, has seen relations much improved.

The composition of the board has changed considerably in the 1980s, especially since the takeover attempt. Sir Herbert Ingram died on 3 July 1980, his place being taken by his brother Michael Warren Ingram. With the English shareholders continuing to hold at least one-eighth of the issued share capital, Article 49 of the Company's Articles still applied, and Michael Crouch continued as their alternate director until 1986, when Christopher Lyndon became the alternate director. Andrew Ingram replaced his father Michael as director in 1991.

Board members essentially represented the various family groups, with the exception of Ken Moore who was appointed in 1980, and Bill Norton, who left the board in 1981 following his appointment to Stanbroke Pastoral Company.

28 Michael Crouch Interview, 12 June 1991.

29 *Australian*, 19 January 1985.

30 *Mercury*, 23 March 1985.

31 *Financial Review*, 27 March 1985.

32 *Financial Review*, 2 April 1985.

33 *Financial Review*, 3 June 1985.

34 There were two Collins males who died relatively young and so did not have sustained influence; Christopher, son of Robert Collins died in 1919 aged 37, and John, son of William Collins died in 1941 aged 35.

Raymond Bruxner, the first former jackaroo of the company to serve on the board, replaced Norton. As the son of Charles Bruxner and grandson of Henry Bruxner, he was the third generation of Bruxners to serve on the board — Henry (1936 to 1957), Charles (1957 to 1976) and Raymond (1981 to 1987) — a total of forty-six years' service to the company. Raymond subsequently left the board in 1987, following the sale of his family's holding in the company. William Foster joined the board in 1986, and Richard Launder in 1987. With the death of Ken Moore in 1990, in late 1991 the board consisted of;

Christopher Lyndon (Chairman)
Michael Crouch (Deputy Chairman)
David Alexander
Henry Foster
William Foster
Bill Fraser
Andrew Ingram
Richard Launder

The expertise of the members is more varied than ever. The strength of the pastoral interests — David Alexander, Henry Foster, Bill Fraser and Ken Moore until his death — has been diluted by the presence on the board of younger members, William Foster and Richard Launder, who have different skills, perceptions and backgrounds. The legal presence has been increased by the addition of solicitor Chris Lyndon. As well as being alternate director for the Ingram family, he was appointed a director in his own right in 1987, and in 1990 was appointed chairman of directors, following the resignation of Michael Crouch after twenty-two years as chairman. Crouch remains on the board as deputy chairman.

The takeover era and Phil Forrest's resignation from the company in 1936 have been the most significant watersheds in the company's history. On each occasion, those shareholders who have wanted to leave have done so, resulting in a more compatible, committed and stable group. Yet the company remains the family-based entity that it has always been. The arrival of Francis Foster following Forrest's departure ensured the company's survival and expansion. The cleansing process forced by the uninvited takeover offers of 1984-85 has been just as significant, and may yet prove to be the stimulus and impetus needed to take the company successfully into the next century.

Such new stability has also been reflected in NAP's recent financial history and profitability. Through bonus issues of shares in 1970 and 1987-88, and a new share issue in 1974, the company's capital base increased from £162,000 in 1939 to a paid up issued capital by 1988 of $14 million. In 1989, for the first time since the very early years of the partnership, shareholders were invited to contribute funds to the company — through a one for five cash issue of shares at $1 each. This raised $2.8 million towards the purchase of Boomarra, and most shareholders accepted the offer — an indication of their confidence in the company's future. Dividends had gradually been improving in the 1980s following losses during the 1970s, and in 1989 annual dividends exceeded $1 million for the first time.

From uncertain beginnings in 1877, NAP has grown to be one of the largest cattle-producing concerns in Australia. Its perspective has changed over the decades, from the early years when it held coastal properties in north Queensland, to a period of twenty-four years after 1910 when the huge property of Alexandria in the Northern Territory was the only holding. From 1934, there has been increasing diversification, with Alexandria used for breeding and Channel properties acquired for fattening. In the last two decades there has been a rapid change of style, by head office, station managers and staff. Increasingly, a greater expertise and sophistication has been required by those succeeding in the pastoral business — and in NAP's case this has been evident in matters such as drought management, breeding programs and marketing strategies.

NAP has shown itself to be more than equal to the challenges that have arisen both in the last century and in this. It has secured for itself a niche in the history of cattle production in Australia; and it has won respect for its achievements from government and from other members of the pastoral industry. Much of this success is attributable to the compact family nature of the company. And, of course, full credit is due to NAP's many loyal and dedicated workers, from the youngest jackaroo to the chairman of the board.

Yet ultimately it has been the elements which have dominated. One lesson which has influenced so many of the endeavours and developments of the company was so poignantly expressed by Phil Forrest in his final days with the company, '. . . but I cannot make it rain'.

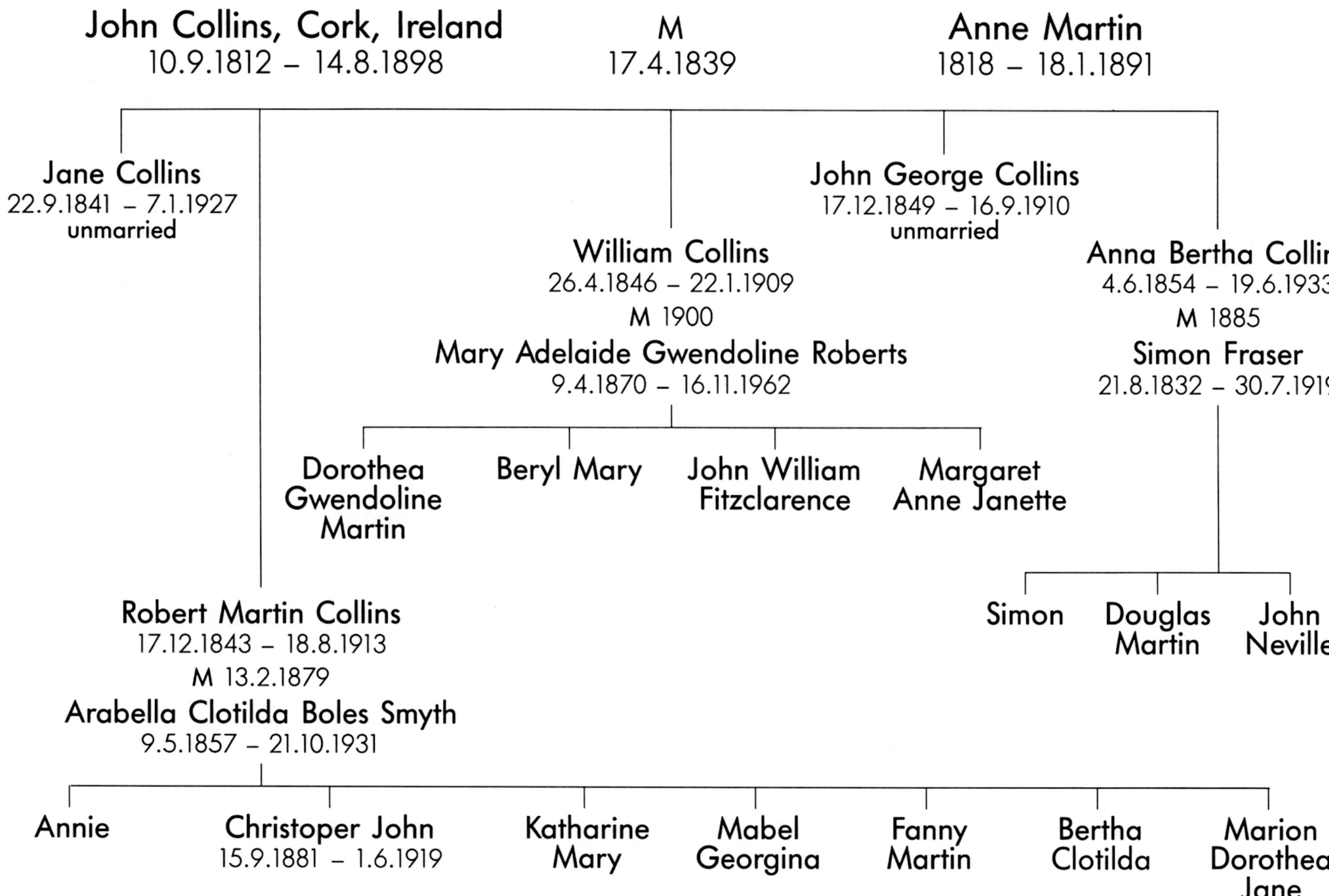
Collins Family
John Collins, Cork, Ireland
10.9.1812 – 14.8.1898
M
17.4.1839
Anne Martin
1818 – 18.1.1891
Jane Collins
22.9.1841 – 7.1.1927
unmarried
John George Collins
17.12.1849 – 16.9.1910
unmarried
William Collins
26.4.1846 – 22.1.1909
M 1900
Mary Adelaide Gwendoline Roberts
9.4.1870 – 16.11.1962
Anna Bertha Collins
4.6.1854 – 19.6.1933
M 1885
Simon Fraser
21.8.1832 – 30.7.1919
Dorothea Gwendoline Martin
Beryl Mary
John William Fitzclarence
Margaret Anne Janette
Robert Martin Collins
17.12.1843 – 18.8.1913
M 13.2.1879
Arabella Clotilda Boles Smyth
9.5.1857 – 21.10.1931
Simon
Douglas Martin
John Neville
Annie
Christoper John
15.9.1881 – 1.6.1919
Katharine Mary
Mabel Georgina
Fanny Martin
Bertha Clotilda
Marion Dorothea Jane

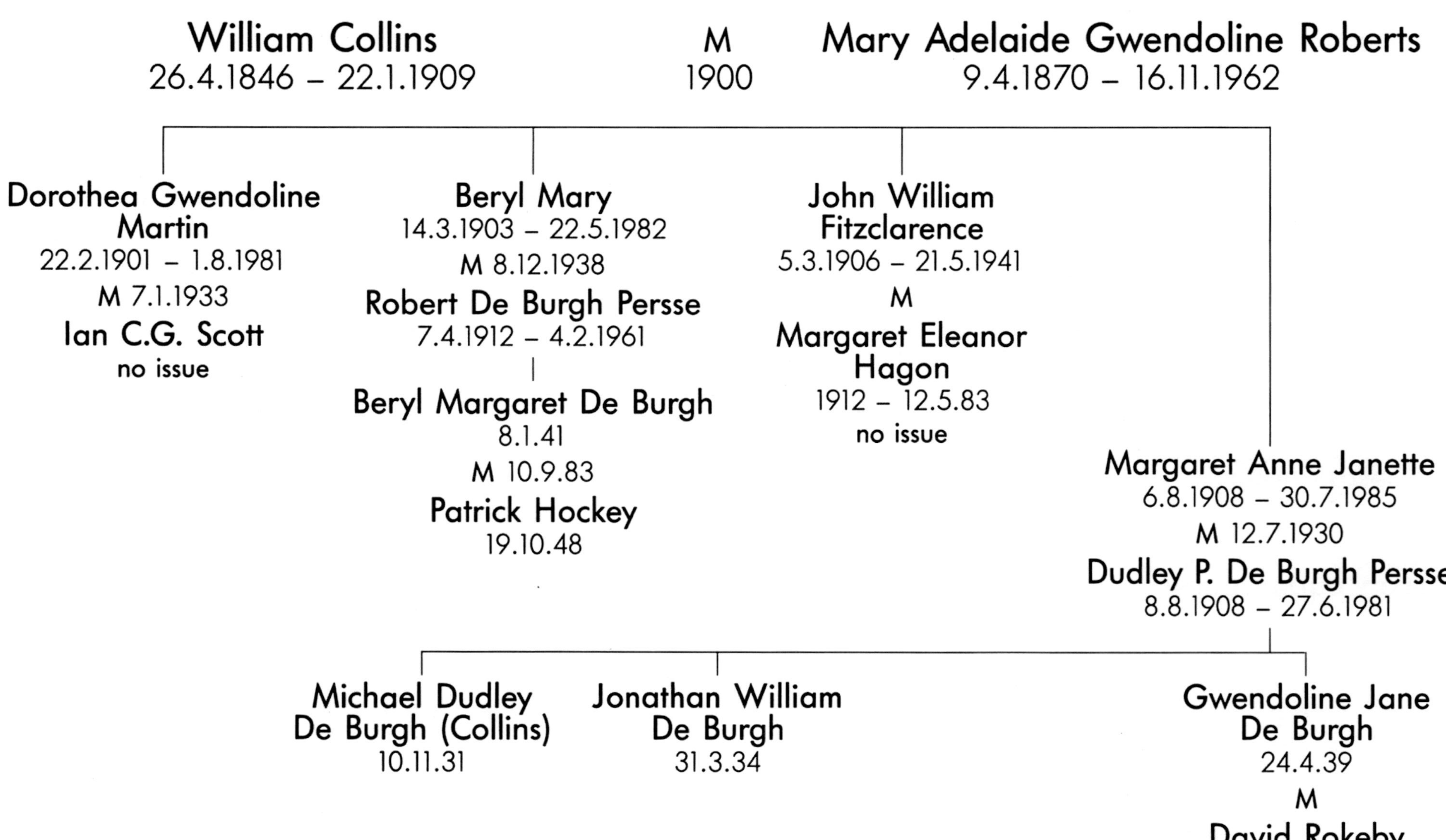
William Collins' Descendants
William Collins
26.4.1846 – 22.1.1909
M
1900
Mary Adelaide Gwendoline Roberts
9.4.1870 – 16.11.1962
Dorothea Gwendoline Martin
22.2.1901 – 1.8.1981
M 7.1.1933
Ian C.G. Scott
no issue
Beryl Mary
14.3.1903 – 22.5.1982
M 8.12.1938
Robert De Burgh Persse
7.4.1912 – 4.2.1961
Beryl Margaret De Burgh
8.1.41
M 10.9.83
Patrick Hockey
19.10.48
John William Fitzclarence
5.3.1906 – 21.5.1941
M
Margaret Eleanor Hagon
1912 – 12.5.83
no issue
Margaret Anne Janette
6.8.1908 – 30.7.1985
M 12.7.1930
Dudley P. De Burgh Persse
8.8.1908 – 27.6.1981
Michael Dudley De Burgh (Collins)
10.11.31
Jonathan William De Burgh
31.3.34
Gwendoline Jane De Burgh
24.4.39
M
David Rokeby Robinson

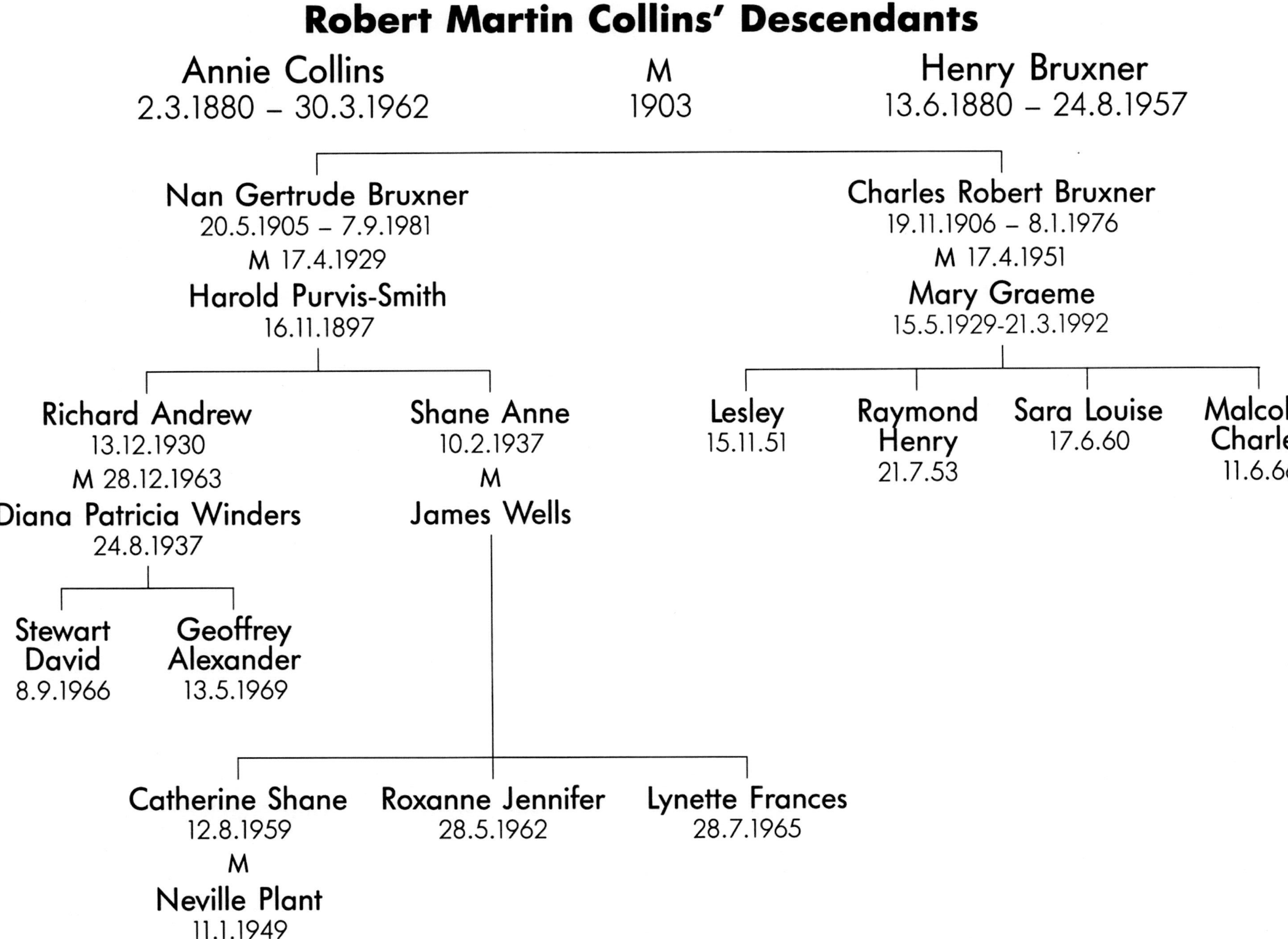
Robert Martin Collins' Descendants
Annie Collins
2.3.1880 – 30.3.1962
M
1903
Henry Bruxner
13.6.1880 – 24.8.1957
Nan Gertrude Bruxner
20.5.1905 – 7.9.1981
M 17.4.1929
Harold Purvis-Smith
16.11.1897
Charles Robert Bruxner
19.11.1906 – 8.1.1976
M 17.4.1951
Mary Graeme
15.5.1929-21.3.1992
Richard Andrew
13.12.1930
M 28.12.1963
Diana Patricia Winders
24.8.1937
Shane Anne
10.2.1937
M
James Wells
Lesley
15.11.51
Raymond Henry
21.7.53
Sara Louise
17.6.60
Malcolm Charles
11.6.66
Stewart David
8.9.1966
Geoffrey Alexander
13.5.1969
Catherine Shane
12.8.1959
M
Neville Plant
11.1.1949
Roxanne Jennifer
28.5.1962
Lynette Frances
28.7.1965

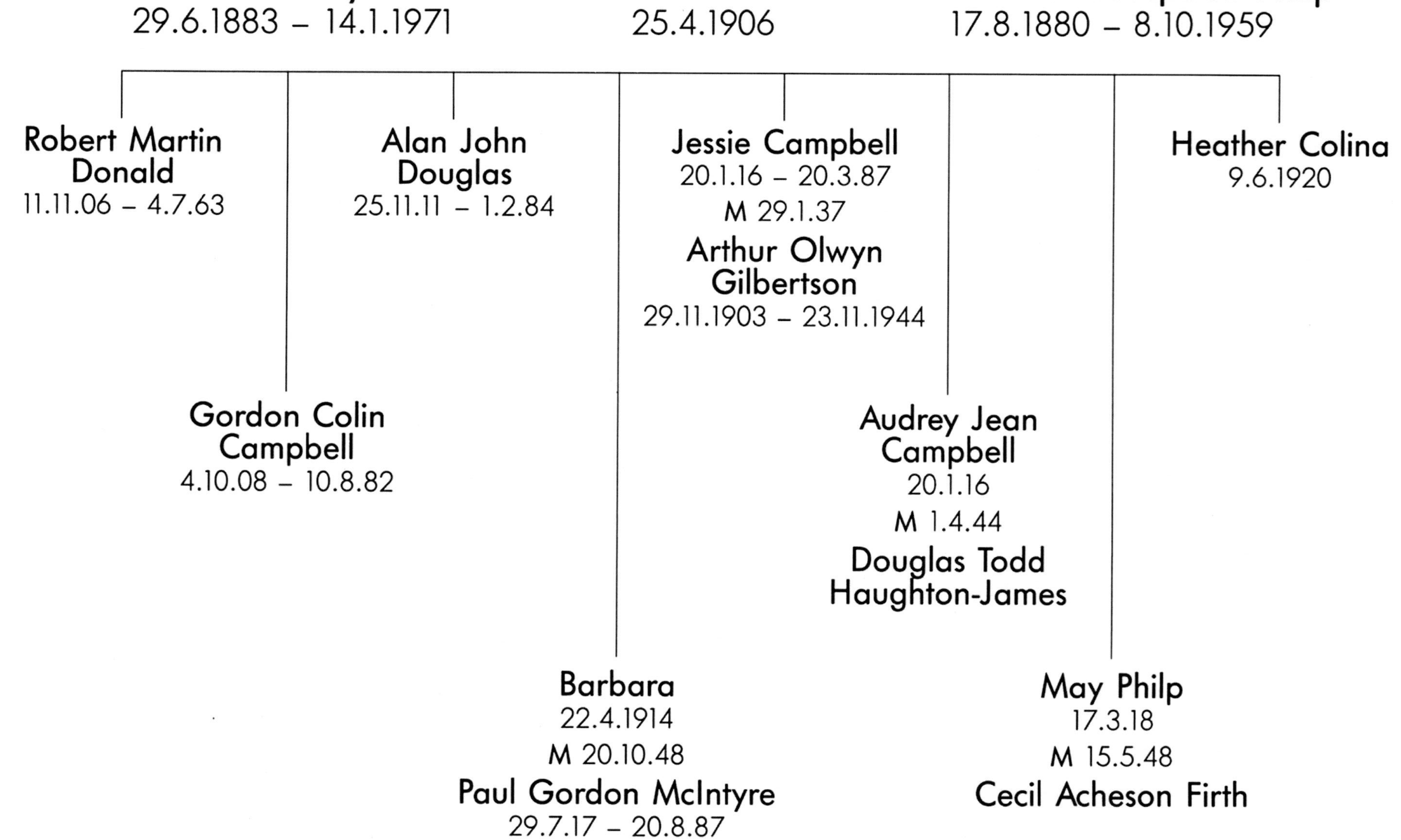
Robert Martin Collins' Descendants
Katharine Mary Collins
29.6.1883 – 14.1.1971
M
25.4.1906
Colin John Campbell Philp
17.8.1880 – 8.10.1959
Robert Martin Donald
11.11.06 – 4.7.63
Gordon Colin Campbell
4.10.08 – 10.8.82
Alan John Douglas
25.11.11 – 1.2.84
Barbara
22.4.1914
M 20.10.48
Paul Gordon McIntyre
29.7.17 – 20.8.87
Jessie Campbell
20.1.16 – 20.3.87
M 29.1.37
Arthur Olwyn Gilbertson
29.11.1903 – 23.11.1944
Audrey Jean Campbell
20.1.16
M 1.4.44
Douglas Todd Haughton-James
May Philp
17.3.18
M 15.5.48
Cecil Acheson Firth
Heather Colina
9.6.1920

Robert Martin Collins' Descendants

Mabel Georgina Collins
17.2.1885 – 10.1.1962

M
1909

Joseph Walter Ralston
17.8.1873 – 2.9.1929

Jean Collins
29.7.1912
M
Douglas Caine

Rosamond Anne Collins
14.12.1914

Arabella Georgina Collins
30.1.1918
M
Frederick Hooper
28.9.1917 – 24.10.1977

Robert John Collins
3.4.1920 – 1.11.1954

Alison Collins
13.9.1927
M
Simon Whiley

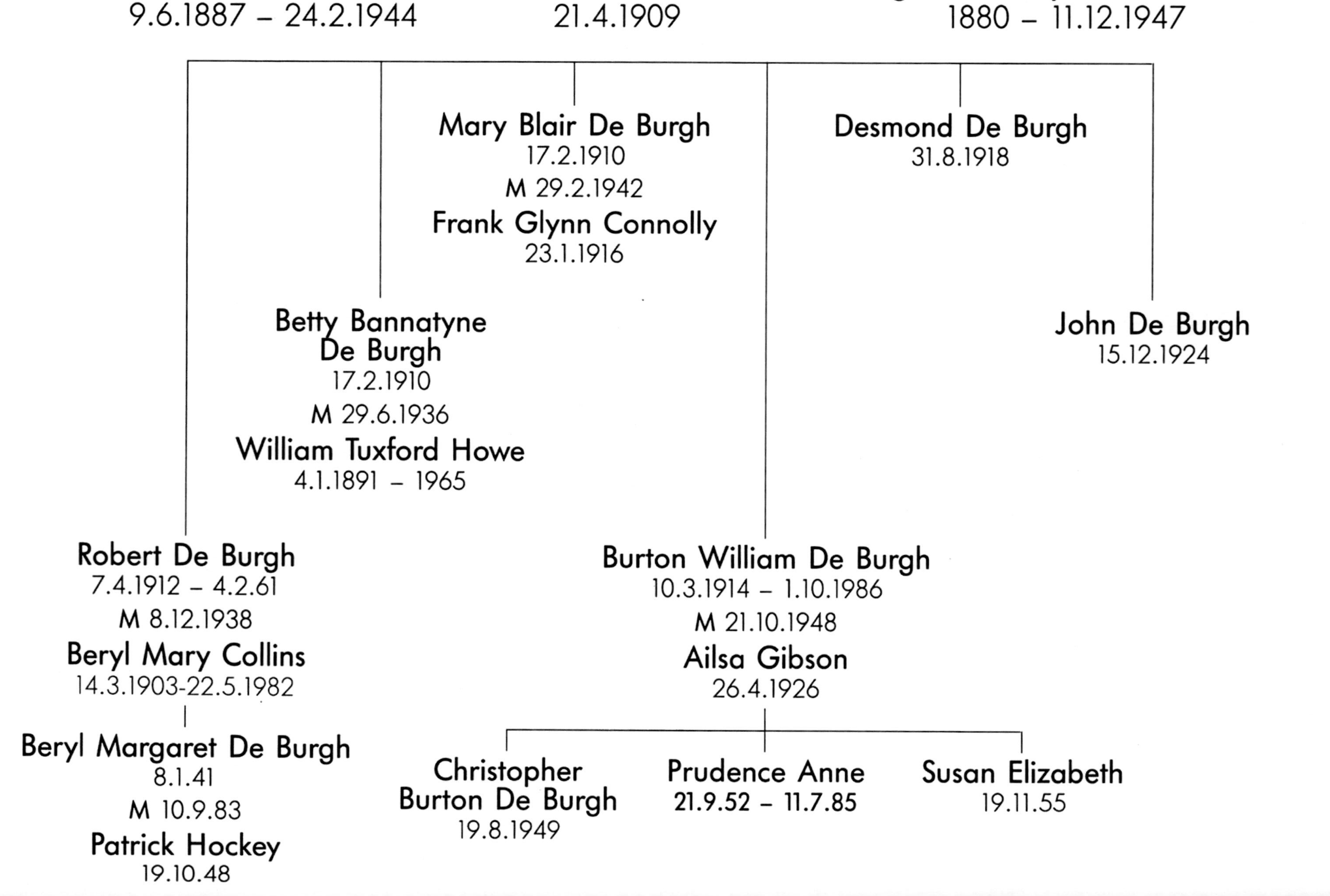
Robert Martin Collins' Descendants
Fanny Martin Collins
9.6.1887 – 24.2.1944
M
21.4.1909
De Burgh Bannatyne Bentinck Persse
1880 – 11.12.1947
Mary Blair De Burgh
17.2.1910
M 29.2.1942
Frank Glynn Connolly
23.1.1916
Desmond De Burgh
31.8.1918
Betty Bannatyne
De Burgh
17.2.1910
M 29.6.1936
William Tuxford Howe
4.1.1891 – 1965
John De Burgh
15.12.1924
Robert De Burgh
7.4.1912 – 4.2.61
M 8.12.1938
Beryl Mary Collins
14.3.1903-22.5.1982
Beryl Margaret De Burgh
8.1.41
M 10.9.83
Patrick Hockey
19.10.48
Burton William De Burgh
10.3.1914 – 1.10.1986
M 21.10.1948
Ailsa Gibson
26.4.1926
Christopher
Burton De Burgh
19.8.1949
Prudence Anne
21.9.52 – 11.7.85
Susan Elizabeth
19.11.55

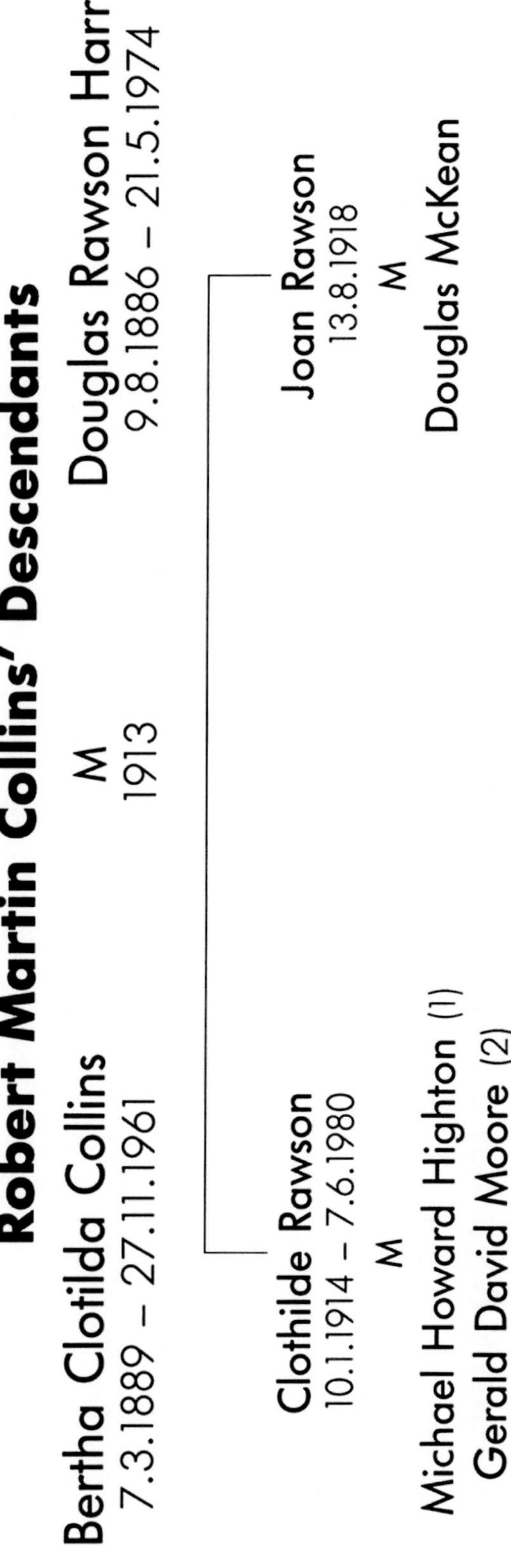
Robert Martin Collins' Descendants
Bertha Clotilda Collins
7.3.1889 – 27.11.1961
M
1913
Douglas Rawson Harris
9.8.1886 – 21.5.1974
Clothilde Rawson
10.1.1914 – 7.6.1980
M
Michael Howard Highton (1)
Gerald David Moore (2)
Joan Rawson
13.8.1918
M
Douglas McKean

Robert Martin Collins'/Anna Bertha Collins' Descendants

Marion Dorothea Jane Collins
18.6.1892 – 9.7.1963

M
ca. 1914

Douglas Martin Fraser
9.12.1888 – 16.5.1968

Douglas Collins	Margaret	William Martin	Anne Dorothea
14.9.1915 – 27.10.1935	9.5.1917	23.2.1922	8.6.1923
	M		M 18.12.48
	Herbert Harris		Alexander H.B. Clarke

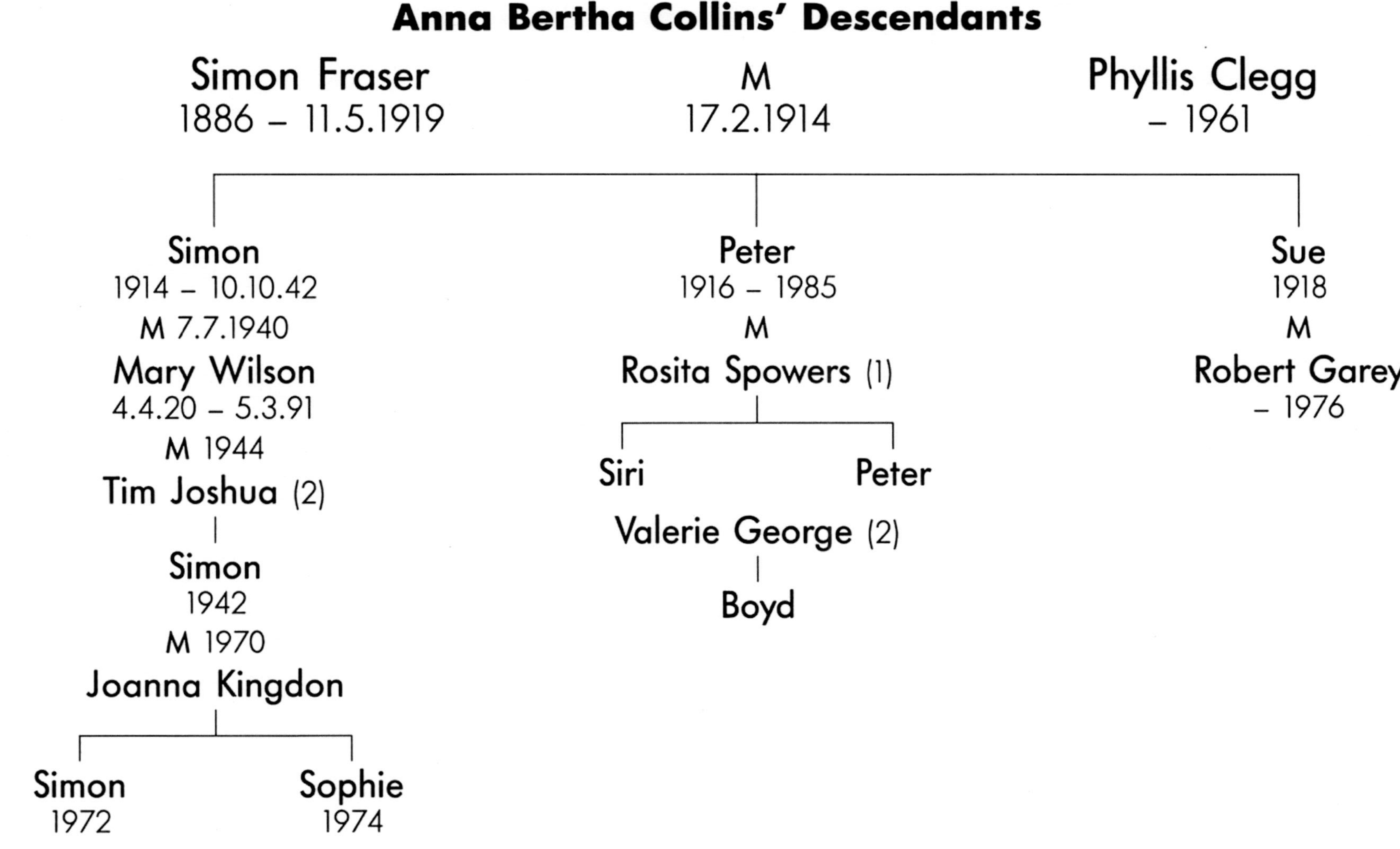
Anna Bertha Collins' Descendants
Simon Fraser
1886 – 11.5.1919
M
17.2.1914
Phyllis Clegg
– 1961
Simon
1914 – 10.10.42
M 7.7.1940
Mary Wilson
4.4.20 – 5.3.91
M 1944
Tim Joshua (2)
Simon
1942
M 1970
Joanna Kingdon
Simon
1972
Sophie
1974
Peter
1916 – 1985
M
Rosita Spowers (1)
Siri
Peter
Valerie George (2)
Boyd
Sue
1918
M
Robert Garey
– 1976

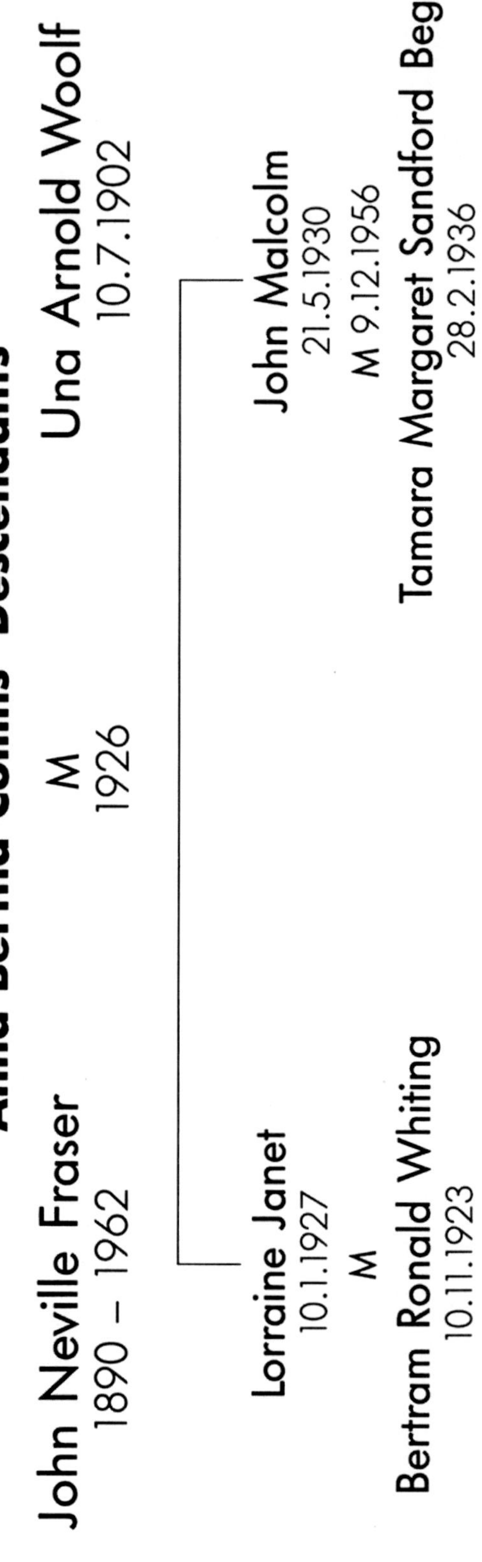
Anna Bertha Collins' Descendants
John Neville Fraser
1890 – 1962
M
1926
Una Arnold Woolf
10.7.1902
Lorraine Janet
10.1.1927
M
Bertram Ronald Whiting
10.11.1923
John Malcolm
21.5.1930
M 9.12.1956
Tamara Margaret Sandford Beggs
28.2.1936

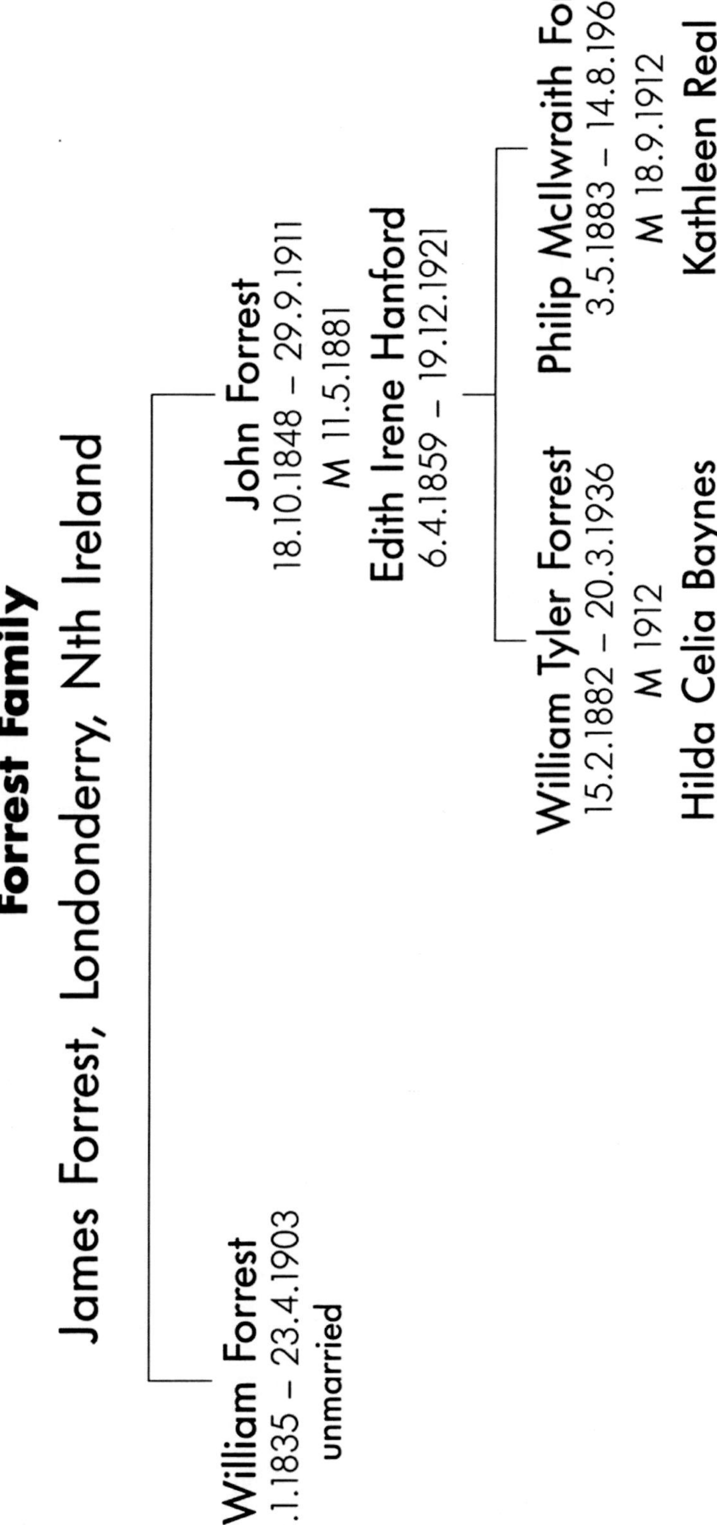
Forrest Family
James Forrest, Londonderry, Nth Ireland
William Forrest
11.1.1835 – 23.4.1903
unmarried
John Forrest
18.10.1848 – 29.9.1911
M 11.5.1881
Edith Irene Hanford
6.4.1859 – 19.12.1921
William Tyler Forrest
15.2.1882 – 20.3.1936
M 1912
Hilda Celia Baynes
Philip McIlwraith Forrest
3.5.1883 – 14.8.1964
M 18.9.1912
Kathleen Real
1.10.1882 – 17.4.1969
Edith Nancy Forrest
5.6.1913 –
M 28.2.63
Edward Cyril
Grover Grace

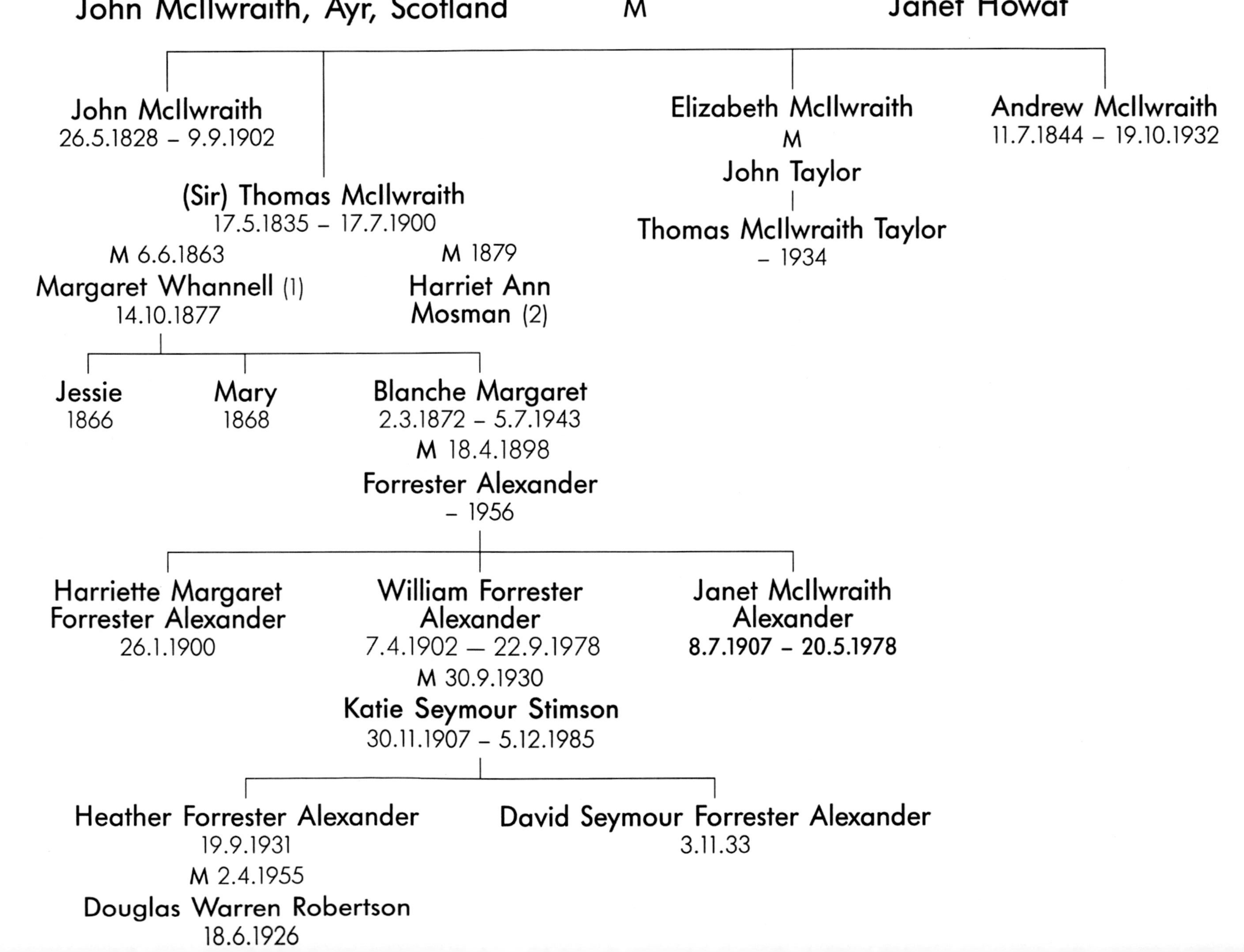
McIlwraith Family
John McIlwraith, Ayr, Scotland
M
Janet Howat
John McIlwraith
26.5.1828 – 9.9.1902
(Sir) Thomas McIlwraith
17.5.1835 – 17.7.1900
M 6.6.1863
Margaret Whannell (1)
14.10.1877
M 1879
Harriet Ann
Mosman (2)
Elizabeth McIlwraith
M
John Taylor
Thomas McIlwraith Taylor
– 1934
Andrew McIlwraith
11.7.1844 – 19.10.1932
Jessie
1866
Mary
1868
Blanche Margaret
2.3.1872 – 5.7.1943
M 18.4.1898
Forrester Alexander
– 1956
Harriette Margaret
Forrester Alexander
26.1.1900
William Forrester
Alexander
7.4.1902 — 22.9.1978
M 30.9.1930
Katie Seymour Stimson
30.11.1907 – 5.12.1985
Janet McIlwraith
Alexander
8.7.1907 – 20.5.1978
Heather Forrester Alexander
19.9.1931
M 2.4.1955
Douglas Warren Robertson
18.6.1926
David Seymour Forrester Alexander
3.11.33

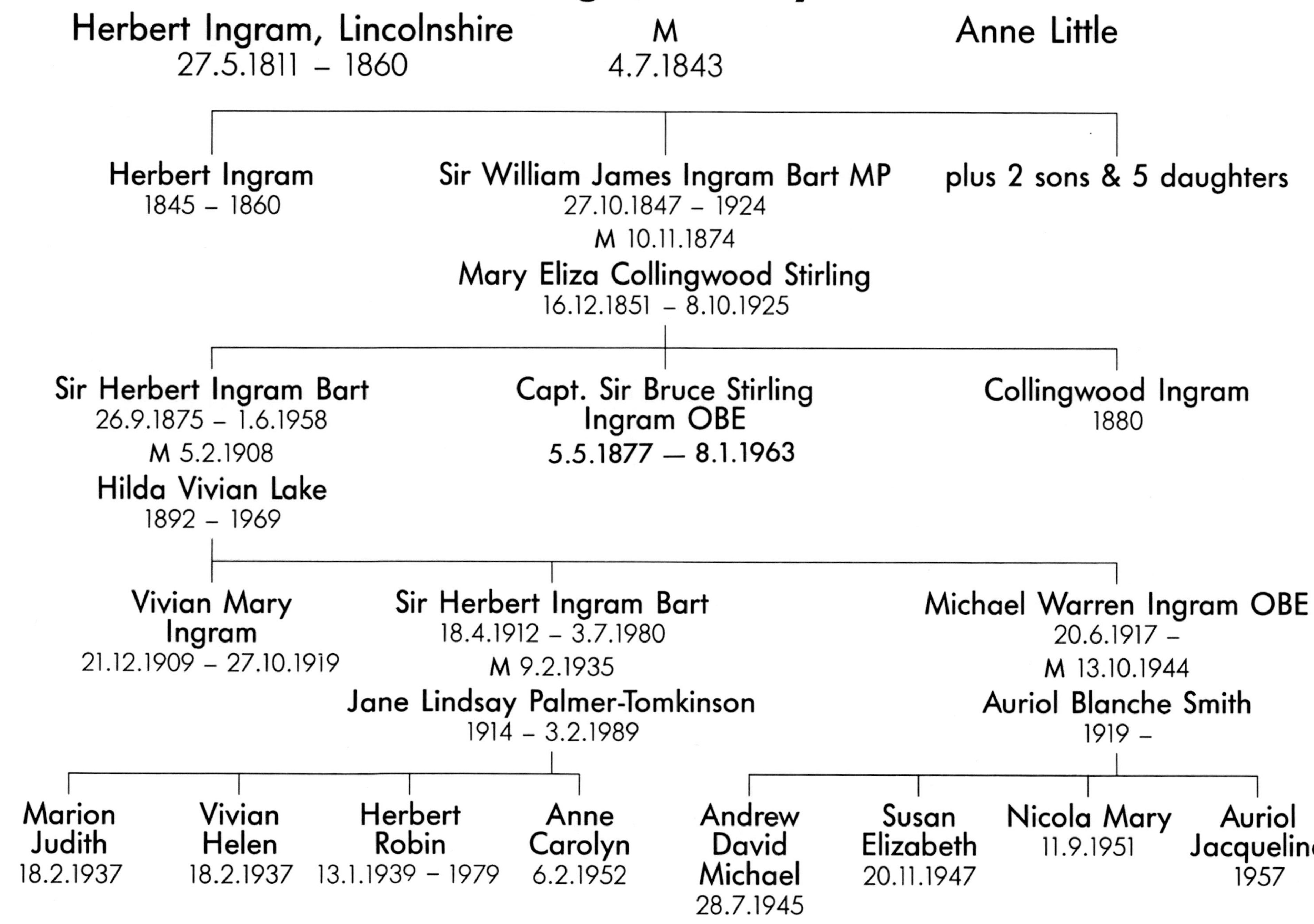
Ingram Family
Herbert Ingram, Lincolnshire
27.5.1811 – 1860
M
4.7.1843
Anne Little
Herbert Ingram
1845 – 1860
Sir William James Ingram Bart MP
27.10.1847 – 1924
M 10.11.1874
Mary Eliza Collingwood Stirling
16.12.1851 – 8.10.1925
plus 2 sons & 5 daughters
Sir Herbert Ingram Bart
26.9.1875 – 1.6.1958
M 5.2.1908
Hilda Vivian Lake
1892 – 1969
Capt. Sir Bruce Stirling
Ingram OBE
5.5.1877 — 8.1.1963
Collingwood Ingram
1880
Vivian Mary
Ingram
21.12.1909 – 27.10.1919
Sir Herbert Ingram Bart
18.4.1912 – 3.7.1980
M 9.2.1935
Jane Lindsay Palmer-Tomkinson
1914 – 3.2.1989
Michael Warren Ingram OBE
20.6.1917 –
M 13.10.1944
Auriol Blanche Smith
1919 –
Marion
Judith
18.2.1937
Vivian
Helen
18.2.1937
Herbert
Robin
13.1.1939 – 1979
Anne
Carolyn
6.2.1952
Andrew
David
Michael
28.7.1945
Susan
Elizabeth
20.11.1947
Nicola Mary
11.9.1951
Auriol
Jacqueline
1957

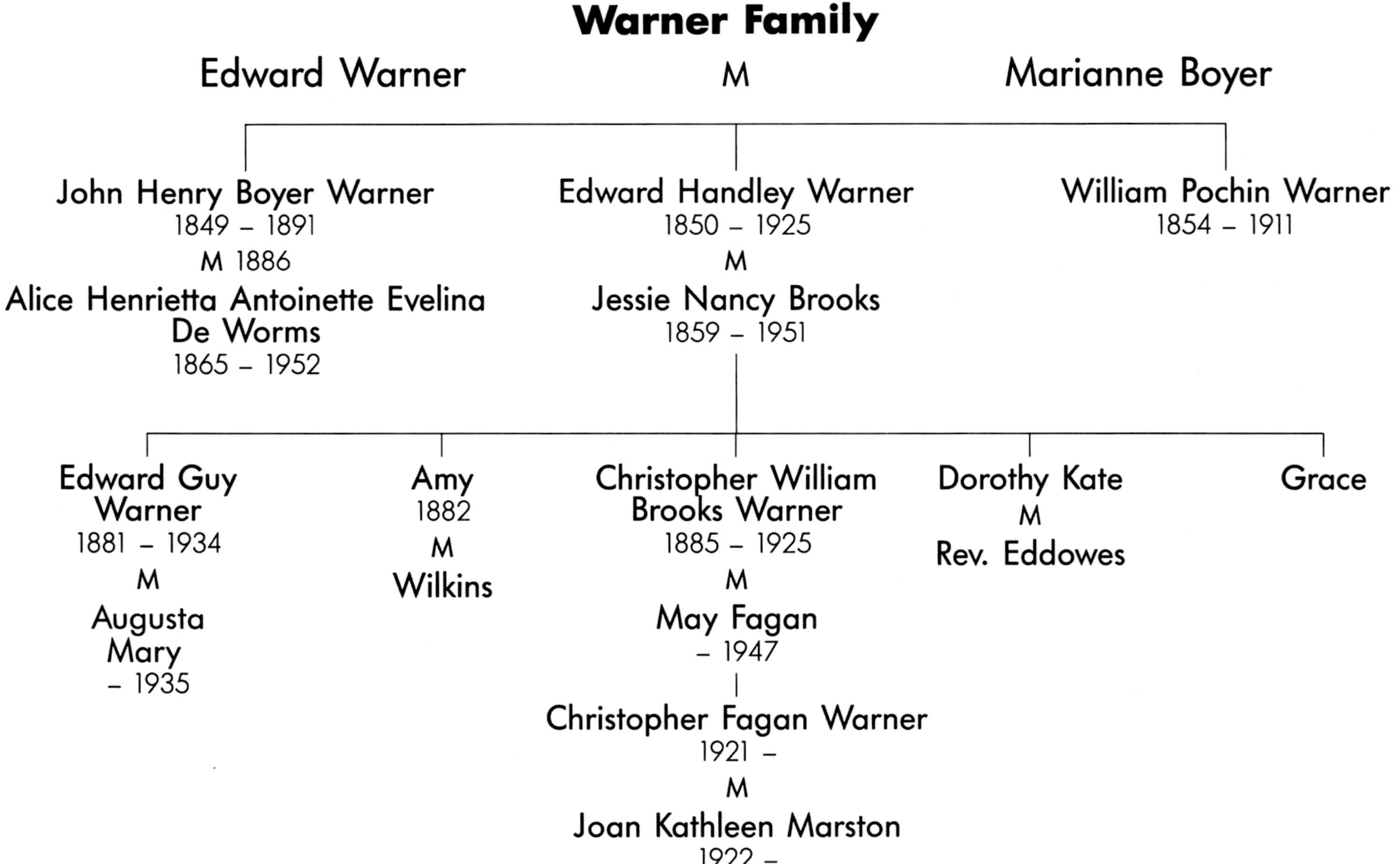
Warner Family
Edward Warner
M
Marianne Boyer
John Henry Boyer Warner
1849 – 1891
M 1886
Alice Henrietta Antoinette Evelina
De Worms
1865 – 1952
Edward Handley Warner
1850 – 1925
M
Jessie Nancy Brooks
1859 – 1951
William Pochin Warner
1854 – 1911
Edward Guy
Warner
1881 – 1934
M
Augusta
Mary
– 1935
Amy
1882
M
Wilkins
Christopher William
Brooks Warner
1885 – 1925
M
May Fagan
– 1947
Christopher Fagan Warner
1921 –
M
Joan Kathleen Marston
1922 –
Dorothy Kate
M
Rev. Eddowes
Grace

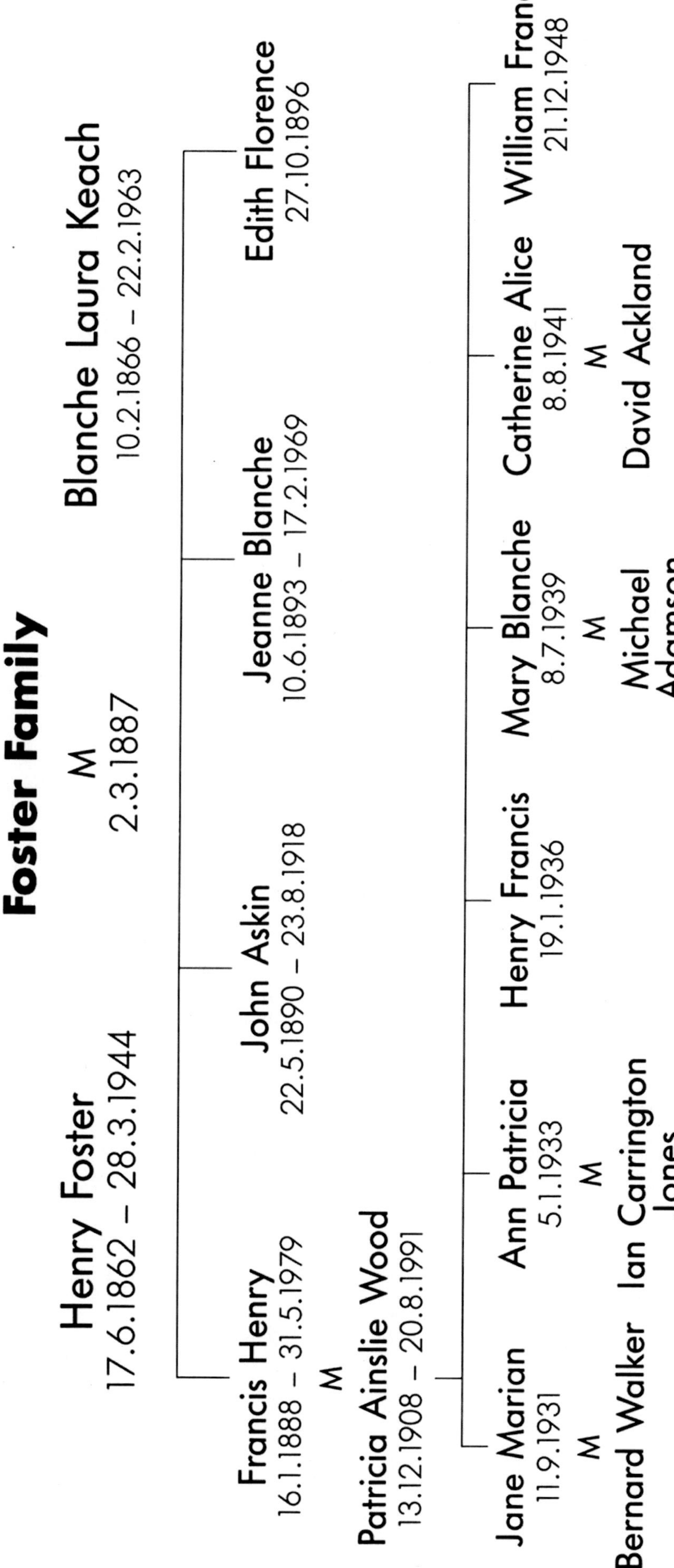
Foster Family
Henry Foster
17.6.1862 – 28.3.1944
M
2.3.1887
Blanche Laura Keach
10.2.1866 – 22.2.1963
Francis Henry
16.1.1888 – 31.5.1979
M
Patricia Ainslie Wood
13.12.1908 – 20.8.1991
John Askin
22.5.1890 – 23.8.1918
Jeanne Blanche
10.6.1893 – 17.2.1969
Edith Florence
27.10.1896
Jane Marian
11.9.1931
M
Bernard Walker
Ann Patricia
5.1.1933
M
Ian Carrington Jones
Henry Francis
19.1.1936
Mary Blanche
8.7.1939
M
Michael Adamson
Catherine Alice
8.8.1941
M
David Ackland
William Francis
21.12.1948

APPENDIX 1

PROPERTIES HELD BY THE NORTH AUSTRALIAN PASTORAL COMPANY PTY. LIMITED DECEMBER 1991

A. PASTORAL PROPERTIES

	square kilometres	square miles
ALEXANDRIA	16,116	6,198
MARION DOWNS	12,700	4,885
COORABULKA	6,494	2,498
GLENORMISTON	6,920	2,662
CONNEMARA	5,025	1,933
MONKIRA	3,730	1,435
VERGEMONT	3,516	1,352
BOOMARRA	1,080	415
KYNUNA	881	339
TOTAL	56,462	21,717

B. LEASED BRISBANE DEPOT — Dunn Road, Rocklea

C. FREEHOLD TITLE

WAINUI AGGREGATION:	Includes Wainui, The Willows and Rangeview - 1,800 hectares (4,446 acres)
BETA STREET, MOUNT ISA:	(Residence)
13 ENTERPRISE ROAD, MOUNT ISA:	(Office)
UNITS 21 and 22, 25 MARY STREET, BRISBANE:	(Head Office)

APPENDIX 2

ADMINISTRATION OF THE NORTH AUSTRALIAN PASTORAL COMPANY PTY. LIMITED

MANAGING PARTNERS

William Forrest and William Collins	1877 - 1902
John Forrest	1902 - 1911
William Tyler Forrest	1911 - 1915
Philip McIlwraith Forrest	1915 - 1931

MANAGING DIRECTORS/GENERAL MANAGERS

Philip McIlwraith Forrest	1931 - 1936
Douglas Martin Fraser	1936 - 1956

MANAGING AND EXECUTIVE DIRECTOR

Edward Michael Crouch	1972 - 1981

CHAIRMAN OF DIRECTORS

Philip McIlwraith Forrest	1931 - 1936
Douglas Martin Fraser	1936 - 1967
Edward Michael Crouch	1967 - 1990
Christopher Brice Lyndon	1990 -

GENERAL MANAGERS

Kenneth Woodhead Moore	1956 - 1982
Ross Murray Brunckhorst	1982 -

APPENDIX 3

BOARD OF DIRECTORS OF THE NORTH AUSTRALIAN PASTORAL COMPANY PTY. LIMITED

DIRECTORS		COMMENCED	DEPARTED
Phillip McI.	FORREST	1931	1936
Edward M.	SHAW	1931	1943
John W.F.	COLLINS	1931	1939
Thomas McI.	TAYLOR	1931	1934
*Walter E.	PUCKERING	1931	1934
William C.	FRITH	1931	1958
Douglas M.	FRASER	1932	1968
*Herbert, Sir Snr	INGRAM	1934	1958
William F.	ALEXANDER	1934	1976
Henry	BRUXNER	1936	1957
Francis H.	FOSTER	1937	1971
William M.	FRASER	1951	
Edward M.	CROUCH	1957	
Charles R.	BRUXNER	1957	1976
*Herbert,Sir Jnr	INGRAM	1958	1980
Henry F.	FOSTER	1971	
Harold W.	CHAMBERS	1971	1975
David S.F.	ALEXANDER	1976	
William S.	NORTON	1976	1980
Kenneth W.	MOORE	1980	1990
*Michael W.	INGRAM	1980	1991
Raymond H.	BRUXNER	1981	1987
William F.	FOSTER	1986	
Christopher B.	LYNDON	1987	
Richard A.	LAUNDER	1987	
*Andrew D.M.	INGRAM	1991	

*Represented by Alternate Directors

Edward R. Crouch	1932 to 1955
Edward M. Crouch	1955 to 1986
Christopher B. Lyndon	1986 -

Appendix 4

SHAREHOLDERS OF THE NORTH AUSTRALIAN PASTORAL COMPANY PTY. LIMITED DECEMBER 1991

		SHAREHOLDERS	FAMILY ORIGIN
1.	ACKLAND	Catherine Alice	Foster
2.	ADAMSON	Mary Blanche	Foster
3.	ALEXANDER	David Seymour Forrester	McIlwraith
4.	ALEXANDER	Harriette Margaret Forrester	McIlwraith
5.	CAINE	Jean Collins	Collins
6.	CARRINGTON-JONES	Ann Patricia	Foster
7.	CAWLA PTY. LTD.		Foster
8.	COLLINS-PERSSE	Michael Dudley de Burgh	Collins
	PERSSE	Jonathon William de Burgh	Collins
9.	CONNOLLY	Mary Blair de Burgh	Collins
10.	CROUCH	Edward Michael	*
11.	DOMA PTY.LTD.		*
12.	FIRTH	May Philp	Collins
13.	FOSTER	Henry Francis	Foster
14.	FOSTER	Patricia Ainslie (dec'd)	Foster
15.	FOSTER	William Francis	Foster
16.	FRASER	Simon	Collins
17.	FRASER	William Martin	Collins
18.	HAUGHTON-JAMES	Audrey Jean Campbell	Collins
19.	HOCKEY	Beryl Margaret de Burgh	Collins
20.	HOOPER	Arabella Georgina Collins	Collins
21.	INGRAM	Estate Lady Jane Lindsay (dec'd) (Michael Warren Ingram and Christopher David Palmer-Tomkinson)	Ingram
22.	INGRAM	Estate Lady Jane Lindsay (dec'd) (Michael Warren Ingram and Christopher David Palmer-Tomkinson) and INGRAM, Michael Warren	Ingram
23.	INGRAM	Michael Warren	Ingram
24.	LAUNDER NOMINEES PROPRIETARY LIMITED		*
25.	LYNDON	Christopher Brice	*

26. McINTYRE	Barbara	Collins
27. McKEAN	Joan	Collins
28. MOORE	Betty Annette MOORE, John Richard MOORE and Susan Louise HENZELL	*
29. PERPETUAL TRUSTEES TASMANIA LIMITED		Foster
30. PERSSE	Christopher Burton de Burgh	Collins
31. PERSSE	Desmond de Burgh	Collins
32. PERSSE	Estate Margaret Anne Janette(dec'd)	Collins
33. PHILP	Heather Colina	Collins
34. RALSTON	Rosamond Anne Collins	Collins
35. ROBERTSON	Heather Forrester	McIlwraith
36. SCOTT	Estate Dorothea Gwendoline Martin (dec'd)	Collins
37. TRUST COMPANY OF AUSTRALIA LIMITED	Harriette Margaret Forrester ALEXANDER and David Seymour Forrester ALEXANDER	McIlwraith
38. WALFOSCO PTY. LTD.		Foster
39. WALKER	Jane Marian	Foster
40. WHILEY	Alison Collins	Collins
41. WIVENHOE PROPRIETARY LIMITED		Foster
42. WIVENHOE NOMINEES PTY. LTD.		Foster

* new shareholders

APPENDIX 5

STATION MANAGERS OF THE NORTH AUSTRALIAN PASTORAL COMPANY PTY. LIMITED

INKERMAN	1877 - 1910	John Townshend
		James Henry Rae
WOODSTOCK	1877 - 1910	John Carr
ALEXANDRIA	1877 - 1882	Unstocked
	1882 - 1898	Thomas Harding
	1898 - 1900	(Tennison H.Robbins
		(Mr Martin
	1900 - 1924	Richard Holt
	1924 - 1938	C.A.Y. Johnston
	1938 - 1946	Harry Barnes
	1946 - 1955	Rowand Murray
	1955 - 1971	Bill Young
	1972 - 1976	Ken McGuire
	1976 - 1981	George MacQueen
	1981 - 1991	John Ohlsen
	1991 -	Ross Peatling
SOUDAN OUT-STATION	1910s- 1915; 1918-30	Jack W. Spratt
	1930 * 1940	*(E.D. Lemon
		*(Tommy Groom
		*(William Gibson
	1940 - 1948	Cliff Nissen
	1948 - 1950	M.A. Marshall
	1951 - 1952	Tommy Groom
	1952 - 1953	George Schultz
	1954 - 1955	Ken McGuire
	1956 - 1957	Bob Napier
	1957	William Gibson
	1958 - 1960	Rod Bellette
	1960 - 1967	Nelson Young
	1967 - 1970	Pat Cullen
	1970 - 1971	Jim Dwyer
	1971	Kevin Gaudie
	1972 - 1975	Graham Stewart
	1975 - 1976	Steve Millard
	1976 - 1980	Arthur Stadhams
	1980	Vivian Chalk
	1980 - 1985	Bob Wales
	1985 - 1989	Aidan Wales
	1989 -	Philip Miller

GALLIPOLI OUT-STATION	1930 - 1938	Jack Crouch
	1938 *	*(Jim Wright
		*(Sandy McIndoe
	1944 - 1946	George Rankine
	1946	Doug Harris
	1947 - 1953	Dick McCullagh
	1953 - 1954	Jim Baker
	1954 - 1955	W.Santowski
	1955	Peter Murray
	1955 - 1958	Rod Bellette
	1958 - 1961	George MacQueen
	1961 - 1970	John Ohlsen
	1971 - 1973	Syd Galvin
	1973 - 1974	Steve Millard
	1975	Graham Stewart
	1975 - 1976	Barry Lowe
	1976 - 1977	Graham Stewart
	1977 - 1982	Marshall Blacklock
	1982 - 1984	Jamie MacFarlane
	1984 - 1985	Peter Wright
	1985 - 1990	Rod Murphy
	1990 -	Glenn Page
ALEXANDRIA OUT-STATION	1977 - 1979	Rodger Johnston
	1981 - 1986	Malcolm Debney
	1986 - 1991	Max Ferris
MARION DOWNS	1889 - 1890s	James Curdie Mackinnon
	1890s- 1895	Gordon McLeod
	1902 - 1934	Joseph R. Coghlan
NAP Ownership from 1934		
	1934 - 1936	Tom Bennett
	1936 - 1939	Mr Macdonald
	1940 - 1955	Thomas Harold Cook
	1955 - 1971	Ken McGuire
	1971 -	Bill Alexander
HERBERT DOWNS OUT-STATION	1890s- 1902	Joseph R.Coghlan
NAP Ownership from 1934		
	1940 - 1945	Michael Naulty
	1945 - 1947	Dick McCullagh
	1947 - 1980	Sam Hill
	1980 - 1982	Don Smith
	1982 - 1989	Peter McLaren
	1990 -	Bob Kirk

MONKIRA	pre 1927	P.D. Edwards
NAP Ownership from 1939		
	1927 - 1961	Bob Gunther
	1961 - 1976	George MacQueen
	1976 - 1985	John Crane
	1985 - 1989	Bob Wales
	1990	Jamie MacFarlane
	1990 -	Peter McLaren
COORABULKA	1909 - 1910	C.S. Delpratt
	1911	H. Afford
	1912 - 1916	Jim Nicholls
	1920s - 1930	Tom Barnes
	1930 - 1934	Eric Barnes
	1934 - 1939	Jack Clanchy
NAP Ownership from 1939		
	1940 - 1942	Les Stretton
	1942 - 1948	Leonard Carrington
	1948 - 1949	Ron Carrington
	1949 - 1951	Tommy Groom
	1951 - 1952	William Kriedemann
	1952 - 1961	Tommy Groom
	1961 - 1963	Harold Cook
	1963 - 1964	Peter Donkin
	1964 - 1971	Bill Alexander
	1972 - 1976	John Crane
	1976 - 1985	Steve Millard
	1985 - 1989	Jamie McFarlane
	1990	Peter McLaren
	1990 - 1991	Dieter Jordan
	1992 -	John Rickertt
GLENORMISTON	1899 - 1910	H.V. Weston
	1910 - 1923	William J. Barnett
	1923 - 1926	F.H. Story
	1926 - 1935	E.H. Hamilton
	1935 - 1968	Martin Hayward
NAP Ownership from 1968		
	1968 - 1971	Peter Donkin
	1971 - 1986	Jim Dwyer
	1986 -	Malcolm Debney
ISLAY PLAINS	1970 - 1980	John Ohlsen
WAINUI FEEDLOT	1985 - 1989	Peter Gilbert
	1989 -	Philip Myers

CONNEMARA	1986 - 1991	Arthur Pryce
	1991 -	Max Ferris
KYNUNA	1900 - 1916	Langton G.C. Reid
	1926 - 1963	Don Selkirk
	1970s	Brian Naughton
NAP Ownership from 1986	1979 -	Nick Murray
BOOMARRA	1989 -	Fred Shephard
VERGEMONT	1990 -	Rod Murphy

* Exact details not known

Appendix 6

VEGETATION: Common and scientific names

Common name	Scientific name
Barley Mitchell grass	*Astrebla pectinata*
Black wattle	*A. leicalyx*
Bluebush	*Chenopodium auricomum*
Brigalow	*Acacia harpophylla*
Buffel	*Cenchrus ciliaris*
Button grass	*Dactyloctenium radulans*
Caustic bush	*Euphorbia spp.*
Channel blue grass	*Dichanthium sericeum*
Coolibah	*Eucalyptus microtheca*
Currant bushes	*Carissa ovata*
Eremophila (Turkey bush)	*Eremophila gilesii*
Flinders	*Iseilema membranaceum*
Gidyea	*Acacia cambagei*
	Acacia georginae
Lantana	*Lantana camara*
Lignum	*Muehlenbeckia cunninghamii*
Mitchell	*Astrebla spp.*
Munyeroo	*Portulaca oleracea*
Native sorghum	*Echinochloa turneriana*
Neverfail	*Eragrostis spp.*
Noogoora burr	*Xanthium pungens*
Parkinsonia	*Parkinsonia aculeata*
Pepper grass	*Panicum laevinode*
	Panicum whitei
Pigweed	*Portulaca filifolia*
Roley-poley	*Bassia quinquecuspis*
Rubber vine	*Cryptostegia grandiflora*
Whitewood	*Atalaya hemiglauca*
Yellow top	*Craspedia spp.*

APPENDIX 7

THE NORTH AUSTRALIAN PASTORAL COMPANY PTY. LIMITED

RECORD OF OPERATING PROFITS AND LOSSES
1901-1990

Year	Operating Profit (Loss)	Year	Operating Profit (Loss)	Year	Operating Profit (Loss)
1901	£9,608	1932	(£13,845)	1963	£66,188
1902	£17,955	1933	£4,680	1964	£157,971
1903	£9,104	1934	£4,787	1965	£241,989
1904	£12,438	1935	(£7,801)	1966	$613,157
1905	£14,405	1936	£5,349	1967	$373,046
1906	£29,594	1937	£14,007	1968	$555,038
1907	£20,323	1938	£1,925	1969	$750,986
1908	£19,407	1939	£2,841	1970	$470,062
1909	£17,119	1940	£48,592	1971	($115,580)
1910	£16,596	1941	£42,112	1972	$473,600
1911	£121,205	1942	£40,939	1973	$272,790
1912	£11,419	1943	(£516)	1974	$945,918
1913	£14,902	1944	£9,299	1975	($303,259)
1914	£20,600	1945	£87,215	1976	($552,676)
1915	£23,598	1946	£16,638	1977	($228,550)
1916	£26,572	1947	£292,694	1978	$158,137
1917	£32,680	1948	£42,215	1979	$673,229
1918	£47,486	1949	£79,431	1980	*$4,301,488
1919	£22,630	1950	£261,523	1981	$1,326,026
1920	£7,731	1951	£537,190	1982	$1,463,310
1921	£52,480	1952	£156,804	1983	$308,139
1922	(£15,078)	1953	(£28,463)	1984	$2,775,592
1923	(£6,968)	1954	£34,477	1985	$2,792,704
1924	(£1,635)	1955	£5,938	1986	$1,029,868
1925	£3,462	1956	(£6,561)	1987	$367,667
1926	(£161)	1957	(£48,607)	1988	$3,180,110
1927	£4,833	1958	£129,559	1989	$2,391,446
1928	(£13,193)	1959	£90,121	1990	$1,081,132
1929	(£2,251)	1960	£281,130		
1930	£16,646	1961	£146,952		
1931	£22,077	1962	(£9,422)		

*17 months

BRANDS and EARMARKS

ALEXANDRIA HERD

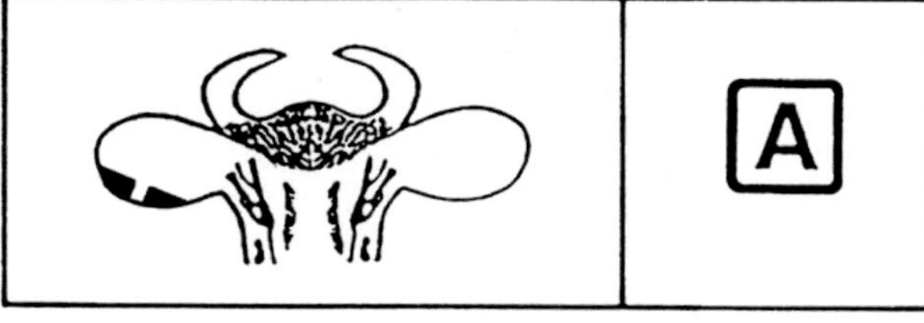

MARION DOWNS

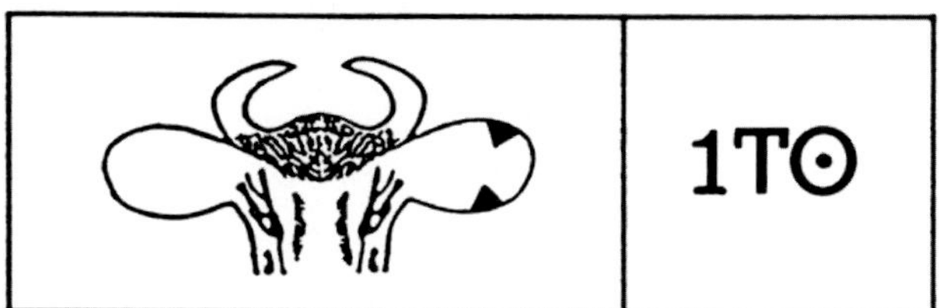

ALEXANDRIA TOP STUD

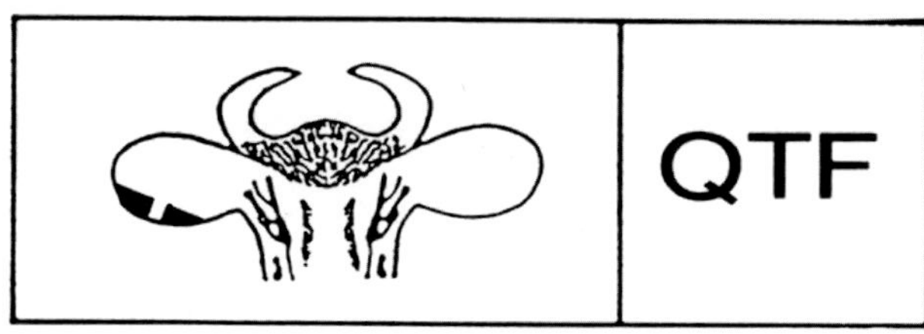

MONKIRA

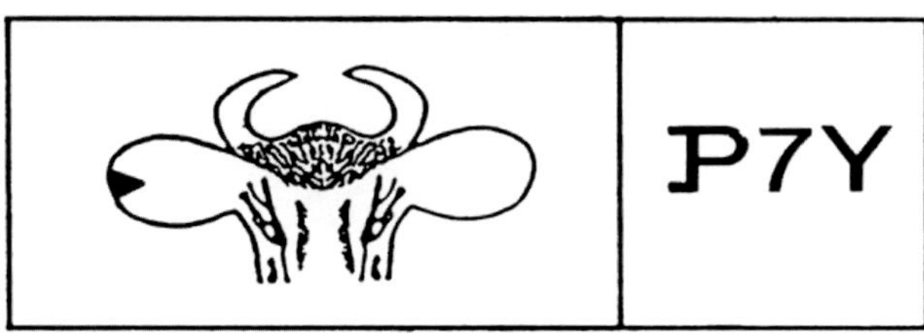

ALEXANDRIA COMMERCIAL STUD

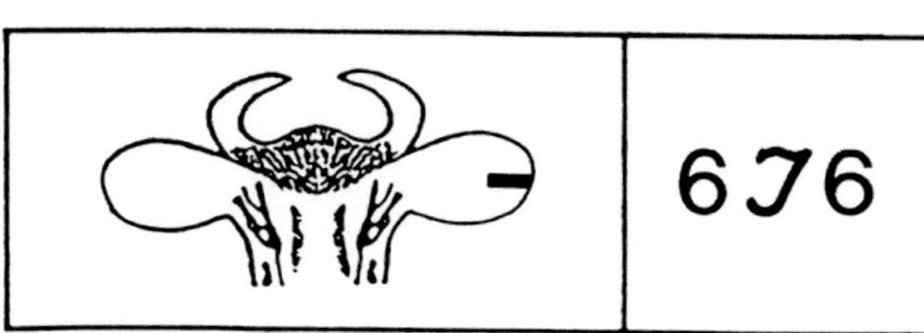

COORABULKA

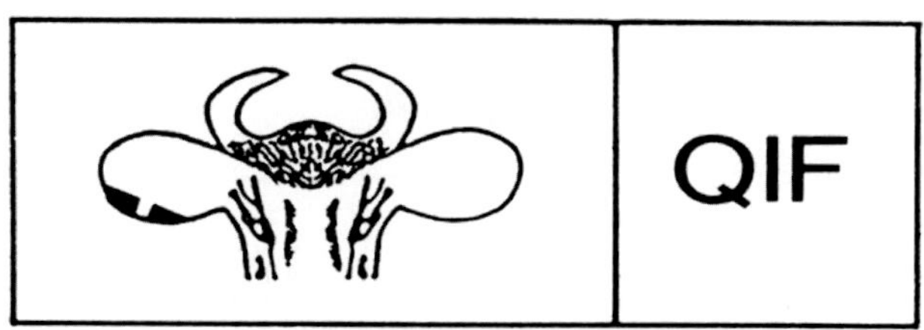

GLENORMISTON

⊙K1

BOOMARRA

IRH
WC.

CONNEMARA

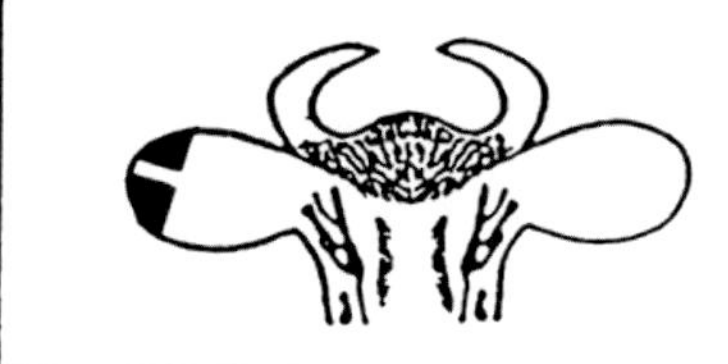

⋖9W

VERGEMONT

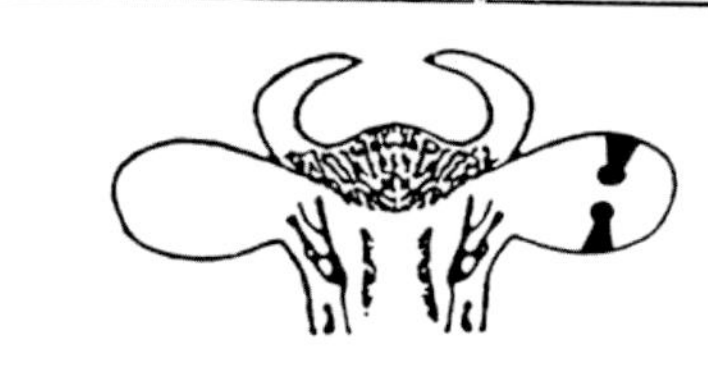

V5C
C.
(F).

CONNEMARA

AN9
Ø.

WAINUI

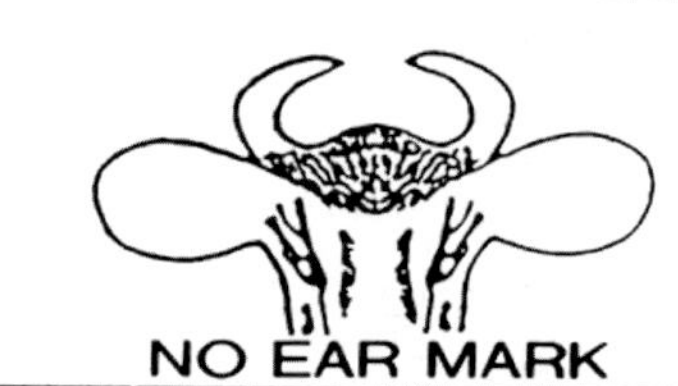

AOm
Æ
~.

KYNUNA

IML
Я.

Bibliography

The bibliography lists only those sources which are cited in the text.

PRIMARY SOURCES

1. ARCHIVAL - GOVERNMENT

AUSTRALIAN ARCHIVES, CANBERRA
Alexandria Bores: A452/1, A1640,

AUSTRALIAN ARCHIVES, DARWIN
Pastoral Lease File: F1

AUSTRALIAN DEPARTMENT OF INDUSTRIAL RELATIONS, BRISBANE
Industrial Relations Commission,
Transcript of Evidence, Cattle Industry Award, (NT), 1965. Evidence by Bill Young at Alexandria, 5 October 1965.

NORTHERN TERRITORY ARCHIVES SERVICE, DARWIN

Lands Files:	NTRS 246, F27.	
Interviews:	Johnston, C.A.Y.	NTRS 226, TS 250
	Johnston, Ellen	NTAS 226, TS 250
	McFarlane, Ted	NTRS 226, TS 273
	Mofflin, Florence	NTAS 226, TS 435
	White, James	NTRS 226, TS 140/1
	Young, Bill	NTRS 226, TS 434

Rankine River Police Journals, 1931-1941, F300.

NORTHERN TERRITORY DEPARTMENT OF LANDS AND HOUSING, DARWIN
Barkly Tableland records held by the Place Names Committee of the Northern Territory.
Maps of the Northern Territory dated 1885, 1907, 1910, 1915.
O'Brien, V.(comp.) 'Kidman Properties in the Northern Territory', Department of Lands and Housing, Darwin, n.d.

QUEENSLAND - JOHN OXLEY LIBRARY, BRISBANE
McIlwraith-Palmer Papers, OM64.19.

Peake, J.H. *A History of the Burdekin*, second edition, Ayr, Ayr Shire Council, 1971.

Stevens, J.F. 'Townsvale and Townsville: the activities of the Hon. Robert Towns MLC in Queensland 1848-1873' Typescript, North Tamborine, 1966.

QUEENSLAND DEPARTMENT OF LANDS, BRISBANE
Lands Files PH 04/364, PH 380, PH 04/5350, PH 29/1201, PH 28/348.

QUEENSLAND STATE ARCHIVES
Register of Runs, CLO/N18-19, CLO/N 33, LAN N42, N43, N44, N45.
Lands Files: LAN/AF 6, LAN/AF 13, LAN/AF 14, LAN/AF 154, LAN/AF 163, LAN/AF 265, LAN/AF 266, LAN/AF 281, LAN/AF 284, LAN/AF 289, LAN/AF 290, LAN/AF 368, LAN/AF 891, LAN/AF 975, LAN/AE 674.
Stock Mortgages SCT/CD 14, Bk 8; SCT/CD 24, Bk 13.

UNIVERSITY OF QUEENSLAND, FRYER LIBRARY

'Joseph Hann and Family, Settlement in North Queensland 1861-1871', Typescript, 15 November 1863.

Reports of two Deputations of Pastoralists in Brisbane, Brisbane, British Australasian Society, 1900.

Fox, M. (comp) *The History of Queensland: its people and industries*, Volume 1, Brisbane, States Publishing, 1919.

Fox, M. (comp) *The History of Queensland: its people and industries*, Volume 2, Brisbane, States Publishing, 1921.

Stirling, A.W. *The Never Never Land*, London, Sampson Low, Marston, Searle and Rivington, 1884.

Sutherland, G. *Pioneering Days: In the early Sixties*, Brisbane, Wendt, 1913.

SOUTH AUSTRALIA - MORTLOCK LIBRARY, ADELAIDE

John Lewis Papers, PRG 247.

Leichhardt, L. *Journal of an Overland Expedition in Australia from Moreton Bay to Port Essington*, Adelaide, Libraries Board of South Australia, 1964.

Map of Frank Scarr's Journey, 1878.

SOUTH AUSTRALIA - PUBLIC RECORD OFFICE OF SOUTH AUSTRALIA

Records pertaining to the Northern Territory of South Australia, 1868-1911: GRS 1, GRS 8/5, GRS 9, GRS 10.

2. ARCHIVAL - NON-GOVERNMENT

RECORDS PRIVATELY HELD

Collins Family — Papers held by Joan McKean, Brisbane.

Collins and Fraser Family — Papers held by William Fraser, Mundoolun.

Crouch, E.M. — Personal Papers held by E.M.Crouch, Brisbane.

Foster Family — Papers held by the Foster family, Tasmania.

Grace, E.N. — Philip Forrest Typescript, Toowoomba, 1989.

MacGregor, Agnes — Papers held by Peg Corbett, Brisbane.

McGuire, Ken — Personal Papers held by Ken McGuire, Bell.

O'Brien, Vern — Personal Papers held by Vern O'Brien, Darwin.

Scarr, Frank — Journal of Expedition to the Northern Territory, 1878, privately held by Bill Kitson, Museum of Mapping and Surveying, Brisbane.

Warner Family — Papers held by C.F. Warner, England.

THE NORTH AUSTRALIAN PASTORAL COMPANY PTY. LIMITED

Company records, 1877-1991, Brisbane.

Barnes, H.R.L. — Typescript, 1990, Brisbane.

Alexander, R. — 'Sam Hill: 50 years of service', Typescript, Brisbane, 31 March 1981.

Alexander, W.F. — 'Ninety years of the N.A.P.Co., 1878-1967', Unpublished manuscript, Brisbane, ca.1968.

MacFarlane, J. — 'Birdsville '89' '.

Moore, Ken — Typescript, 1990, Brisbane.

3. NEWSPAPERS/PERIODICALS

Adelaide Chronicle 1948.

Australian Sugar Journal 1910.

Brisbane Courier 1884, 1911.

Cleveland Bay Express 1869.

Courier-Mail 1948, 1963, 1964, 1965.

Mimag, 1976.

North Queensland Herald 1910.

North Queensland Register 1949, 1964, 1970, 1979.

Northern Standard 1933.

Northern Territory Gazette 1892, 1920.

N.T.Times 1909.

N.T. Times and Gazette 1886, 1892, 1901.

Pastoral Review 1894-95, 1896, 1904-05.

Pastoralist's Review 1906.

Port Denison Times 1867, 1877, 1879.
Queensland Agricultural Journal 1907.
Queensland Country Life 1959, 1961, 1962, 1965, 1966.
Queenslander 1879, 1884.
South Australian Register 1892.
Sunday Sun 1988.
Townsville Bulletin 1948, 1965.
Walkabout 1938.

4. PUBLISHED

GOVERNMENT

Administrator's Annual Reports, Northern Territory, *Commonwealth Parliamentary Papers*, 1911-1949.

Australian Bureau of Agricultural Economics, *The Australian Beef Cattle Industry Survey: 1962-63 to 1964-65*, Canberra, Bureau of Agricultural Economics, 1970.

Australian Bureau of Agricultural Economics, *The Economics of Road Transport of Beef Cattle*, Canberra, Bureau of Agricultural Economics, 1959.

Australian Dictionary of Biography, 1969, 1974, 1981.

Australian Meat Board, *An Outline of the Australian Export Meat Industry*, Sydney, Australian Meat Board, 1969.

Bishop, F.A.C. *Report on an inspection of the pastoral holdings, stock route, bores, and dips on the Barkly Tableland*, Northern Territory, Melbourne, Government Printer, 1923.

Commonwealth Government Gazette, 1942.

Commonwealth Law Reports, 1945, 1946.

Dictionary of National Biography, 1917.

Government Resident's Quarterly, Half-yearly and Annual Reports, *South Australian Parliamentary Papers*, 1880-1910.

N.T. Government Gazette, 1967.

Payne-Fletcher Report, *Report of the Board of Inquiry appointed to inquire into the Land and Land Industries of the Northern Territory of Australia*, Canberra, Commonwealth Printer, 1937; also in *CPP*, 4/1937, Volume 3, 1937, pp.813-926.

Queensland Parliamentary Papers, 1907, 1911, 1912, 1913.

Queensland Votes and Proceedings, 1884, 1887.

South Australian Government Gazette, 1903.

South Australian Parliamentary Debates, 1891, 1893.

NON-GOVERNMENT

Bourne, G. *Journal of Landsborough's Expedition from Carpentaria in search of Burke and Wills*, Melbourne, Dwight, 1962.

Bundaberg and District Historical Museum, Bundaburg. Newspaper Cutting Book, Book A1, 1890s-1900s.

Ingram, C. 'On the birds of the Alexandra District of Northern Territory of South Australia', *The Ibis*, July 1907, pp.387-415.

Ingram, C. 'Supplementary list of the birds of the Alexandra District, Northern Territory of South Australia', *The Ibis*, October 1909, pp.613-8.

Ingram, W. 'On the Display of the King Bird-of-Paradise', *The Ibis*, April 1907, pp.225-9.

Muir, P. *Pituri Pete: a saga of cattle land and desert sand*, Leonara, Muir, 1985.

Spencer Browne, R. *A Journalist's Memories*, Brisbane, Read, 1927.

5. INTERVIEWS

Alexander, R. Marion Downs, 15 November 1990.
Alexander, W. Marion Downs, 15 November 1990.
Brunckhorst, R. Brisbane, 26 July 1991.
Crouch, M. Brisbane, 12 June 1991.

Debney, M.	Bedourie, 17 November 1990.
Doglione, C.	Brisbane, 17 October 1990.
Dwyer, J.	Mount Isa, 12 November 1990.
Foster, W.	Brisbane, 11 June 1991.
Fraser, W.	Mundoolun, 15 September 1990.
Fraser, M.	Mundoolun, 15 September 1990.
Kirk, R.	Herbert Downs, 14 November 1990.
MacQueen, G.	Alphadale, 26 March 1991.
McGuire, K.	Bell, 5 February 1990.
McLaren, L.	Monkira, 18 November 1990.
McNicholl, L.	Dulacca, 16 July 1991.
Young, W.	Casino, 26 January 1986.

SECONDARY SOURCES

1. PUBLISHED

BOOKS

Barrett, R.	*Principles of Income Taxation*, Sydney, Butterworths, 1975.
Bernays, C.A.	*Queensland politics during sixty years: 1859-1919*, Brisbane, Government Printer, n.d.
Bolton, G.C.	*A Thousand Miles Away*, Brisbane, Jacaranda, 1963.
Bowen, J.	*Kidman: The Forgotten King*, North Ryde, Angus and Robertson, 1987.
Buchanan, G.	*Packhorse and Waterhole*, Sydney, Angus and Robertson, 1933.
Burke, Sir B.	*History of the Landed Gentry of Great Britain and Ireland*, sixth edition, Volume 1, London, Herrison, 1879.
Challoner, N.E. and Greenwood, J.M.	*Income Tax Law and Practice (Commonwealth)*, second edition, Sydney, Law Book Company, 1962.
Connolly, R.	*John Drysdale of the Burdekin*, Sydney, Ure Smith, 1964.
Cotton, A.J.	*With the big herds in Australia*, Hidden Vale, A.J.Cotton, 1931.
Critchell, J.T. and Raymond, J.	*A History of the Frozen Meat Trade*, London, Constable, 1912.
Donovan, P.F.	*A Land Full of Possibilities*, St.Lucia, University of Queensland Press, 1981.
Duncan, R.	*The Northern Territory Pastoral Industry 1863-1910*, Melbourne, Melbourne University Press, 1967.
Durack, M.	*Sons in the Saddle*, London, Corgim, 1985.
Dymock, J.	*Nicholson River (Waanyi-Garawa) Land Claim*, Darwin, Northern Land Council, 1982.
Favenc, E.	*The History of Australian Exploration from 1788 to 1888*, Sydney, Turner and Henderson, 1888.
Fysh, H.	*Taming the North*, Sydney, Angus and Robertson, 1950.
Guinness Publishing.	*The Guinness Book of Records, 1992*, England, Guinness Publishing, 1991.
Hall, V.C.	*Outback Policeman*, Melbourne, Rigby, 1970.
Hill, E.	*The Territory*, Sydney, Angus and Robertson, 1951.
Holmes, J.M.	*Australia's Open North*, Sydney, Angus and Robertson, 1963.
Hungerford, T.G.	*Diseases of Stock*, 8th revised edition, Sydney, McGraw Hill, 1975.
Kelly, J.H.	*Report on the Beef Cattle Industry in Northern Australia*, Canberra, Bureau of Agricultural Economics, 1952.
Kerr, J.	*Triumph of Narrow Gauge: A History of Queensland Railways*, Brisbane, Boolarong, 1990.
Lilley, G.W.	*Story of Lansdowne*, Melbourne, Lansdowne Pastoral Company, 1973.
Mason, H.H. and Priddle, L.G.	*Case Companion to Ryan's Income Tax Manual*, Sydney, Law Book Company, 1976.
May, D.	*From Bush to Station*, Townsville, James Cook University, 1983.
McGrath, A.	*Born in the Cattle*, Sydney, Allen and Unwin, 1987.
McKnight, T.L.	*The Long Paddock*, Armidale, University of New England, 1977.
Mahood, M.	*The Australian Stockman*, Sydney, Lansdowne, 1988.
Miller, L.A.	*The Border and Beyond: Camooweal 1884-1984*, Toowoomba, Harrison Printing, 1984.
O'Loghlen, F.(ed.)	*Beef Cattle in Australia 1956*, Sydney, Johnston, 1956.
Neal, J.C.	*Beyond the Burdekin*, Charters Towers, Mimosa, 1984.
Perry, H.C.	*Pioneering: the life of R.M.Collins MLC*, Brisbane, Watson Ferguson, 1923.
Powell, A.	*Far Country: a short history of the NT*, Melbourne, Melbourne University Press, 1982.
Ryan, K.W.	*Manual of the Law of Income Tax in Australia*, third edition, Sydney, Law Book Company, 1972.

Schmidt, P.J. and Yeates, N.T.M. *Beef Cattle Production*, (second edition), Sydney, Butterworths, 1985.

Seddon, H.R. *Diseases of Domestic Animals in Australia*, Part 3, Canberra, Department of Health, 1951.

Stevens, F. *Equal Wages for Aborigines*, Sydney, Aura Press, 1968.

Tinsdale, B. *Aboriginal Tribes of Australia*, Berkeley, University of California Press, 1974.

Troughton, E. *Furred Animals of Australia*, second edition, Sydney, Angus and Robertson, 1943.

Wagstaff, C. *Avon Downs NT 1882-1982*, Armidale, Australian Agricultural Company, 1982.

Webster, M.S. *John McDouall Stuart*, Melbourne, Melbourne University Press, 1958.

MONOGRAPHS AND JOURNAL ARTICLES

Bauer, F.H. 'Significant factors in the white settlement of northern Australia', *Australian Geographical Studies*, Volume 1, No.1, April 1963, pp.39-48.

Beardmore, G.O. 'Glimpses of early Australia', Part V, *Queensland Government Mining Journal*, 65, November 1964, pp. 390-4; 441-7; 475-9; 517-20; 525; 596-601.

Borrow, K.T. 'The Northern Territory of Australia, North Australia and Alexandra Land (1862-1900)', *The Australian Law Journal*, Volume 28, 22 July 1954, pp.148-51.

Camm, J.C.A. 'The Queensland frozen beef industry 1890-1914', *Australian Geographer*, Volume 16, No.1, May 1984.

Charlton, J. 'Moleskin-moulded in the Murranji', *Stockman's Hall of Fame Newspaper*, Volume 42, March 1992.

Cilento, R. 'Medicine in Queensland', *Journal Royal Historical Society Queensland*, Volume 6, 1961-62.

Duncan, R. 'South Australia's contribution to the development of the Northern Territory cattle industry', *Historical Studies in Australia and New Zealand*, Volume 11, No.43, October 1964, pp.324-42.

Gough, Tony 'Tom McIlwraith, Ted Drury, Hugh Nelson and the Queensland National Bank 1896-1897', *Queensland Heritage*, Volume 5, No. 9, November 1978, pp.3-13.

Hare, W.T. *The Early History of Animal Industry in the N.T.*, N.T.Conservation Commission, 1985.

Holt, R.M. and Bertram, J.D. *The Barkly Tableland Beef Industry 1980*, N.T. Department of Primary Production, Technical Bulletin No. 14, June 1981.

Keen, I. 'The Alligator Rivers Aborigines - Retrospect and Prospect', R.Jones (ed.), *Northern Australia: Options and Implications*, Canberra, A.N.U., 1980.

Kingston, B. 'The origins of Queensland's Comprehensive Land Policy', *Queensland Heritage*, Volume 1, No. 2, May 1965, pp.3-9.

May, D. 'The North Queensland Beef Cattle Industry: an historical overview', *Lectures on North Queensland History*, No.4, James Cook University, History Department, 1984.

O'Brien, V. 'The Nation Builder; Pioneer Pastoral Attempts in the Territory', V.Dixon, (ed.) *Looking Back: The Northern Territory in 1888*, Casuarina, Historical Society of N.T. 1988.

Parsons, J.L. *Proceedings Royal Geographical Society of Australasia*, S.A.Branch, Volume 5, 1901-02, Appendix 1, p.11.

Waterson, D. 'Pastoral capitalism and the politician Thomas McIlwraith and two land companies, 1877-1900', *Royal Historical Society Queensland Journal*, 12,6, November 1986.

Waterson, D. *Personality, Profit and Politics: Thomas McIlwraith in Queensland 1866-1894*, The John Murtagh Macrossan Lecture, St. Lucia, University of Queensland Press, 1984.

Williams, W.W. 'The Payne-Fletcher Report on the Northern Territory', *Australian Quarterly*, March 1938, pp.53-60.

2. UNPUBLISHED

Corfield, N. The Development of the Cattle Industry in Queensland, 1840-1890, B.A. (Hons) Thesis, University of Queensland, 1959.

Griggs, P.D. Plantation to small farm: A historical geography of the lower Burdekin Sugar Industry, 1880-1930, PhD Thesis, University of Queensland, 1990.

May, D. 'The Impact of the 15 Year Meat Agreement on the Development of Northern Australia', Unpublished Paper, Australian Historical Association Conference, Darwin, 1991.

McCulloch, K.E. Cattle Tick in the Northern Territory in the Nineteenth Century, B.A. (Hons) Thesis, University of Adelaide, 1962.

INDEX

ILLUSTRATIONS ARE IN BOLD

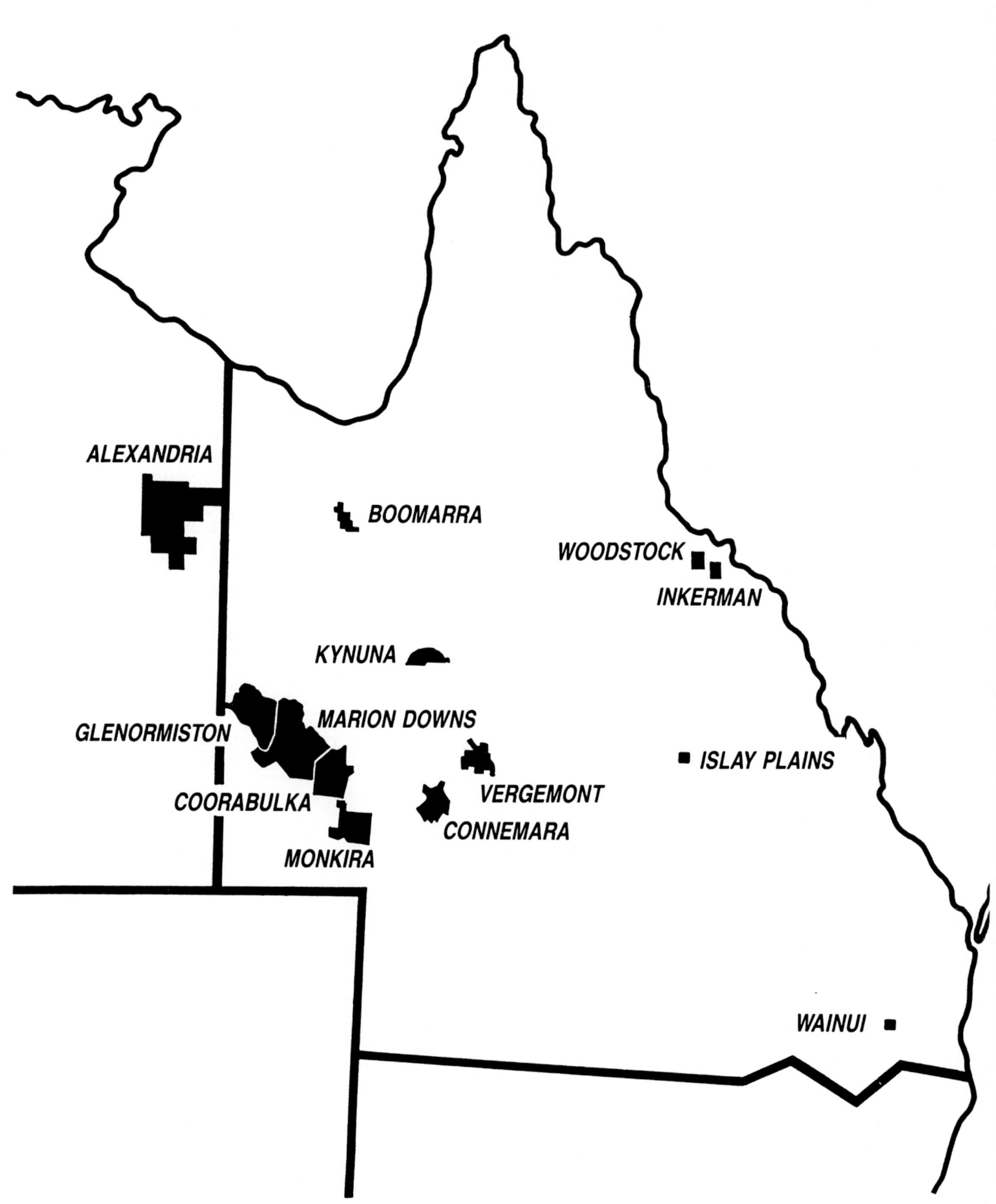

NAP properties, past and present.

THE NORTH AUSTRALIAN PASTORAL COMPA

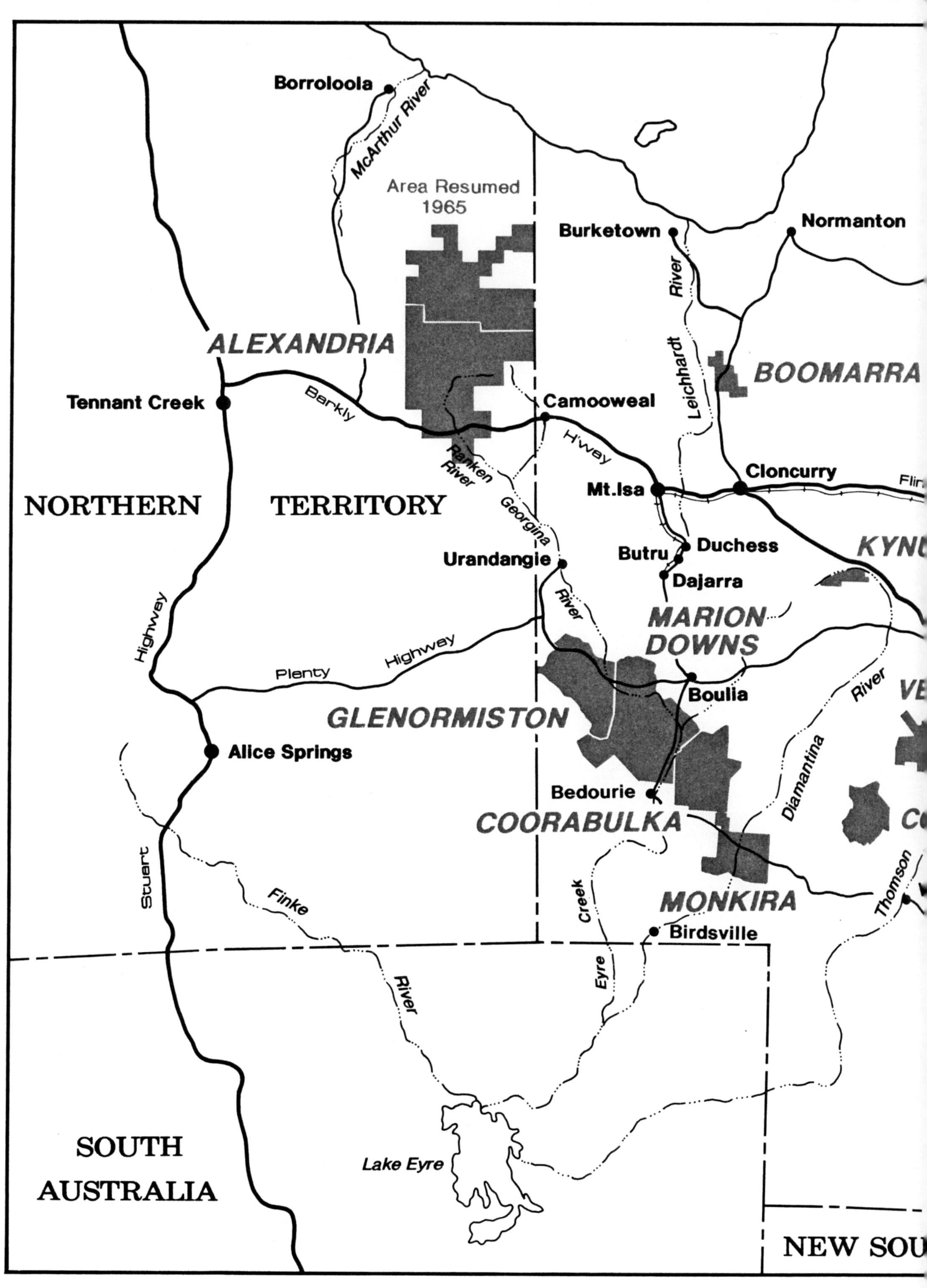

TY. LIMITED – PROPERTY LOCATION MAP

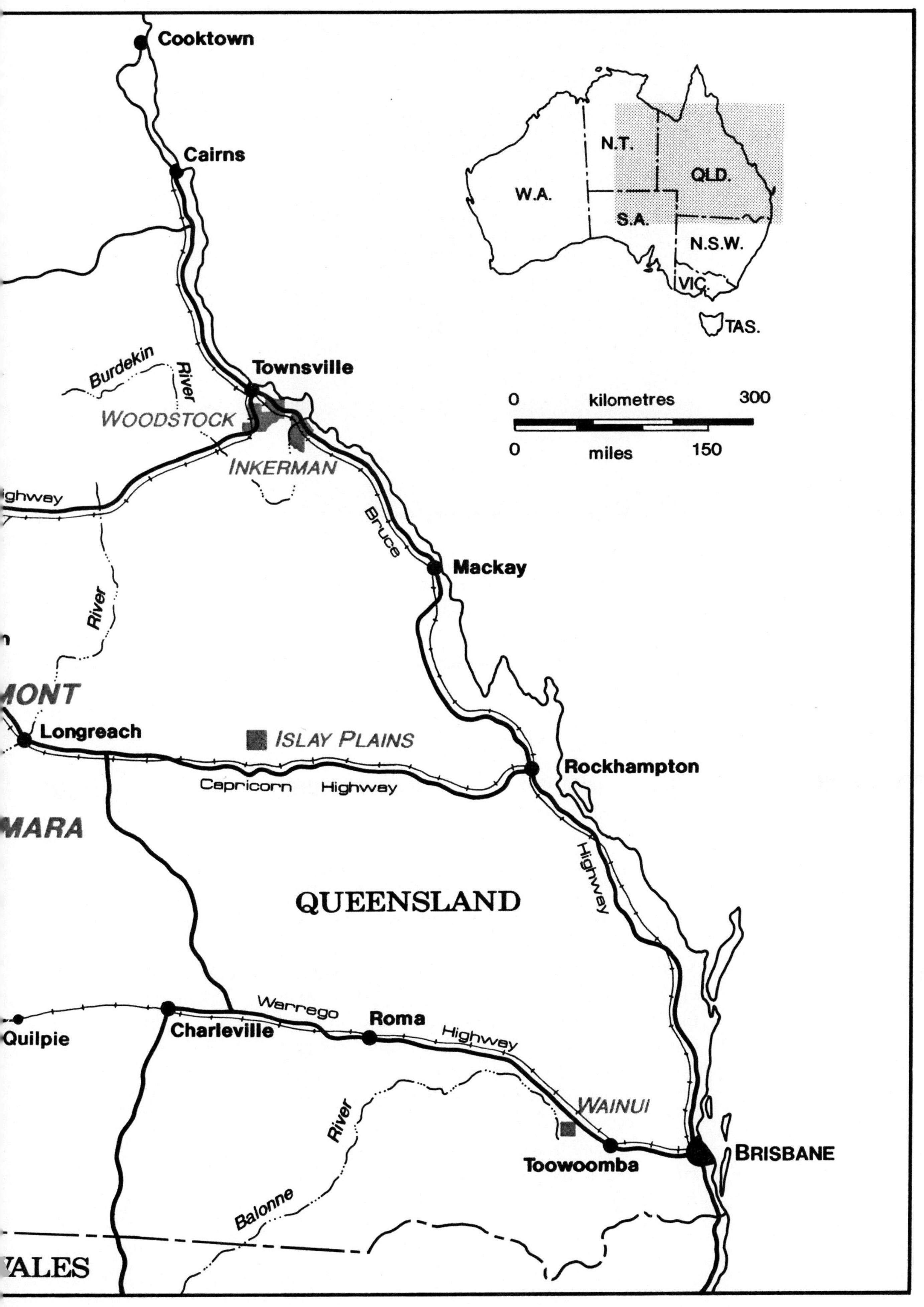

Margaret Kowald B.A., B.Ed.Stud., M.A. was born in Cairns and after spending her childhood in north Queensland, she taught for ten years in secondary schools in Queensland, Canberra and Darwin. Her formal interest in history began in 1987 as a researcher for Dr Ross Johnston in his history of the Queensland Police Force. She is currently studying the Queensland Pastoral Industry 1890-1940 for her Ph.D. at The University of Queensland.

William Ross Johnston, B.A.(Syd.) LL.B., M.A., Ph.D.(Duke) is an Associate Professor in History at The University of Queensland. He has specialised in British Commonwealth History, South Pacific History and Queensland History. His numerous publications reflect the breadth of this interest; recent titles on Queensland history include *A Bibliography of Queensland History, A History of the Queensland Bar, The Call of the Land, Brisbane: the first thirty years, A Documentary History of Queensland, The Long Blue Line,* and three local studies - *Bauhinia; Hyne-sight, a story of a Timber Family;* and a *History of Monto.*